Fatigue and Fracture Mechanics of High Risk Parts

Join Us on the Internet

WWW: http://www.thomson.com
EMAIL: findit@kiosk.thomson.com

thomson.com is the on-line portal for the products, services and resources available from International Thomson Publishing (ITP).

This Internet kiosk gives users immediate access to more than 34 ITP publishers and over 20,000 products. Through *thomson.com* Internet users can search catalogs, examine subject-specific resource centers and subscribe to electronic discussion lists. You can purchase ITP products from your local bookseller, or directly through *thomson.com*.

Visit Chapman & Hall's Internet Resource Center for information on our new publications, links to useful sites on the World Wide Web and an opportunity to join our e-mail mailing list. Point your browser to: http://www.chaphall.com or
http://www.thomson.com/chaphall/mecheng.html for Mechanical Engineering

A service of

Fatigue and Fracture Mechanics of High Risk Parts

Join Us on the Internet

WWW: http://www.thomson.com
EMAIL: findit@kiosk.thomson.com

thomson.com is the on-line portal for the products, services and resources available from International Thomson Publishing (ITP).

This Internet kiosk gives users immediate access to more than 34 ITP publishers and over 20,000 products. Through *thomson.com* Internet users can search catalogs, examine subject-specific resource centers and subscribe to electronic discussion lists. You can purchase ITP products from your local bookseller, or directly through *thomson.com*.

Visit Chapman & Hall's Internet Resource Center for information on our new publications, links to useful sites on the World Wide Web and an opportunity to join our e-mail mailing list. Point your browser to: http://www.chaphall.com or
http://www.thomson.com/chaphall/mecheng.html for Mechanical Engineering

A service of

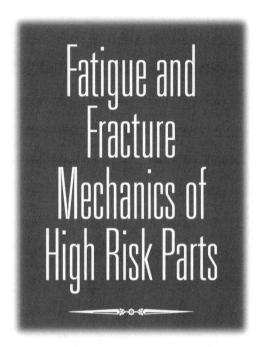

Fatigue and Fracture Mechanics of High Risk Parts

Application of LEFM & FMDM Theory

Bahram Farahmand
Senior Specialist and MDC Fellow
McDonnell Douglas Corporation
Huntington Beach, CA

**with
George Bockrath
James Glassco**

CHAPMAN & HALL

INTERNATIONAL THOMSON PUBLISHING

New York • Albany • Bonn • Boston • Cincinnati • Detroit • London
Madrid • Melbourne • Mexico City • Pacific Grove • Paris • San Francisco
Singapore • Tokyo • Toronto • Washington

Cover design: Curtis Tow Graphics

Copyright © 1997 by Chapman & Hall

Printed in the United States of America

Chapman & Hall
115 Fifth Avenue
New York, NY 10003

Chapman & Hall
2-6 Boundary Row
London SE1 8HN
England

Thomas Nelson Australia
102 Dodds Street
South Melbourne, 3205
Victoria, Australia

Chapman & Hall GmbH
Postfach 100 263
D-69442 Weinheim
Germany

International Thomson Editores
Campos Eliseos 385, Piso 7
Col. Polanco
11560 Mexico D.F
Mexico

International Thomson Publishing–Japan
Hirakawacho-cho Kyowa Building, 3F
1-2-1 Hirakawacho-cho
Chiyoda-ku, 102 Tokyo
Japan

International Thomson Publishing Asia
221 Henderson Road #05-10
Henderson Building
Singapore 0315

All rights reserved. No part of this book covered by the copyright hereon may be reproduced or used in any form or by any means—graphic, electronic, or mechanical, including photocopying, recording, taping, or information storage and retrieval systems—without the written permission of the publisher.

1 2 3 4 5 6 7 8 9 10 XXX 01 00 99 98 97

Library of Congress Cataloging-in-Publication Data

Farahmand, Bahram.
   Fatigue and fracture mechanics of high risk parts : application of
LEFM & FMDM theory / Bahram Farahmand, George Bockrath, James
Glassco.
     p. cm.
   Includes bibliographical references and index.
   ISBN 0-412-12991-4 (hb : alk. paper)
   1. Metals -- Fatigue. 2. Fracture mechanics. 3. Metals — Ductility.
4. Service life (Engineering). I. Bockrath, George. II. Glassco,
James. III. Title.
TA460.F337   1997
620.1'66--dc21   97-6586
                                                                                    CIP

British Library Cataloguing in Publication Data available

"Fatigue and Fracture Mechanics of High Risk Parts" is intended to present technically accurate and authoritative information from highly regarded sources. The publisher, editors, authors, advisors, and contributors have made every reasonable effort to ensure the accuracy of the information, but cannot assume responsibility for the accuracy of all information, or for the consequences of its use.

To order this or any other Chapman & Hall book, please contact **International Thomson Publishing, 7625 Empire Drive, Florence, KY 41042**. Phone: (606) 525-6600 or 1-800-842-3636. Fax: (606) 525-7778. e-mail: order@chaphall.com.

For a complete listing of Chapman & Hall titles, send your request to **Chapman & Hall, Dept. BC, 115 Fifth Avenue, New York, NY 10003**.

This book is dedicated to the memory of my loving father, Hadi, who instilled in me his enormous respect for higher education, and the values of patience and discipline.

Contents

Contents vii

Preface xiii

Chapter 1 A Brief Introduction to Fatigue and Fracture Mechanics 1

1.1 Failures in Riveted Joints   1
1.2 Failures in Welded Joints   2
1.3 Fatigue and Fracture Mechanics   3
     References   11

Chapter 2 Conventional Fatigue (High- and Low-Cycle Fatigue) 13

2.1 Background   13
2.2 Cyclic or Fluctuating Load   15
2.3 Fatigue Spectrum   18
    2.3.1 Load Spectrum for Space Structures Using the Space Transportation System (STS) or Shuttle   28
        2.3.1.1 Flight Cycles   31
2.4 The S–N Diagram   32
    2.4.1 Factors Influencing the Endurance Limit   34
        2.4.1.1 Brief Description of Constant Amplitude Axial Fatigue Tests   34
    2.4.2 Empirical Representation of the S–N Curve   38
    2.4.3 Parameters Affecting the S–N Curve   40
2.5 Constant-Life Diagrams   41
    2.5.1 Development of a Constant-Life Diagram   42

Contents

- 2.5.2 Development of a Constant-Life Diagram by Experiment 43
- 2.6 Linear Cumulative Damage 47
  - 2.6.1 Comments Regarding the Palmgren–Miner Rule 50
- 2.7 Crack Initiation (Stage I) and Stable Crack Growth (Stage II) 51
  - 2.7.1 Fracture Surface Examination 51
  - 2.7.2 Introduction to Crack Initiation 57
    - 2.7.2.1 Fracture Due to Monotonic Applied Load 57
    - 2.7.2.2 Crystal Structures of Metals and Plastic Deformation 59
  - 2.7.3 Crack Initiation Concept (Intrusion and Extrusion) 63
- 2.8 Low-Cycle Fatigue and the Strain-Controlled Approach 67
  - 2.8.1 Crack Initiation Life 68
- 2.9 The Conventional or Engineering Stress–Strain Curve 69
- 2.10 The Natural or True Stress–Strain Curve and Hysteresis Loop 70
  - 2.10.1 The Natural or True Stress–Strain Curve 70
  - 2.10.2 Hysteresis Loop 71
- 2.11 Cyclic Stress–Strain Curve 74
  - 2.11.1 The Power Law Representation of the Cyclic Stress–Strain Curve 78
- 2.12 Strain-Life Prediction Models 82
  - 2.12.1 Transition Point, N_T 86
  - 2.12.2 The Mean Stress Effect 88
  - 2.12.3 Fatigue Notch Factor, K_f 88
  - 2.12.4 Stress and Strain at Notch (Neuber Relationship) 93
- 2.13 Universal Slope Method 98
- References 99

Chapter 3 Linear Elastic Fracture Mechanics 103

- 3.1 Energy Balance Approach (the Griffith Theory of Fracture) 103
- 3.2 The Stress Intensity Factor Approach 105
  - 3.2.1 General 105

	3.2.2	Crack Tip Modes of Deformation 106
	3.2.3	Derivation of Mode I Stress Intensity Factor 107
		3.2.3.1 Airy Stress Function, ϕ 108
	3.2.4	Combined Loading 114
	3.2.5	Stress Intensity Factor Equations for Several Through Cracks 117
	3.2.6	Critical Stress Intensity Factor 122
3.3	Fracture Toughness 124	
	3.3.1	Fracture Toughness and Material Anisotropy 128
	3.3.2	Other Factors Affecting Fracture Toughness 130
3.4	Residual Strength Capability of a Cracked Structure 134	
	3.4.1	Residual Strength Diagram for Material with Abrupt Failure 136
	3.4.2	The Apparent Fracture Toughness 138
	3.4.3	Development of the Resistance curve (R-curve) and K_R 139
3.5	Plasticity at the Crack Tip 146	
3.6	Plastic Zone Shape Based on the Von Mises Yield Criterion 148	
3.7	Surface or Part Through Cracks 152	
	3.7.1	Stress Intensity Factor Solution for Surface Cracks (Part Through Cracks) 153
	3.7.2	Longitudinal Surface Crack in a Pressurized Pipe 157
	3.7.3	Part Through Corner Crack Emanating from a Hole 158
	3.7.4	Part Through Fracture Toughness, K_{Ie} 159
	3.7.5	The Leak-Before-Burst (LBB) Concept 160
3.8	A Brief Discription of ASTM Fracture Toughness Determination 164	
	3.8.1	Plane Strain Fracture Toughness (K_{Ic}) Test 164
		3.8.1.1 Standard K_{Ic} Test Specimen and Fatigue Cracking 165
	3.8.2	Plane Stress Fracture Toughness (K_c) Test 169
		3.8.2.1 M(T) Specimen for Testing K_c 171
		3.8.2.2 Grip Fixture Apparatus, Buckling Restraint, and Fatigue Cracking 172
	References 174	

Chapter 4 Fatigue Crack Growth   177

4.1 Introduction   177
    4.1.1 Stress Intensity Factor and Crack Growth Rate   179
4.2 Crack Growth Rate Empirical Descriptions   180
    4.2.1 Brief Review of Fatigue Crack Growth Testing   188
4.3 Stress Ratio and Crack Closure Effect   194
    4.3.1 Elber Crack Closure Phenomenon   196
        4.3.1.1 Threshold Stress Intensity Factor, ΔK_{th} 199
    4.3.2 Newman Crack Closure Approach   203
4.4 Variable Amplitude Stress and the Retardation Phenomenon   216
    4.4.1 Wheeler Retardation Model   218
    4.4.2 Willenborg Retardation Model   219
4.5 Cycle by Cycle Fatigue Crack Growth Analysis   227
4.6 Structural Integrity Analysis of Bolted Joints Under Cyclic Loading   230
    4.6.1 Preloaded Bolt Subjected to Cyclic Loading   230
    4.6.2 Fatigue Crack Growth Analysis of Pads in a Bolted Joint   237
4.7 Material Anisotropy and Its Applications in Beam Analysis   248
    References   249

Chapter 5 Fracture Control Program and Nondestructive Inspection   253

5.1 Introduction   253
    5.1.1 Design Philosophy   253
        5.1.1.1 Slow Crack Growth Design   254
        5.1.1.2 Fail-Safe Design and Low Released Mass   256
        5.1.1.3 Material Selection, Testing, Manufacture, and Inspection   257
5.2. Fracture Control Plan   258
    5.2.1 Purpose   258
    5.2.2 Causes of Unstable Crack Growth   259

	5.2.3	Fracture Control in the Stages of Hardware Development 259

- 5.2.3 Fracture Control in the Stages of Hardware Development 259
- 5.2.4 Typical Fracture Control Activities 260
- 5.2.5 Data Required for Fracture Control 261
- 5.2.6 Contents of a Fracture Control Plan 261
- 5.2.7 Fracture Control Classification of Components 261
- 5.2.8 Analysis and/or Testing to Determine Fracture Control Acceptability of Hardware 263
- 5.2.9 Traceability and Design Annotation for Drawings 265
- 5.2.10 Overall Review and Assessment of Fracture Control Activities and Results 267
- 5.2.11 Fracture Control Board 267

5.3 Nondestructive Inspection Techniques 269
- 5.3.1 Introduction 269
- 5.3.2 Liquid Penetrant Inspection 269
- 5.3.3 Magnetic Particle Inspection 270
- 5.3.4 Eddy Current Inspection 270
- 5.3.5 Ultrasonic Inspection 271
- 5.3.6 Radiographic Inspection 272
- 5.3.7 Flaw Size Verification 274
- 5.3.8 Probability of Detection Statistics 276
- 5.3.9 Variables Affecting NDI Flaw Detectability Testing 278
  - 5.3.9.1 Variables Affecting the Liquid Penetrant Inspection 279
  - 5.3.9.2 Variables Affecting the Magnetic Particle Inspection 283
  - 5.3.9.3 Variables Affecting the Eddy Current Inspection 284
  - 5.3.9.4 Variables Affecting the Ultrasonic Inspection 284
  - 5.3.9.5 Variables Affecting the Radiographic Inspection 286
- 5.3.10 Critical Initial Flaw Sizes for Standard NDI 286
- 5.3.11 Special Level Inspection 288

References 288

Chapter 6 The Fracture Mechanics of Ductile Metals Theory   289

6.1 Introduction   289
6.2 Fracture Mechanics of Ductile Metals   290
6.3 Determination of the $\frac{\partial U_F}{\partial c}$ Term   292
6.4 Determination of the $\frac{\partial U_U}{\partial c}$ Term   294
    6.4.1 Octahedral Shear Stress Theory (Plane Stress Conditions)   295
6.5 Octahedral Shear Stress Theory (Plane Strain Conditions)   303
6.6 Applied Stress, σ, and Half Crack Length, c, Relationship   305
6.7 Mixed Mode Fracture and Thickness Parameters   307
6.8 The Stress–Strain Curve   309
6.9 Verification of FMDM Results with the Experimental Data   309
6.10 Fracture Toughness Computation by the FMDM Theory   312
    6.10.1 Introduction   312
    6.10.2 Fracture Toughness, K_c, Evaluation for 2219-T87 Aluminum Alloy   313
    6.10.3 Fracture Toughness, K_c, Evaluation for 7075-T73 Aluminum Alloy   316
    References   322

Appendix A NASA/FLAGRO 2.0 Materials Constants   323

Appendix B Equations for Uniaxial True Stress and True Strain   358

Index   367

Preface

In the preliminary stage of designing new structural hardware that must perform a given mission in a fluctuating load environment, there are several factors the designers should consider. Trade studies for different design configurations should be performed and, based on strength and weight considerations, among others, an optimum configuration selected. The selected design must be able to withstand the environment in question without failure. Therefore, a comprehensive structural analysis that consists of static, dynamic, fatigue, and fracture is necessary to ensure the integrity of the structure. During the past few decades, fracture mechanics has become a necessary discipline for the solution of many structural problems. These problems include the prevention of failures resulting from preexisting cracks in the parent material, welds or that develop under cyclic loading environment during the life of the structure.

The importance of fatigue and fracture in nuclear, pressure vessel, aircraft, and aerospace structural hardware cannot be overemphasized where safety is of utmost concern. This book is written for the designer and strength analyst, as well as for the material and process engineer who is concerned with the integrity of the structural hardware under load-varying environments in which fatigue and fracture must be given special attention. The book is a result of years of both academic and industrial experiences that the principal author and co-authors have accumulated through their work with aircraft and aerospace structures. However, the material contained in this book can also be applied to other industries, such as nuclear, pressure vessel, and shipbuilding, where fracture and fatigue are equally important. Moreover, the scope and contents of the book are adequate for use as a textbook for both graduate and undergraduate level courses in the mechanical, material, and aerospace engineering departments. Each chapter has several example problems that have been handpicked from industrial experiences that the authors have accumulated throughout the years in the field of fracture mechanics.

This book addresses the conventional fatigue approach to life evaluation of a structural part where it is assumed that the structure is initially free from cracks and, after N number of load cycles, the crack will initiate in some highly localized stressed areas. In contrast to the conventional fatigue approach, Linear Elastic Fracture Mechanics (LEFM) assumes the existence of a crack in the structural part in the most unfavorable location perpendicular to the applied load. This book covers in detail both the conventional fatigue (the stress to life, S–N, and the strain to life, ε–N) and the LEFM approaches to determining the life of structural hardware in a load-varying environment. In using the LEFM approach to evaluate the life of a part, an initial crack size provided by the standard nondestructive inspection method and the fracture toughness data for the material under study must be available to the analyst. Obtaining fracture toughness data through laboratory testing is costly and time consuming. A new approach, called Fracture Mechanics of Ductile Metals (FMDM), can satisfactorily correlate the fracture toughness data obtained through the analysis with the test data and was first developed in the late 1960s by the authors of this book.

In the late 1960s, G. E. Bockrath and J. B. Glassco, while working at McDonnell Douglas Aerospace Company (formerly known as Douglas Aircraft Company) as principal engineers, were researching high fracture stress under low-cycle fatigue of high-strength material. It was noticed that some Ti-6Al-4V titanium alloy (STA condition) test specimens containing small part through cracks fractured at the ultimate stress. This led to the question of how large can a crack be in a specimen that will fracture at the ultimate stress. For metals that have a large amount of necking, under normal rate of loading, the critical crack size was large, and for a metal with no necking (such as beryllium) it was very small. The maximum flaw size that has a failure stress equal to the ultimate stress can be obtained through the FMDM computer program.

Another important observation as a result of this study was made. While studying the fracture surface of test specimens, the authors noticed that metals whose full-range uniaxial tensile engineering stress–strain curve showed local plastic deformation beyond the ultimate stress (necking) had a slope of 45 degrees over part, or all, of the fracture surface. This indicated that ductile metals with observed necking absorb more energy at the crack tip than metals whose stress–strain curves do not show plastic deformation beyond the ultimate stress.

The crack tip plastic deformation defined by the FMDM theory is composed of two distinct regions: (1) the local strainability at the crack tip (the region of highly plastic deformation) and (2) the uniform strainability near the crack tip. The energy absorption rate for these two regions was calculated (see Chapter 6) and used to extend the Griffith theory of fracture that originally was developed for brittle materials. In contrast to LEFM, the FMDM theory was shown to accurately correlate with test data for commonly used structural metals over a wide range of crack sizes at stresses above as well as below the yield stress. The FMDM computer program is capable of generating the variation of fracture toughness as a function of the material thickness for ductile metals and requires only the stress–strain curve as an input.

The contents of this book represent a complete overview of the field of fatigue and fracture mechanics, a field that is continuously being advanced by many investigators.

The author is very thankful to the following individuals: Mr. David Ollodort and Mr. Kerry Michaels of McDonnell Douglas Aerospace for contributing a portion of the material in Chapter 5; to Mr. Robert Muller for his constructive suggestions pertaining to sections of the fracture control plan. Also my great appreciation goes to Dr. Bijan Irani Nejad for his contribution to the sections of load spectrum material. Special thanks also goes to Mr. Vector Kerlins and Mr. Ray Toosky of the metallurgy department of McDonnell Douglas Aerospace. The author wishes to thank Professor Ares Rosakis of the California Institute of Technology for his constructive comments regarding Chapter 6. As a final note, the author would like to acknowledge the co-authors, Mr. George Bockrath and Mr. Jim Glassco, for their contribution to the development of the FMDM theory and to Mr. David Ollodort for his editorial assistance with the manuscript.

I would also like to acknowledge the support of my loving and devoted wife, children, and especially my dear mother whose sacrifices made it possible for me to complete this book.

Fatigue and Fracture Mechanics of High Risk Parts

Chapter 1

A Brief Introduction to Fatigue and Fracture Mechanics

During the past few decades fracture mechanics has become a necessary discipline for the prevention of many structural failures. These are problems resulting primarily from structures containing part through or through cracks. Such cracks can originate in many ways. For example, they may be introduced as cracks or as incipient cracks during manufacture of structural parts; they may grow from defects in the parent metal, from incomplete welds, from shrink cracks or other imperfections in weldments; or they may nucleate and grow in structure under fatigue loading. It is the role of fracture mechanics to determine when these cracks become critical, that is, when they will reach a size at which the crack will grow catastrophically at an operational stress well below the yield strength.

The following are descriptions of a few documented fatigue-related failures that had occurred during the service life of the structure and are directly extracted from Reference [1] by Parker. Extensive scientific investigation on the nature of these structural failures indicated that in almost all cases, flat (cleavage) or brittle fracture occurred in a catastrophic manner, with high velocities and little or no shear lips. Defects in the welded parts, induced residual stresses in the weld and during the structural assembly, steel composition, cyclic and corrosion environment, transition temperature effect (ductile-brittle failure) poor design, and lack of inspection can all be responsible for these failures. Of the sources of failures reported in Reference [1], seven started in riveted structures as the source of stress concentration and 14 in welded parts.

1.1 Failures in Riveted Joints

- Water Standpipe, Gravesend, Long Island, New York, October 7, 1886
- Gasholder, Brooklyn, New York, December 23, 1898
- Water Standpipe, Sanford, Maine, November 17, 1904

- Molasses Tank, Boston, Massachusetts, January 15, 1919. The tank contained 2,300,000 gallons of molasses that caused 12 deaths and 40 injuries.
- Crude Oil Storage Tank, Ponca City, Oklahoma, December 19
- Eight Riveted Crude Oil Tanks, south and midwest United States, 1930-1940
- Oil Storage Tank, midwest United States, December 14, 1943

1.2 Failures in Welded Joints

- Three Vierendeel Truss Bridges, Albert Canal (in Hasselt, Herenthals–Oolen, and Kaulille), Belgium, 1938 through 1940
- Duplessis Bridge, Three Rivers, Quebec, Canada, January 31, 1951
- Spherical hydrogen storage Tank, Schenectady, New York, February 1943.
- Spherical ammonia tank, Pennsylvaia, March 1943
- Spherical pressure vessel, Morgantown, West Virginia, January 1944
- Cylindrical gas pressure vessel, Cleveland, Ohio, October 1944
- Five oil storage tanks in Russia, December 1947
- Crude oil storage tank, midwest area, United States, February 1947
- Two oil storage tanks, Fawley, England, February and March 1952
- Failures of three empty storage tanks in Europe during 1952
- Penstank at the Anderson Ranch Dam, Boise, Idaho, January 1950.

The total cost associated with material fracture and failure in the United States is estimated to be $88 billion dollars per year (based on 1978 dollars) [2]. Research directed toward fracture-related problems could reduce the costs of structural failure in the United States by almost 30%; that is, an estimated $21 billion dollars per year (based on 1978 dollars). Extensive study and research on these failures indicated that potential savings can be obtained by focusing attention on two major areas: material and structural. The summary of research in these areas concluded that (1) a reduction in material variability (tensile and yield stress, as well as fracture properties) which can contribute to prevention of structural failures; (2) better use of fracture mechanics in evaluating the life of the component; (3) increased use of nondestructive inspection and improvement in NDE techniques; and (4) implementation of a fracture control plan can all reduce the costs of fracture related failure in the United States [2].

1.3 Fatigue and Fracture Mechanics

Fracture mechanics theory began with A. A. Griffith (1893-1963) at the Royal Aircraft Establishment (RAE) in the United Kingdom with the help of the mathematical work of Professor C. E. Inglis [3, 4], but the major development took place in the United States at the Naval Research Laboratory (NRL) in 1950 by George R. Irwin [5]. Griffith's theory applies to brittle materials, that is, materials that, under normal rates of loading (monotonically increasing tensile loads as compared to fluctuating or cyclic loads) and at room temperature, do not undergo plastic deformation before fracture in a tensile test.

The Griffith theory as described in "The Phenomenon of Rupture and Flow in Solids" assumes material contains crack-like defects and that work must be performed on the material to supply the energy needed to propagate the crack by creating two new crack surfaces. For brittle materials (such as high-strength steels), the Griffith theory gives the correct relationship between the fracture strength, σ_F, and critical flaw size, c. However, this theory is not satisfactory as a means for determining the fracture strength of ductile material, because cracked structural metals undergo considerable plastic deformations in the region at the crack tip prior to final fracture. Almost 30 years after Griffith's theory, Irwin and Orowan [6, 7] pointed out that for ductile material to fracture, the stored strain energy is consumed for both the formation of two new cracked surfaces and the work done in plastic deformation. This energy balance approach can be stated as:

$$\text{Energy Input} \geq \text{Surface Energy} + \text{Energy of Plastic Deformation}$$

Irwin realized that for ductile materials, the energy required to form the new crack surfaces is negligible compared to the work done in plastic deformation at the crack tip. Generally in metallic materials, the energy required for plastic deformation is approximately 10^3 times larger than the surface energy. This is the reason why metallic materials have much greater resistance to fracture than brittle material. Irwin further proposed that the elastic energy release rate can be regarded as a force called crack driving force that is denoted by G (after Griffith), where its critical value, G_c, corresponds to crack instability. The quantity G is equal to the derivitive of elastic energy and its magnitude is the same for cracked plate subjected to constant load or fixed ends conditions.

Fracture mechanics became an engineering discipline with the development of Linear Elastic Fracture Mechanics (LEFM). Griffith's energy balance approach is applicable only to the crack analysis cases where the condition of instability under static load is required. However, this approach is not applicable to other important cases, such as stable crack growth that occurs when a crack is subjected to cyclic loading. The stress intensity parameter, K, can address important cracking problems (e.g., stable crack growth under fluctuating load, as well as the condition of instability) that are of interest in the aircraft, space vehicle, pressure vessel, nuclear, and shipbuilding industries where safety of people and structures is of great concern.

LEFM is an extension of the Griffith theory, modified to account to a limited extent for the presence of the plastic zone at the crack tip. To be able to correlate the LEFM results with the test data, the energy consumed at the crack tip for the plastic deformation must be considered. Irwin and Dugdale [8, 9] estimated the size of the plastic zone formed at the crack tip by assuming that local stresses are equal to the yield stress of the material. In Dugdale's model, the shape of the plastic zone formed at the crack tip was considered to be in the form of a strip in front of the physical crack tip (a region that carries yield stress tending to close the crack) called the strip yield model, whereas Irwin assumed that the plastic zone ahead of the crack tip is a circle. In both models, the effective crack length, rather than the physical crack length, was used to calculate the stress intensity factor.

The critical value of the crack tip stress intensity factor is associated with the fracture toughness of the material. Fracture toughness can be defined as the material's ability to resist unstable cracking in a non-corrosive environment. It can be described in terms of the critical stress intensity factor under the condition of plane strain, K_{Ic}, or plane stress, K_c, under slow rate of loading and normal temperature. Ductile materials under normal rate of loading have the ability to absorb energy and deform plastically prior to rupture, hence possessing high fracture toughness. Brittle material on the other hand fails at low stresses in a sudden manner with high velocities and over great distances, hence possessing low fracture toughness and tending to have cleavage or low energy fracture.

To be able to resolve any cracking problem, the fracture toughness value must be available to the analyst. This is analogous to the strength analysis requirement that the limit stress,[1] σ, must always be compared with both the yield, σ_{Yield}, and the ultimate, σ_{Ult}, strength of the material. That is, the analysis would be incomplete without both material allowables. With the material fracture toughness value available from ASTM Standard specimen testing, the residual strength capability of a structural part containing a crack can be determined through a relationship that describes the critical stress intensity factor, fracture stress, and the crack geometry. The load-carrying capability or residual strength of a structure is significantly affected by the presence of cracks and it is substantially lower than the strength of the undamaged structure. The residual strength diagram (the fracture stress as a function of crack length or time) can be used for the purpose of planning the inspection interval prior to the total loss of the structural part during its service life. For single load path structure, the residual strength capability of the structure or the ability of the structure to tolerate damage can be determined with ease. For fail-safe or multiple load path structures or structures with a crack arrest design, such as stiffened structure, the construction of a residual strength diagram requires calculation of a failure stress at which partial failure occurs when a load path fails and determination of whether the redistributed load can be tolerated by the remaining structure.

Almost all low-to-medium strength structural steels that are used in constructing bridges, ships, pressure vessels, etc. and many aluminum alloys used in the design of aircraft and space structure have sufficient ductility and under normal rate of loading form a large plastic zone at the crack tip. The R-curve method [10],

---

[1] The terms "limit stress", "applied stress", "gross stress", and "operating stress" all have the same meaning and are used interchangeably throughout the literature. No factor of safety is associated with these values.

an extension of LEFM into elastic–plastic fracture mechanics, can be utilized to describe the crack instability when the plastic zone size is large and substantial stable crack growth is expected prior to catastrophic failure (tearing type fracture). The construction of the R-curve for material with considerable plasticity at the crack tip is well documented in ASTM E-561. It is shown that the R-curve can be described as a plot of material crack growth resistance (expressed in units corresponding to K_c associated with the plane stress fracture toughness) as a function of physical crack movement, ΔC. Just as in plane strain, K_{Ic}, fracture toughness testing, the procedure for conducting K_R tests are complex and tedious. The experimental values of the fracture toughness for all thickness ranges are essential in evaluating the life expectancy of the cracked structure. The crack tip stress intensity concept and its derivation, an in-depth discussion of fracture toughness for both surface and through cracks, crack tip plasticity, and plastic zone shape based on different yield criteria are covered in Chapter 3.

The presence of surface or embedded cracks in a pressure vessel containing hazardous gases or fluids can produce catastrophic failure when the induced stress due to applied pressure overcomes the residual strength capability of a vessel during its service life, or while it is subjected to proof test. The failure can occur when the critical stress intensity factor, K, of the surface crack exceeds the surface fracture toughness of the material, K_{Ie}. When the depth of the growing surface crack becomes equal to the thickness of the vessel prior to catastrophic burst failure, the surface crack becomes a through crack and the container will leak. The leakage of a vessel prior to its final failure (known as leak-before-burst failure mode) is desirable, since it can be a warning signal for needed inspection and consequently may prevent catastrophic failure by repairing the container. Chapter 3 provides a complete description of the leak-before-burst failure mode (when $K < K_{Ie}$ and $K \geqslant K_c$) and the failure criteria defining catastrophic fracture when $K \geqslant K_{Ie}$.

A new approach, called the Theory of Fracture Mechanics of Ductile Metals (FMDM), that could satisfactorily correlate the fracture test data with the analysis was first developed in the 1960s by the authors of this book [11]. The theory assumes the existence of two distinct plastic zones (Fig. 1.1) at the crack tip that absorb energy as the crack advances. That is, the localized plastic deformation is composed of: (1) the local strainability at the crack tip (the region of highly plastic deformation) and (2) the uniform strainability near the crack tip. In other words, the fracture characteristics of a metal are directly related to the strainability of the metal.

In the FMDM theory, the total energy per unit thickness absorbed in plastically straining the material around the crack tip, U_P, was described in terms of plastically straining the material beyond the ultimate at the crack tip, U_F, and below the ultimate of the material near the crack tip, U_U, where $U_P = U_F + U_U$. Thus, fracture can be characterized by two parameters shown to be determinable from the uniaxial stress–strain curve.

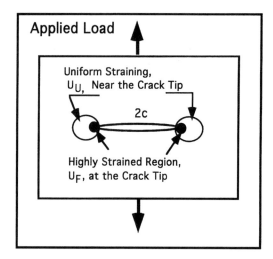

Figure 1.1 The Crack tip plastic zones as viewed by the FMDM theory.

The FMDM theory was later used to develop analytically the plane strain and plane stress fracture toughness for different isotropic materials [12]. The conventional approach for obtaining the K_c variation with respect to thickness is to conduct several tests with cracked plates of different thicknesses and crack lengths according to ASTM procedures. These tests are costly and time consuming. The fracture toughness value provided by the FMDM theory depends on a stress–strain curve that is readily available in MIL-HDBK-5 and other reliable sources or can be generated in the laboratory for the material under consideration. The results of the computed fracture toughness generated by the FMDM approach for several aerospace alloys were shown to be in excellent agreement with the test data. A detail description pertaining to the FMDM approach is given in Chapter 6.

Failure of structures under working environments seldom occurs by static loading. The majority of unexpected structural failures are due to fluctuating or cyclic loads where the preexisting flaw will grow undetected in a stable manner and finally reach its critical length. Generally speaking, fatigue can be defined as an accumulative failure of a part under repeated or fluctuating loads. While fatigue failures start from microscopic cracks, the presence of tool or grinding marks left on the surface of the part make the formation of fatigue cracks easier. Fatigue failure occurs at calculated applied nominal stresses considerably below the tensile strength of the materials involved, and although the materials have sufficient ductility, the failure generally shows little or no ductility. The sequence in which fatigue failure occurs consists of three parts: microscopic crack nucleation/initiation along the slip surfaces (stage I), stable crack growth perpendicular to the applied load (stage II), and final rupture, where the crack will propagate in an

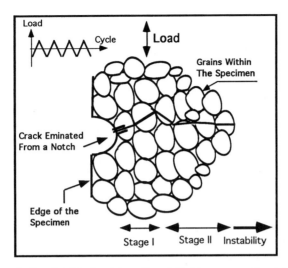

Figure 1.2 Crack initiation (stage I) and Stable propagation (stage II).

unstable manner. Figure 1.2 illustrates the crack initiation and crack growth stages where the crack is nucleated from a notch (a source of stress concentration).

August Wöhler (1819-1914), the railway engineer who became director of the German Imperial Railway (1847-1889), was the first investigator to address fatigue failure by conducting cyclic loading tests on full-scale railway axles, as well as on small-scale specimens [13]. He plotted his test data in terms of applied stress versus the number of cycles to failure. This type of plot became known as the S–N diagram or Wöhler line. Later, the Wöhler S–N diagram was used in other applications, such as bridges, ships, aircraft, and machinery equipment, that are also subjected to fluctuating loads.

The S–N curve is a useful tool for assessing the total life of a structural part when failure occurs under a relatively large number of cycles, and stresses and strains in the bulk of the material are within the elastic range of the material. The results of fatigue tests conducted in the laboratory in accordance with the ASTM E-466 procedure can be presented in tabular form or in the form of S–N curves. From the collected S–N data, the total life expectancy of the structural part can be determined and compared with the total number of cycles that the structure is exposed to during its service life. When the structural part is subjected to several load environments having cycles of different stress magnitude, the partial damage due to each loading environment must be calculated for the total life evaluation. The ratio of the sum of all cycles, Σn_i, to the total cycles to failure, N_{fi}, are then compared ($\Sigma n_i / N_{fi} = n_1/N_{f1} + n_2/N_{f2}$ ...) using the classical linear damage theory proposed by Palmgren and Miner [14, 15]. The total life assessment of a machinery part, using the S–N diagram is an acceptable approach, provided that the service conditions of the part under study are parallel to the test conditions conducted in the laboratory. This is known as the similitude law. That is, the life of a structural part is the same as the life of a test specimen if both have undergone the same loading environment (Chapter 2).

If the magnitude of the fluctuating stress is no longer in the elastic range of the material, significant plastic straining occurs throughout the body, especially in the highly localized areas at stress concentration regions and the number of cycles to failure is expected to be relatively low. Low cycle fatigue failure, sometimes referred to as the strain-life approach (ε–N), can no longer be characterized by an S–N curve. The number of cycles to crack failure in the region of plastic deformation immediately adjacent to the notch can be estimated by a strain-life prediction model using the Neuber relationship [16] and the cyclic stress-strain curve conducted under strain-controlled conditions. Chapter 2 contains a comprehensive review of conventional fatigue, including high and low cycle fatigue, together with several example problems.

To prevent fatigue failure of a structural part in a load-varying environment, it is important to have a good understanding of (1) all loading events that a component will experience and the number of times that each event will occur (the fatigue spectrum); (2) an empirical equation that can relate the fatigue crack growth rate, da/dN, with the crack tip stress intensity, ΔK; (3) the material fracture toughness; and (4) some estimate of the initial flaw size. The function f (relating da/dN with ΔK) can be obtained as the result of laboratory test data and can then be utilized to solve crack growth problems where the structural part has undergone the same loading conditions. Having the above information available to the analyst, the remaining life cycles (number of cycles for a crack to grow from its initial length, a_i, to its final length, a_f) can be calculated.

The earliest and most well known relation between the crack growth rate (da/dN) and the stress intensity factor range (ΔK) was given by Paris, Gomez, and Anderson [17, 18]. The Paris "law" (an empirical equation based on the experimental data obtained through laboratory testing) was later modified to account for other parameters, such as stress ratio, threshold and critical stress intensity factors, retardation effect, etc. Probably one of the most useful crack growth rate equations is given in the NASA/FLAGRO computer program [19], the current state-of-the-art computer code that is being used in the aerospace industry for the safe-life analysis of space structural parts. The constant amplitude crack growth rate relationship (da/dN, ΔK) for all regions of the crack growth rate curve (including the threshold stress intensity factor, ΔK_{th}, and the critical regions, K_c) are presented in the NASA/FLAGRO computer code. The Newman's crack closure phenomenon approach [20] is implemented in the computer code to account for different stress ratio, R, under constant amplitude loading. Chapter 4 provides a detailed description of the crack growth rate concept, including a discussion on the NASA/FLAGRO computer code, together with several example problems pertaining to crack growth rate prediction for aerospace structural parts.

In constant amplitude loading conditions where there is no interaction among cycles (baseline fatigue crack growth condition when $dR/dt = 0$, and $R = \sigma_{min}/\sigma_{max}$), the crack growth rate analysis is relatively simple to assess. However, the actual cyclic loading that most structures experience during their service life is by no means of constant amplitude and the crack growth rate delay or acceleration due to load interactions between low–high or high–low cycles must be taken into consideration (Fig. 1.3). For example, when aircraft wings are subjected to gust and maneuver loads, the tensile overload or peak load forms a tensile plastic zone at the crack tip larger than the subsequent constant amplitude cycles (from high gust load to low load). Upon load release, the overload-induced plastic zone causes a crack tip compressive stress and delay in the crack growth rate, da/dN. Use of the constant amplitude crack growth rate equations to express variable amplitude loading will give conservative results when the number of cycles to failure of the structural part is of interest. Two well-known mathematical closure models, called the Wheeler and Willenborg models, based on yield zone [21, 22], are presently available and are both discussed in Chapter 4.

Many structural parts used in assembling aircraft and space vehicles are either welded together or mechanically fastened as bolted or riveted joints. A joint can be viewed as a source of stress concentration that can shorten the life of the jointed parts unless preventative measures are taken to minimize structural failure. Fatigue failure of bolts usually occurs at the threaded location where the bolt and nut are engaged. In some cases (less probable), bolt failure is observed in the shank to bolt head area. Experimental investigations have shown that thread processing done by rolling the threads improves the fatigue life of the part. Rolling

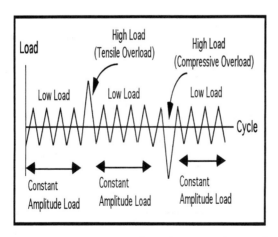

Figure 1.3 Low-high-low tensile and compressive overloads.

threads will create compressive residual stresses at thread roots that help to prevent fatigue failure of bolts in a joint. A tensile preloaded bolt is also considered to be a desirable practice in reducing the fatigue failure and increasing the life of a bolt when it is exposed to cyclic loading. Another area of concern when dealing with a bolted joint is fatigue cracking that may initiate from the hole where the two plates are bolted together (a source of stress concentration, Fig. 1.4). The initiated crack from one of the holes in the joint may grow to its critical size due to the fluctuating load environment and cause the complete separation of the two structural parts (Chapter 4).

To maintain trouble-free damage tolerant structural hardware, a multidisciplinary process is required that starts in the early stage of design and continues through manufacturing and into the operational phase. The implementation of this process requires the execution of a fracture control plan that can control and prevent damage due to the preexisting flaws in the structure. The required fracture control procedures are based heavily upon the good engineering and manufacturing practices already embedded in the hardware development process. Fracture control imposes additional engineering and product assurance requirements needed to ensure the structural integrity of high-risk or fracture critical structures throughout all phases of the component's lifetime. The design philosophy, material selection, analysis, testing, inspection, and manufacturing are all elements of the fracture control plan that will contribute to building and maintaining a trouble-free space structure. Section 1 of Chapter 5 briefly describes the implementation approach to a fracture control plan for a man-rated space structure. The fracture control plan methodology used in this chapter is also applicable to other industries where safety is the primary concern.

For manned aircraft or space vehicles, it is commonly required to assume cracks preexist in all structural parts. These cracks shall not grow to their critical

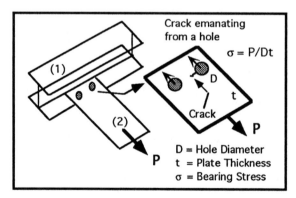

Figure 1.4 Crack emanating from a hole where the two structures (1 and 2) are jointed together.

size when subjected to crack growth analysis at a specified load during their usage period. In general, the material as received from the vendor will contain defects of small size, such as porosity, scratches, inclusions, microcracks, and machine marks. These inherent flaws are considerably smaller than the Nondestructive Inspection (NDI) capability to detect them and will not grow appreciably in service. There are numerous NDI methods utilized for flaw detection in structural components. The most prevalent of these NDI techniques, commonly used in the detection of flaws in aerospace components, are liquid penetrant, magnetic particle, eddy current, ultrasonic, and radiography. Section 2 of Chapter 5 provides a detail description of different NDI techniques that are currently used in the aerospace and aircraft industries.

References

1. E. R. Parker, "Brittle Behavior of Engineering Structures," John Wiley & Sons, 1957, pp. 253–271.
2. R. P. Reed, J. H. Smith, and B. W. Christ, "The Economic Effects of Fracture in the United States," SP 647-1, NBS, March 1983, p. 3.
3. A. A. Griffith, "The Phenomena of Rupture and Flow in Solids," Philos. Trans., R. Soc. Lond., Ser. A., Vol. 221, 1920, p. 163.
4. C. E. Inglis, "Stresses in a Plate due to the Presence of Cracks and Sharp Corners," Proc. Inst. Naval Architects, March 14, 1913.
5. G. R. Irwin, "Analysis of Stresses and Strains Near the End of a Crack Traversing a Plate," Trans. ASME, J. Appl. Mech., Vol. 24, 1957, p. 361.
6. G. R. Irwin, "Fracture Dynamics," Fracture of Metals, ASM, 1948, pp. 147–166.
7. E. Orowan, "Fracture and Strength of Solids," Rep. Prog. Physics, Vol. 12, 1949, pp. 185–232.
8. G. R. Irwin, Fracture Handbuch der Physik, Springer-Verlag, Heidelberg, VI, 1958, pp. 551–590.
9. D. S. Dugdale, "Yielding of Steel Sheets Containing Slits," J. Mech. Phys. Solids, Vol. 8, 1960, p. 100.
10. Fracture Toughness Evaluation by R-Curve Method, ASTM STP 527, edited by D. E. McCabe, American Society for Testing and Materials, 1973.
11. G. E. Bockrath and J. B. Glassco, Fracture Mechanics of Ductile Metals. California State University, Long Beach, revised 1985.
12. B. Farahmand and G. E. Bockrath, "A Theoretical Approach for Evaluating the Plane Strain Fracture Toughness of Ductile Metals," Engin. Fract. Mech., Vol. 53, No. 6, March 1996.
13. A. Wöhler, "Wöhler's Experiments on the Strength of Metals," Engineering, August 23, 1967, p. 160.
14. M. A. Miner, "Cumulative Damage in Fatigue," Trans. ASME, J. Appl. Mech., Vol. 67, September 1945, p. A159.

15. H. Neuber, Kerbspannungslehre, Springer, (Berlin), 1958: Translation Theory of Notch Stresses, U.S. Office of Technical Services, 1961.
16. A. Palmgren, "Ball and Roller Bearing Engineering," translated by G. Palmgren and B. Ruley, SKF Industries, Inc., Philodelphia, 1945, pp. 82–83.
17. P. C. Paris, M. P. Gomez, and W. E. Anderson, "A Rational Analytic Theory of Fatigue," the Trend Engin., Univ. Wash., 13 (1), 1961, pp. 9–14.
18. P. C. Paris, "The Fracture Mechanics Approach to Fatigue" in Syracuse University Press, 1964, pp. 107–132.
19. Fatigue Crack Growth Computer Program "NASA/FLAGRO," Version 2.0, Document Number JSC-22267A, January 1993.
20. J. C. Newman, Jr. "A Crack Opening Stress Equation for Fatigue Crack Growth," Int. J. Fract., Vol. 24, No. 3, March 1984, pp. R131–R135.
21. O. E. Wheeler, "Spectrum Loading and Crack Growth," J. Basic Engin., 94D, 1972, pp. 181–183.
22 J. D. Willenborg, R. M. Engle, and H. A. Wood, "A Crack-Growth-Retardation Model Using an Effective Stress Concept," AFFDL-TM-71-1 FBR, 1971.

Chapter 2
Conventional Fatigue (High- and Low-Cycle Fatigue)

2.1 Background

It is known that when metals are subjected to fluctuating load, the failure occurs at a stress level much lower than the fracture stress corresponding to a monotonic tension load.[1] With the development of the railway in the nineteenth century, the fatigue failure of railway axles became a problem and much attention was given to the understanding of the fatigue failure phenomenon. To understand fatigue failure mechanism induced by repeated loading, full-scale as well as small-scale fatigue tests were conducted in the laboratory. In 1852, the German railway engineer August Wöhler (Director of Imperial Railways in Germany from 1847 to 1889), conducted several constant amplitude fatigue tests on full and small-scale railway axles. The results of this work [1] were presented in the form of plots of the failure stress as a function of the number of cycles to failure. This plot is a useful tool for the total life prediction of a part subjected to constant amplitude cyclic loading and is known as the Wöhler S–N diagram. The Wöhler approach was extended to other areas of concern, such as bridges, ships, and machinery equipment that undergo repeated loading. The S–N approach is still a useful tool to assess fatigue failure of many modern structures that are subjected to repeated loading, where the applied stress is under the elastic limit of the material and the number of cycles to failure is large. When material failure occurs under a relatively large number of cycles, and stresses and strains are within the elastic range of the material, the failure mechanism is called high-cycle fatigue. If the magni-

---

[1] Monotonic tension load is defined as the application of a single load to failure during the lifetime of the member.

tude of the fluctuating stress is no longer in the elastic range of the material, significant plastic straining occurs throughout the body, especially in the highly localized areas at stress concentration sites, and the number of cycles to failure is expected to be relatively short. This failure mechanism is referred to as low-cycle fatigue. Low-cycle fatigue failure, sometimes referred to as the strain-controlled or strain-life (ε-N) approach, can no longer be characterized by an S-N curve. Low-cycle fatigue life is usually associated with a number of cycles to failure between 100 and 10,000 cycles (depending on material strength and ductility) and for high-cycle fatigue the number is above 10,000 cycles. The results of low cycle fatigue tests can be important in the design and failure analysis of industrial hardware when they are subjected to mechanically or thermally induced repeated strain where failure occurs in relatively short cycles (less than 10,000 cycles). For example, aircraft components can be subjected to high mechanically induced cyclic strain during severe gust and maneuvering load environment. Jet engines, nuclear reactor parts, and pressure vessels are example of thermally induced cyclic strain.

In the preliminary stage of design, when a new vehicle or structure is under consideration to perform a given mission, there are several factors that the designers should consider. Trade studies for different design configurations should be performed and, based on strength and weight considerations among others, an optimum configuration is selected. The selected design must withstand the environment in question without failure. Therefore, a comprehensive failure analysis (static, dynamic, and fatigue) is necessary to ensure the integrity of the structure under study. There are two primary groups of information that are necessary as an input for a comprehensive fatigue analysis. One group of information is the data related to the material behavior when subjected to cyclic loading, such as the laboratory tests for constructing the S-N diagram, modified Goodman or Gerber diagrams, and other factors that would help to evaluate the life of the structure. The laboratory tests must simulate the stress environment that the structural component experience. The stress environment could be inertia, thermal, pressure, sonic, or other environmentally induced stresses. The second group of information is the determination of the total number of cycles that the structure will undergo throughout its life (the life cycle or service history of the structure). This is usually presented in terms of stress environment versus time. Determination of life cycle is discussed in Section 2.3. With these two groups of information provided to the engineer, a complete fatigue analysis is possible. If the analysis reveals that the structure does not have sufficient life, a redesign might be considered. Alternatively, it is often helpful to reexamine the analysis as a whole to see if it is possible to reduce the degree of conservatism in the assumptions made.

The S-N curve, in addition to estimation of the structural life, is also useful for evaluating the following cases:

1. To study the differences in fatigue behavior between two or more materials when subjected to a given fluctuating load environment
2. For a given alloy, to select the heat treatment conditions that would give the best fatigue results when subjected to a given fluctuating load environment
3. To study the fatigue behavior of a given material when it is subjected to mechanical working
4. To study the fatigue behavior of a given material with respect to its material orientation
5. To study the effect of stress concentration on the fatigue behavior.

Having the S–N curve and load spectrum available, a third concept known as cumulative damage theory must be introduced that relates cycles of different magnitudes to the S–N curve in order to predict the total life of the structural component. Section 2.6 describes the classical linear damage theory proposed by Palmgren and Miner [2, 3] to estimate failure when summation of the cycles' ratio for each event becomes equal to 1.

The number of cycles to failure in the region of high plastic deformation, immediately adjacent to the notch, can be estimated by the strain-life prediction model via the Neuber relationship [4] and cyclic stress–strain curve conducted under strain-controlled conditions. Sections 2.8 through 2.11 discuss the development of the cyclic stress–strain curve by connecting the tips of stable hysteresis loops obtained through testing. Each hysteresis loop represents one complete load-unload cycle of a constant strain amplitude. The strain-life (ε-N) prediction model and relating local stress and strains to the far field applied stress and strains by the Neuber relationship, are discussed in Section 2.12.

In Section 2.2, different types of cyclic loading occurring in real structures are defined. These fluctuating loads may have constant amplitude or may vary randomly throughout the service life of the structure (as shown in Fig. 2.1).

2.2 Cyclic or Fluctuating Load

Typical forms of cyclic loading that occur in real structures are almost random in nature and vary in magnitude during their service life (Fig. 2.1a). In constant amplitude cyclic loading, where the amplitude and mean stress stay constant, there are five stress parameters that can define the loading characteristics (Fig. 2.1b). These characteristics are: cyclic stress amplitude S_a; mean stress S_M; maximum stress S_{max}; minimum stress, S_{min}; and the stress ratio, R. Any two of the above quantities are sufficient to completely define the cyclic loading.

The mean or steady-state stress (S_M) is the average algebraic sum of the maximum (S_{max}) and minimum (S_{min}) cyclic stresses. The alternating or variable stress amplitude, S_a, is defined as:

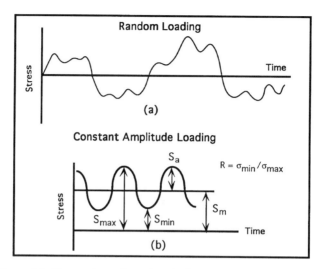

Figure 2.1 Illustrations of random and constant amplitude loading

$$S_a = (S_{max} - S_{min})/2 \qquad (2.1)$$

The stress ratio, R, is also an important parameter and is defined as the algebraic ratio of the minimum to maximum cyclic stresses. R is widely used to distinguish different constant amplitude cyclic loading conditions in fatigue analysis. Another two less commonly used parameters in fatigue loading are the algebraic ratio of the stress amplitude to the mean stress ($A = S_a/S_M$) and the strain ratio $R_\varepsilon = \varepsilon_{min}/\varepsilon_{max}$ during a complete cycle. The strain ratio, R_ε, is used when dealing with strain-controlled low cycle fatigue where total strain $\Delta\varepsilon$ is controlled throughout the cycle.

The load case presented in Fig. 2.2a is a fully reversed sinusoidal shape stress cycle and has a stress ratio $R = -1$. In this case, the maximum and minimum cyclic stresses are equal. For the same range, the case of $R = -1$ is considered to be a less damaging cyclic load case when evaluating the fatigue life of a structure. An example of the $R = -1$ load case is a rotating–bending test that uses four–point loading to apply a constant moment to a rotating cylindrical specimen, where the maximum and minimum stresses are equal but of opposite sign.

Figure 2.2b presents the cyclic stress case for the stress ratio $R = 0$, where the minimum stress is zero. An example of this case is the pressurization and depressurization cycle of a pressurized tank, where the maximum and minimum induced stresses are related to the maximum and minimum pressure. Figure 2.2c is for the loading condition where both the maximum and minimum cyclic stresses are positive ($0 < R < 1$). A preloaded bolt subjected to cyclic tensile stresses such that the maximum and minimum fatigue stresses are positive represents this case. For

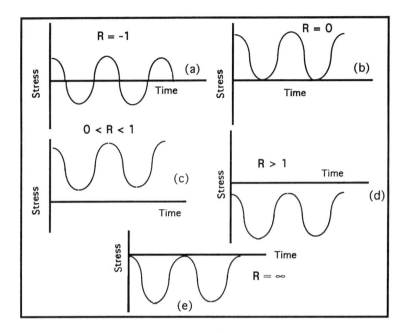

Figure 2.2 Typical constant amplitude loading cycles

the same stress range value, the case of $R > 0$ is considered to be the most damaging cyclic load case when evaluating the fatigue life of a structure.

Figure 2.2d shows the loading condition for the case of $R > 1$. A plate with a hole that has undergone a sleeve cold expansion or mandrelizing process [5] and subjected to fluctuating load is an example of this case. The mandrelizing process creates a massive zone of compressive residual stress field ($S_M < 0$) around the hole up to the compressive yield strength of the material (see Fig. 2.3). The compressive stresses are a result of applied tensile load (when it is above the tensile

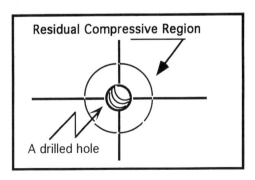

Figure 2.3 Residual compressive stress caused by the cold sleeve process

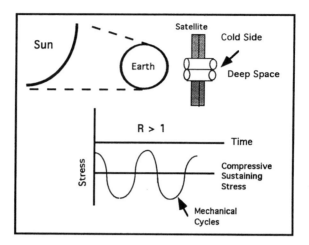

Figure 2.4 Illustration of the case of $R > 1$ for a satellite orbiting the earth

yield of the material) that causes the surrounding elastic material upon unloading to induce compressive residual stress in the plastic region local to the hole. The residual compressive stresses extend fatigue and crack growth lives of the part. Note that the static load case represents the condition of stress ratio $R = 1$.

The case of a satellite orbiting around the earth is another example illustrating the case of $R > 1$. For a satellite orbiting around the earth, at a certain position in its orbit (Fig. 2.4) not exposed to the sun, the induced thermal stress is compressive. The induced compressive stress can be thought of as a compressive sustained load. The mechanical cycling that occurs at the same time (having much smaller wavelength) together with the sustaining compressive load ($S_M < 0$) will create a load cycling that would be similar to the case shown in Fig. 2.2d. The case of $R = \infty$ is a variation of $R > 1$, where $S_{max} = 0$, as shown in Fig. 2.2e.

The random loading case shown in Fig. 2.1a is complex and it may contain any combination of the above mentioned cyclic stresses. The random loadings that involve more than one amplitude and mean stress are more representative of what a structure may be subjected to. An example of this case is the loading environment that an aircraft or space structure will experience during its lifetime. Table 2.1 shows the stress ratio and the corresponding maximum and minimum stresses, together with an example that represents the pertinent loading condition.

2.3 Fatigue Spectrum

Load, stress, or fatigue spectrum is the engineering definition of the fatigue environment that a component experiences throughout its design life and is defined by the load (or stress) amplitude versus the number of cycles. The fatigue spec-

2.3 Fatigue Spectrum

Table 2.1 Stress ratio and representative load description

Stress ratio R	Loading case	Load description
$R = -1$	S_{max} is positive, S_{min} is negative with $S_M = 0$ (fully reversed)	Rotating shaft without the overload
$R = 1$	Static loading	Static loading
$R = 0$	S_{max} is positive and $S_{min} = 0$	Pressurization and depressurization of a tank
$0 < R < 1$	S_{max} and S_{min} are both positive (with positive S_M)	Preloaded bolt subjected to fully reversed load
$R > 1$	S_{max} and S_{min} are both negative (with negative S_M)	Mandrelized hole subjected to fluctuating load
$R = \infty$	S_{max} is equal to zero and S_{min} is negative	Mandrelized hole under fluctuating load with $S_{max} = 0$

trum should be available early in the engineering phase to permit initiation of design and analysis activities. This section provides an overview of how a fatigue spectrum is developed.

The most important step in developing a fatigue spectrum is the definition of all loading events that a component will experience and the number of times that each event will occur. Completion of this step requires a comprehensive understanding of the component under consideration, its role in the system being designed, and the design requirements. For example, an aircraft may be required to perform 1000 missions during its service life. For an aircraft, each mission profile may be divided into distinct events, such as taxi to takeoff, ascent, cruise, descent, landing, and taxi after landing (Fig. 2.5a). The fatigue spectrum for each event may be a function of several variables. For example, the wing load magnitude and its cyclic behavior during ascent depends on speed, weight, gust factor, etc. Figure 2.5b shows a simplified version of a complex load spectrum that the wings of a transport aircraft are exposed to during each mission.

Contractual or top level system requirements usually specify operations and/or conditions to which the system is exposed and the frequency of this exposure. It is usually the analyst's responsibility to identify the relevant system level requirements for each component and to establish the variation of load or stress versus time (time history). If the component responds statically to a loading event, the load/stress time history can be established with relative ease. For example, consider a spur gear that is designed to actuate the wing flaps of an aircraft during ascent and would undergo a fixed number of complete revolutions under a constant torque. In this case, the stress time history for a typical gear tooth during each ascent event can be calculated by using the equations of static equilibrium.

In many cases where flexible structures and rapidly changing loads are involved, however, the component/system responds dynamically to the loading

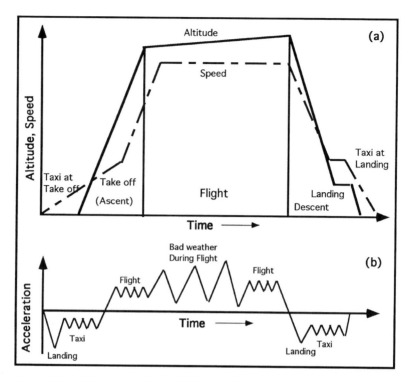

Figure 2.5 (a) **Mission profile and (b) simplified load spectrum for an aircraft**

environment. In these cases, establishing the time history would require dynamic analysis and the cyclic behavior of the time history would be a function of both the load variation and the dynamics of the structure. For example, in the absence of a shock absorber, the landing gear of an aircraft would experience large load vibrations during touchdown (Fig. 2.6).

Once the time history for each event is established using a cycle counting procedure, each time history is converted to a fatigue spectrum for that event consisting of load range/mean range versus number of cycles (range is defined as the algebraic difference between successive valley and peak loads; see Fig. 2.7). A variety of cycle counting procedures are available. One of the most commonly recognized and widely used cycle counting approaches is the rain flow method. Several variations on the rain flow method for specialized applications have been discussed in the literature [6, 7]. Other cycle counting methods are peak counting, level crossing, and range-pair counting. In all of the aforementioned cases, the irregular load sequence can be converted to a sum of cycles, N_i, with different stress amplitudes, S_i, to assess the total damage induced on a given part. It is important to note that the prediction of fatigue life by one technique (for example,

2.3 Fatigue Spectrum

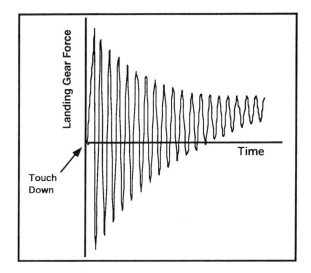

Figure 2.6 An undamped landing gear force time history during touch down

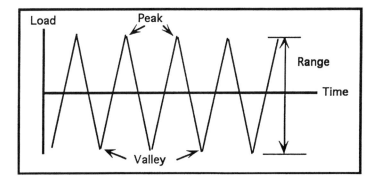

Figure 2.7 A typical fatigue cycle with fatigue loading parameters (range, peak, valley)

the level crossing method) may produce results that differ by an order of magnitude from the others.

The following 6 steps are the step-by-step procedures for using a standard rain flow method technique that are directly extracted from the ASTM E-1049. Figure 2.8 is used here to clarify the procedures defined in steps 1 through 6. Note that this procedure should be preceded by identifying the local peaks and valleys in the time history. Let X denote the range under consideration; Y, the previous range (see Fig. 2.7 for the definition of range) adjacent to X; and S, the starting point in the history. The starting point, S, in the load history shown in Fig. 2.8a is A and the ranges $X = |B - C|$ and $Y = |A - B|$.

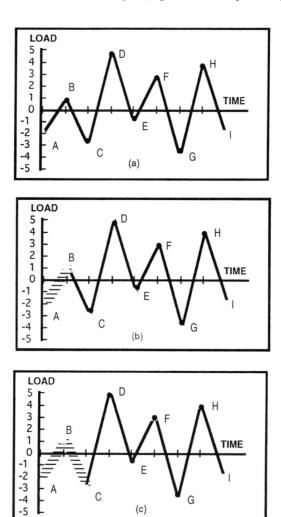

Figure 2.8 Rain flow cycle counting technique

1. Read the next peak or valley. If out of data, go to Step 6.
2. If there are less than three points, go to Step 1. Form ranges X and Y using the three most recent peaks and valleys that have not been discarded.
3. Compare the absolute values of ranges X and Y.
   (a) If $X < Y$, go to Step 1.
   (b) If $X \geq Y$, go to Step 4.
4. If range Y contains the starting point S, go to Step 5; otherwise, count range Y as one cycle; discard the peak and valley of Y; and go to Step 2.

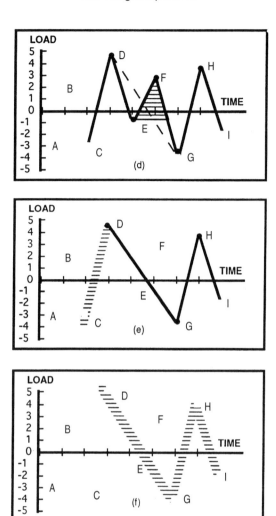

Figure 2.8 *(cont.)*

5. Count range Y as one-half cycle; discard the first point (peak or valley) in range Y; move the starting point to the second point in range Y; and go to Step 2.
6. Count each range that has not been previously counted as one-half cycle.

Example 2.1

By following rain flow cycle count process described by Steps 1 through 6, calculate the fatigue spectrum for the load time history shown in Fig. 2.8.

Solution

Step 1. The starting point is $S = A$ as shown in Fig. 2.8a and the ranges are $X = |B - C|$ and $Y = |A - B|$, where $X > Y$ (Y contains the starting point $S = A$). Count $Y = |A - B|$ as one-half cycle. The new starting point is $S = B$. Note that in Fig. 2.8b the discarded quantity $Y = |A - B|$ is shown by a dashed line which indicates that it was counted as one-half cycle.

Step 2. The new starting point is $S = B$, and the ranges $X = |C - D|$ and $Y = |B - C|$, where $X > Y$, (see Fig. 2.8b). Count $Y = |B - C|$ as one-half cycle. In Fig. 2.8c, the discarded quantity $Y = |B - C|$ is shown by a dashed line.

Step 3. The starting point is $S = C$. The new ranges are $X = |D - E|$ and $Y = |C - D|$ where $X < Y$, go to the next range.

Step 4. The next ranges are $X = |E - F|$ and $Y = |D - E|$, where $X < Y$, go to the next range.

Step 5. The next ranges are $X = |F - G|$ and $Y = |E - F|$, where $X > Y$. In this case, count $Y = |E - F|$ as one full cycle, as shown in Fig. 2.8d. Points E and F are discarded and they are shown by the dashed area illustrated in Fig. 2.8d.

Step 6. The new starting point is the same as in step 3, $S = C$, and $X = |D - G|$ and $Y = |C - D|$, where $X > Y$. Count $Y = |C - D|$ as one-half cycle. The new starting point is $S = D$ and the next ranges are $X = |G - H|$ and $Y = |D - G|$ where $X < Y$. Count $Y = |D - G|$ as one-half cycle (Fig. 2.8e).

Step 7. The new starting point is $S = G$ and $X = |H - I|$ and $Y = |G - H|$, where $X < Y$. Count $|G - H|$ as one half cycle and $|H - I|$ as one-half cycle (end of counting), Fig. 2.8f. The resulting fatigue load spectrum is shown in Table 2.2. In many engineering applications, the spectrum for a typical (or in some cases the worst) loading event is calculated and the off nominal events are developed by applying statistical distributions to the range and mean portion of the calculated nominal spectrum. Such an approach results in the redistribution of cycles among load ranges and means while maintaining the total number of cycles the same.

The cycle count for each individual event is multiplied by the number of expected occurrences of that event in the life cycle of the component and is added to all other expected events to form the fatigue spectrum for the component.

Example 2.2

(a) Use the rain flow technique to calculate the fatigue spectrum for the load time history shown in Fig. 2.9 and the associated Table 2.3. (b) Calculate the fatigue spectrum for a component that would experience 100 events characterized by Fig. 2.8 and 200 events characterized by Fig. 2.9 during its life cycle. (c) Calculate the fatigue spectrum for a component that would experience 200 events that have a uniform distribution between 2 and 5 for its positive peak load and is characterized by a typical time history shown in Fig. 2.9.

2.3 Fatigue Spectrum

Table 2.2 Fatigue spectrum for time history of Example 2.1

Load range	Mean load						
	−3.0	−2.0	−1.0	0.0	1.0	2.0	3.0
10							
9				0.5 (DG)			
8				0.5 (GH)	0.5 (CD)		
7							
6					0.5 (HI)		
5							
4			0.5 (BC)		1.0 (EF)		
3				0.5 (AB)			
2							
1							

Solution

Part a. Prior to applying the rain flow procedure the local peaks and valleys are identified and labeled A through S in Table 2.3. The rain flow procedure is then applied to the peaks and valleys. A summary of each step is given here:

1. $S = A$, $\quad X = |B - C|, Y = |A - B|, X > Y, Y$ includes S ∴ eliminate A, $AB = .5$ cycle.
2. $S = B$, add D, $X = |C - D|, Y = |B - C|, X < Y$.
3. $S = B$, add E, $X = |D - E|, Y = |C - D|, X > Y, Y$ excludes S ∴ eliminate CD, $CD = 1$ cycle.
4. $S = B$, add F, $X = |E - F|, Y = |B - E|, X > Y, Y$ includes S ∴ eliminate B, $BE = .5$ cycle.
5. $S = E$, add G, $X = |F - G|, Y = |E - F|, X < Y$.
6. $S = E$, add H, $X = |G - H|, Y = |F - G|, X < Y$.
7. $S = E$, add I, $\ X = |H - I|, Y = |G - H|, X > Y, Y$ excludes S ∴ eliminate GH, $GH = 1$ cycle.
8. $S = E$, $\quad X = |F - I|, Y = |E - F|, X < Y$.
9. $S = E$, add J, $X = |I - J|, Y = |F - I|, X < Y$.
10. $S = E$, add K, $X = |J - K|, Y = |I - J|, X < Y$.
11. $S = E$, add L, $X = |K - L|, Y = |J - K|, X < Y$.
12. $S = E$, add $M, X = |L - M|, Y = |K - L|, X > Y, Y$ excludes S ∴ eliminate KL, $KL = 1$ cycle.

26   Chap. 2 Conventional Fatigue (High- and Low-Cycle Fatigue)

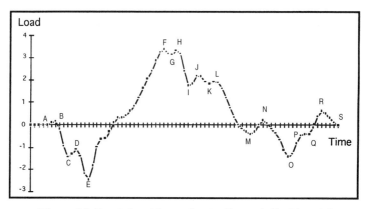

Figure 2.9 Load time history for the Example Problem 2.2

Table 2.3 Component load time history for Example 2.2

	Time	Force		Time	Force		Time	Force
(A)	0.04	0.00		1.04	0.85		2.04	0.03
	0.08	0.00		1.08	1.20		2.08	−0.17
	0.12	0.00		1.12	1.64		2.12	−0.29
	0.16	0.00		1.16	2.04	(M)	2.16	−0.41
	0.20	0.01		1.20	2.41		2.20	−0.30
	0.24	0.11		1.24	2.86		2.24	0.02
(B)	0.28	0.17		1.28	3.26	(N)	2.28	0.22
	0.32	−0.19	(F)	1.32	3.38		2.32	0.09
	0.36	−0.94		1.36	3.19		2.36	−0.15
(C)	0.40	−1.44	(G)	1.40	3.14		2.40	−0.34
	0.44	−1.32	(H)	1.44	3.34		2.44	−0.66
(D)	0.48	−1.08		1.48	3.18		2.48	−1.10
	0.52	−1.40		1.52	2.40	(O)	2.52	−1.41
	0.56	−2.17	(I)	1.56	1.75		2.56	−1.28
(E)	0.60	−2.45		1.60	1.84		2.60	−0.82
	0.64	−1.83	(J)	1.64	2.18		2.64	−0.45
	0.68	−0.98		1.68	2.17	(P)	2.68	−0.41
	0.72	−0.63		1.72	1.91	(Q)	2.72	−0.42
	7.76	−0.58	(K)	1.76	1.82		2.76	−0.10
	0.80	−0.29	(L)	1.80	1.91		2.80	0.39
	0.84	0.12		1.84	1.87	(R)	2.84	0.62
	0.88	0.32		1.88	1.61		2.88	0.53
	0.92	0.33		1.92	1.20		2.92	0.33
	0.96	0.45		1.96	0.77		2.96	0.14
	1.00	0.63		2.00	0.35	(S)	3.00	0.01

[1] local peaks and valleys are Labeled by A through S

13. $S = E$, $X = |J - M|$, $Y = |I - J|$, $X > Y$, Y excludes S ∴ eliminate IJ, $IJ = 1$ cycle.
14. $S = E$, $X = |F - M|$, $Y = |E - F|$, $X < Y$.
15. $S = E$, add N, $X = |M - N|$, $Y = |F - M|$, $X < Y$.
16. $S = E$, add O, $X = |N - O|$, $Y = |M - N|$, $X > Y$, Y excludes S ∴ eliminate MN, $MN = 1$ cycle.
17. $S = E$, $X = |F - O|$, $Y = |E - F|$, $X < Y$.
18. $S = E$, add P, $X = |O - P|$, $Y = |F - O|$, $X < Y$.
19. $S = E$, add Q, $X = |P - Q|$, $Y = |O - P|$, $X < Y$.
20. $S = E$, add R, $X = |Q - R|$, $Y = |P - Q|$, $X > Y$, Y excludes S ∴ eliminate PQ, $PQ = 1$ cycle.
21. $S = E$, $X = |O - R|$, $Y = |F - O|$, $X < Y$.
22. $S = E$, add S, $X = |R - S|$, $Y = |O - R|$, $X < Y$.
23. Count the remaining peaks and valleys ($EF = .5$ cycles, $FO = .5$ cycles, $OR = .5$ cycles, $RS = .5$ cycles).

Results are summarized in Table 2.4.

Part b. The solution to part b is simply obtained by multiplying the cycle counts from the spectrum in Table 2.2 by 100 and adding it to the spectrum of Table 2.4 multiplied by 200; the result is shown in Table 2.5.

Part c. The detailed solution is discussed for segment BE of the time history. The BE segment represents [200 events × 0.5 cycles/event] = 100 cycles with a

Table 2.4 Fatigue spectrum for part a of Example 2.2

Load range	Mean load						
	−3.0	−2.0	−1.0	0.0	1.0	2.0	3.0
10							
9							
8							
7							
6				0.5 [EF]			
5		0.5 [FO]					
4							
3			0.5 [BE]	0.5 [OR]			
2							
1			1 [CD]	3 [AB, MN, PQ, RS]		2 [KL, IJ]	1 [GH]

Table 2.5 Fatigue spectrum for part b of Example 2.2

Load range	Mean load						
	−3.0	−2.0	−1.0	0.0	1.0	2.0	3.0
10							
9				50			
8				50	50		
7							
6				100	50		
5		100					
4			50		100		
3			100	150			
2							
1			200	600		400	200

range value of $[0.17 + 2.45] = 2.62$ and mean value of $\left[\frac{0.17 - 2.45}{2}\right] - 1.14$. Since the maximum positive peak (point F at 3.38) has uniform distribution between 2 and 5, the range for segment BE would have a uniform distribution between $\left[\frac{2.62 \times 2}{3.38}\right] 1.55$ and $\left[\frac{2.62 \times 5}{3.38}\right] 3.87$. The mean value would have a uniform distribution between $\left[\frac{-1.14 \times 2}{3.38}\right] - 0.67$ and $\left[\frac{-1.14 \times 5}{3.38}\right] - 1.68$. Therefore, $\left[\frac{1.50 - 0.67}{1.68 - 0.67}\right] = 82\%$ of the cycles fall in the -1.0 mean category and the remaining 18% are in the -2.0 mean category. For the range, however, $\left[\frac{2.00 - 1.55}{3.87 - 1.55}\right] = 19.4\%$ are in the 2 load range which translates to $[100 \times 0.194] = 19$ cycles in the "2 range and -1 mean" category. $\left[\frac{3.00 - 2.00}{3.87 - 1.55}\right] = 43.1\%$ are in the 3 load range which translates to 43 cycles in "3 range and -1 mean" category. The remaining $\left[\frac{3.87 - 3.00}{3.87 - 1.55}\right] = 37.5\%$ are in the 4 load range. Since only the lower 82% of the cycles fall in the -1 mean load category, however, there will be $[100 \times (82\% - 19.4\% - 43.1\%)] = 19$ cycles in the "4 load range and -2 mean load" category. The remaining $[100 \times 18\%] = 18$ cycles will be in the "5 load range and -2 mean load" category. A similar procedure is followed in analyzing all cycles identified in part a, the complete solution is shown in Table 2.6.

2.3.1 Load Spectrum for Space Structures Using the Space Transportation System (STS) or Shuttle

To evaluate the fatigue life of a space structure it is important to consider the total number of cycles associated with each event that the part is exposed to during its service life. The initial step is to break the load history associated with each event into its constituent events, and then the damage effect of each of those events are added together. For example, the cyclic load environments that a space structure will experience during its mission can be broken down as follows (see Fig. 2.10):

Table 2.6 Fatigue spectrum for part c of Example 2.2

Load range	Mean load									
	−4.0	−3.0	−2.0	−1.0	0.0	1.0	2.0	3.0	4.0	5.0
10										
9										
8	2 [FO]									
7		24 [FO]			5 [EF]	12 [EF]				
6		24 [FO]			19 [EF]	19 [EF]				
5			24 [FO]		19 [EF]	14 [EF]				
4			18 [BE] + 24 [FO]	19 [BE]	11 [EF]					
3				43 [BE] + 4 [FO] + 24 [OR]	32 [OR]					
2			65 [CD]	19 [BE]						
1				135 [CD] + 62 [PQ]	44 [OR] + 100 [AB] + 200 [MN] + 138 [PQ] + 100 [RS]	39 [IJ] + 48 [KL]	40 [GH] + 115 [IJ] + 121 [KL]	70 [GH] + 46 [IJ] + 31 [KL]	70 [GH]	20 [GH]

30 Chap. 2 Conventional Fatigue (High- and Low-Cycle Fatigue)

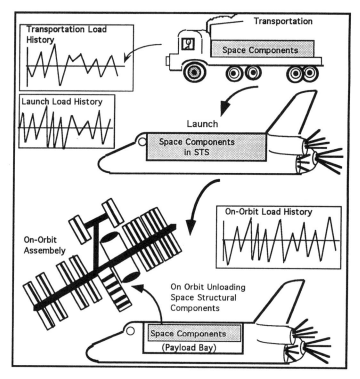

Figure 2.10 The load environment that space componenets experience during their service life

- Prelaunch cycles (acceptance, proof testing, etc.)
- Transportation (including ferry flight) cycles prior to flight
- Flight (liftoff/ascent) cycles, including abort landing
- On-orbit cycles due to unloading and any other on-orbit activities
- Thermal cycles
- Descent/landing cycles
- Postlanding event cycles

The Ferry flight is referred to the transportation of the shuttle structure (by the Boeing 747) from one location to its final distination prior to or after its mission. When the structural hardware is assembled in the shuttle, while it is being transported to another location for launch, the flight load cycles (as the result of transportation) must be included in the total life analysis of parts.

A brief description of flight environment and the corresponding number of cycles are provided in Section 2.3.1.1.

2.3.1.1 Flight Cycles

Components of space structures experience cyclic loadings of different stress levels as a result of launch and landing of the space shuttle. The payload random loading spectrum for these events is shown in Table 2.7 and is called the Goddard load spectrum because it was developed at Goddard Space Flight Center (GSFC) [8]. There are a total of 13 load steps associated with the Goddard load spectrum and the percent of limit load for each stress level, S_i, associated with the number of cycles, N_i, is given in Table 2.7. The Goddard spectrum has been developed for the main load carrying payload structure in the shuttle orbiter payload bay. Moreover, the payload structural parts major natural frequency must be below 50 Hz. If the payload major frequency is above 50 Hz, the number of cycles in the Goddard spectrum (see Table 2.7) should be multiplied by the factors shown in Table 2.8 [9].

When the load environment is completely evaluated in terms of the number of cycles, N_i, and the associated load amplitude, S_i, the total fatigue life of the structural part can be predicted. However, the $S-N$ curve representing the load environment for the material must be well established. Section 2.4 provides a comprehensive overview of the constant load amplitude $S-N$ diagram which can be established by plotting test data from the load controlled method.

Table 2.7 Launch and landing load spectrum (called Goddard fatigue spectra), extracted from [9]

Load step number	Cycles (N_i)/flight			Cyclic stress (% limit stress)	
	Launch	Landing	Total	Minimum	Maximum
1	1	1	2	−100	100
2	3	1	4	−90	90
3	5	3	8	−80	80
4	12	3	15	−70	70
5	46	3	49	−60	60
6	78	3	81	−50	50
7	165	13	178	−40	40
8	493	148	641	−30	30
9	2229	891	3120	−20	20
10	2132	1273	3405	−10	10
11	2920	2099	5019	−7	7
12	22272	6581	28853	−5	5
13	82954	8701	91655	−3	3

Table 2.8 Major frequencies and the multiplication factor for the shuttle payload structural parts

Major frequency	Multiplication factor
0–50 Hz	1
50–100 Hz	2
100–200 Hz	4
200–300 Hz	6

2.4 The S–N Diagram

In evaluating the number of cycles to failure for a given structure subjected to in-service fluctuating loads, fatigue test data representing the load environment must be available. The concept of similitude states that the life of a structural part is the same as the life of a test specimen if both have undergone the same nominal stress. Figure 2.11 shows that the service life of a bridge part exposed to a fluctuating load environment can be evaluated by conducting a laboratory fatigue test that simulates the same environment.

Fatigue test data can be provided to the analyst in tabular form or in the form of an S–N diagram. The S–N (stress-life) diagram is a plot of stress amplitude, stress range, or the maximum cyclic stresses, S (selected as the controlled or independent variable), versus the number of cycles to failure, N (the dependent variable). There are two methods of plotting S–N curves: (1) The S–N diagram is plotted as the actual stress, S, versus the logarithmic scale of cycles, N (semilogarithmic

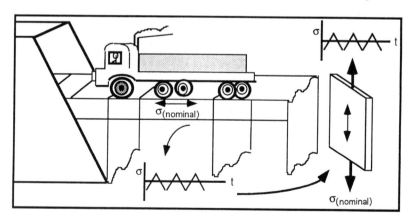

Figure 2.11 The concept of similitude: the life of a bridge part can be determined by obtaining fatigue test data on a Laboratory specimen subjected to the same nominal stress.

plotting). (2) Both S and N are plotted in the form of a log–log plot of S versus N (logarithmic plot). For most materials, the logarithmic plot of the S–N curve is approximated by a straight line. For some alloys, including the ferrous metals, the logarithmic plotting method will generate an additional straight line (horizontal) to account for the endurance limit (also called the fatigue limit) (Fig. 2.12). The semilogarithmic method is the most widely used in engineering applications.

Several types of machines and specimens are available to develop fatigue data. Test data are generated under axial loading, plate bending, rotating bending, and torsion tests. For aerospace use, none of these methods, except for the axial loading tests according to ASTM E-466 (Conducting Constant Amplitude Axial Fatigue Tests of Metallic Materials) and ASTM E-606 (Standard Practice for Strain-Controlled Fatigue Testing), are now conducted for developing fatigue data. A brief description of the constant amplitude axial fatigue tests is provided in Section 2.4.1.1. The usual laboratory procedure for determining the S–N curve for a given material is to use about 18 specimens [10]; see Fig. 2.14 for the specimen geometry. The results of testing are expected to sometimes have wide scatter, so that statistical analysis (the best fit curve method by regression analysis discussed in ASTM E-739) is needed to establish a meaningful S–N diagram [11]. The first few specimens are used where the applied cyclic stress is equal to about 70% of the static tensile strength of the material (the induced cyclic stress magnitude in the test specimen must be below elastic limit). For this region of the S–N curve, the number of cycles to failure is expected to be about 10^3 to 10^4 cycles. The remainder of the test specimens are utilized for other regions of the S–N curve, where the test stress decreases and the number of cycles to failure is expected to increase from 10^5 to 10^7 cycles. Usually six to eight stress levels are selected and for each stress level two to three test specimens are used to develop an S–N curve. When the number of cycles to failure exceeds 10^7 without failure (runout condition), the fatigue limit has been reached and the S–N curve becomes

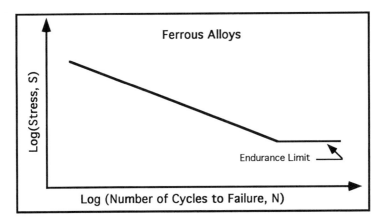

Figure 2.12 *S–N* **curve for ferrous alloys**

asymptotic to a horizontal line. The stress associated with this limit is called the endurance limit (or fatigue limit), see Fig. 2.12. The endurance limit is an important parameter when designing a part to have infinite life.

When prior information regarding the shape of the S–N curve is available, it is unnecessary to select several stress levels to trace out the S–N curve point by point. Not more than four or five test specimens are needed to establish the shape of the S–N curve.

2.4.1 Factors Influencing the Endurance Limit

For most materials, the endurance limit or fatigue limit is not a constant and varies with the stress ratio, R. For a given material, the endurance limit can be influenced by the type of cyclic loading. Experimental data obtained in the laboratory show that the endurance limit of a material tested in uniaxial loading is lower than the endurance limit tested in reverse bending, provided that the two loading cases are subjected to the same stress ratio, R. Under axial loading, the stresses are uniform throughout the part, as compared to a nonuniform stress distribution where bending load is applied.

Other factors affecting the endurance limit are: degree of surface finish, heat treatment, stress concentration, and corrosive environment. Therefore, it is expected that the endurance limit will have a wide range of values depending on the conditions described above. Consider the case of high strength steel with little ductility exposed to a corrosive environment and containing stress concentration. In this case, it is possible to have an endurance limit value as low as 15% of its ultimate tensile strength. On the other hand, consider the case of stainless steel in the annealed condition and subjected to a noncorrosive environment. The endurance limit can be as high as 70% of its ultimate tensile strength.

Experimental data have shown that certain alloys, such as ferrous material, exhibit a clear fatigue limit. For the maximum applied stress below this limit, failure will not occur and, therefore, the material has infinite life. Typical fatigue data (maximum stress, S, versus N) for 4130 alloy steel is plotted in the semilog plot shown in Fig. 2.13a [12]. The S–N diagram for this alloy exhibits an endurance limit of 43.3 ksi with stress ratio of $R = -1$. In comparison with ferrous alloys, aluminum alloys do not show a well-defined endurance limit. Fig. 2.13b shows the S–N curve for 2024-T4 aluminum alloy where the fatigue limit is not well defined [12].

The S–N curves shown in Figs. 2.13a and 2.13b can be represented by an empirical relationship that describes the variation of stress amplitude, S, and fatigue cycles, N. See Section 2.4.2 for a more detailed description related to this topic.

2.4.1.1 Brief Description of Constant Amplitude Axial Fatigue Tests

In determining the fatigue life of a metallic part subjected to constant amplitude axial cyclic load (where the strains are predominantly elastic and the number of

2.4 The S–N Diagram

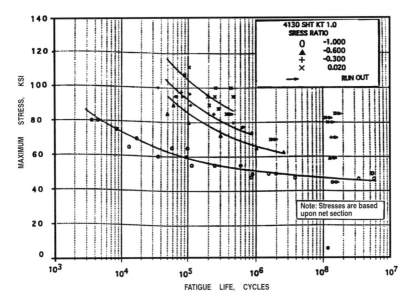

Figure 2.13a The *S–N* diagram for 4340 alloy steel (from MIL-HDBK-5E)

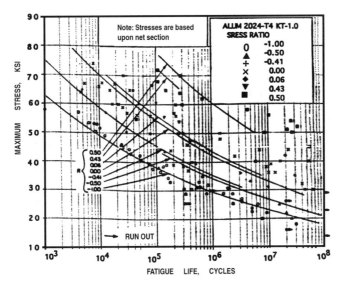

Figure 2.13b The *S–N* diagram for 2024-T4 aluminum alloy (from MIL-HDBK-5E)

cycles to failure is expected to be large), laboratory fatigue tests in accordance with the ASTM E-466 Standard must be performed. Small-scale laboratory fatigue tests, closely resembling or simulating the load environment, are often necessary in order to replace the costly and time consuming full-scale fatigue tests.

The number of cycles to failure is defined as either complete separation of the test specimen or when a crack of some specified dimension has been reached. The results of fatigue tests can be presented in tabular form or in the form of *S–N* curves. These data are useful for design of machinery parts where the service conditions are parallel to the test conditions. Four types of specimens are recommended by the ASTM E-466 Standard and they are shown in Fig. 2.14. Attention should be given to ensure that the size of the test specimen is compatible with the capability of the loading machine employed to generate the *S–N* data. These specimens are designed such that failure as a result of fluctuating load is expected to occur in the middle region, where the stresses are uniform and cross-sectional area is reduced. The end threads of the test specimen which are used to fasten the specimen to the testing machine must be designed such that failure occurs in the reduced region and not in the threads area. Specimens with circular cross-sections are of two types: (1) a type in which the test specimen has tangentially blending fillets between the reduced section and the ends (Fig. 2.14a) and (2) a type where the test specimen has a continuous radius between ends (Figs. 2.14b). Circular specimens should have a minimum of 0.2 in. and a maximum of 1.0 in. diameter

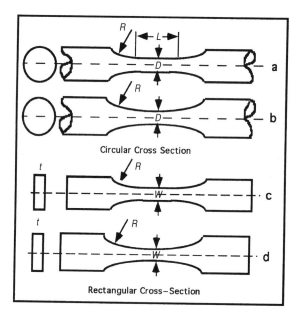

Figure 2.14 Four types of fatigue specimens recommended by the ASTM E-466

in the reduced section. A minimum and maximum grip cross sectional area of 1.5 to 4 times the reduced cross-sectional area is recommended by the ASTM-E466. Furthermore, the ratio of the length of the reduced section, L, to its diameter, D, should be greater than 3. To minimize the stress concentration between the grip and the reduced section, the fillet radius, R, of the blended section should be at least eight times the test section. The fillet radius stress concentration criterion mentioned above is also applicable to test specimens with rectangular cross-sections as shown in Figs. 2.14c and d.

Test specimens with rectangular cross-section must have a minimum and maximum reduced area of 0.03 and 1.0 in.^{2}, respectively. The width-to-thickness ratio, W/t, in the reduced section should be between 2 and 6. For example, a test specimen with rectangular cross-section, in which the width and thickness dimensions are 0.5 in. and 0.15 in., respectively, should have a cross-sectional length $L = 1.5$ in., fillet radius $R = 4.0$ in., and $W_{grip} = 0.75$ in. For a circular cross-sectional test specimen, with $D = 0.25$ in., the cross-sectional length $L = 1.0$ in., $R = 2.0$ in., and $D_{grip} = 0.75$ in.

When calculating the applied load, the width and the thickness dimensions from which the area is calculated must be measured to the nearest 0.001 in. For test specimens with dimensions smaller than 0.2 in., the dimensional measurements should be to the nearest 0.0005 in. (ASTM-E466). The same dimensional requirements are also applicable to the circular test specimens. Consider a rectangular test specimen that is prepared from 2219-T851 aluminum alloy plate (ultimate and yield stress of 62,000 and 45,000 psi, respectively) and is axially loaded. It is necessary to obtain the number of cycles to failure for a stress equal to 60% of the material ultimate allowable (37,200 psi) with a stress ratio of $R = 0$. The calculated applied load, based on a measured thickness, t, of 0.1 in. and width, W, of 0.5 in. is 1860 lb. If the dimensional measurements of thickness and width differ by 0.004 and 0.002 in. (not to the nearest 0.001 and 0.0005 in. as mentioned above) the calculated applied load will be 1778 lb. This would result in a predicted fatigue life considerably higher or lower than the expected value.

Axially loaded test specimens must be free from any induced rotation or bending stress introduced as a result of misalignment or rotation of the grips when mounting the specimen. To eliminate misalignment and to ensure axiality of the applied load, strain gages can be used to measure the bending strains. The value of bending strain should be compared with the axially measured strain due to applied load. The calculated percent bending strain should not exceed 5% of total strain measured by the strain gages (ASTM-E-466). Care must be taken to ensure that the installation of strain gages does not cause damage to the surface (resulting in a stress riser) of the test specimen.

Upon completion of constant amplitude axial fatigue testing, a description of the parameters that can significantly influence the test results must be provided to the user of the data. The most important parameters are:

Material

Grade designation, heat number, melting practice, last mechanical and heat treatment, chemical composition, tensile and yield strength, elongation and reduction of area

Fatigue specimen

Shape size, stress concentration factor, preparation, forming, heat treatment

Fatigue tests

Fatigue testing machine, type of test, frequency, dynamic load verification, dynamic load mounting procedures, failure criterion, number of specimens tested, laboratory temperature and relative humidity

Fatigue data

Fatigue test data should include a table that contains dynamic stresses, fatigue life, test sequence, specimen mark, and remarks related to the nature of failure.

2.4.2 Empirical Representation of the S–N Curve

In some cases, fatigue data are presented in the form of an empirical equation that describes the variation between the stress amplitude, S, and fatigue cycles, N, for different stress ratio, R. For example, in Fig. 2.15 the equation that describes the variation of S versus N for 4130 alloy steel with stress ratios of -1 to 0.02 is given [12] as:

$$\log N = 9.65 - 2.85 \log [S_{max}(1 - R)^{0.41} - 61.3] \quad (2.2)$$

For stress ratios of $R = -1$ and $R = 0$, equation (2.2) becomes:

$$\log N = 9.27 - 3.57 \log (S_{max} - 43.3) \quad (2.3)$$

$$\log N = 9.27 - 2.85 \log (S_{max} - 61.3) \quad (2.4)$$

The relationship described by Eq. (2.2) is very useful for the cases where the test data for the intermediate values of R are not available. However, it is impor-

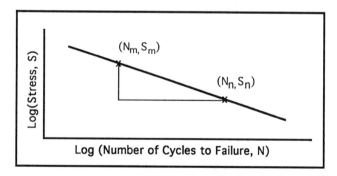

Figure 2.15 Idealized S–N curve shown in a log–log plot

tant to note that the use of Eq. (2.2) is not recommended in making life predictions for the conditions beyond the range given by the test data. Finally, Eq. (2.2) is a convenient way of evaluating the number of cycles to failure when it is programmed in the computer for cases where the load environment contains stress amplitudes of different magnitude (see example 2.5).

The fatigue life equation described above is formulated by plotting the stress data, S, versus the number of cycles to failure, N, in log–log form. For an idealized case, shown in Fig. 2.15, the slope of a straight line can be written by establishing the coordinates of two points along the line:

$$\log S_m - \log S_n = -b (\log N_n - \log N_m) \quad (2.5)$$

where b is the slope of the line (after Basquin [13], who proposed the idea) and (N_m, S_m) and (N_n, S_n) are two points taken along the line. By knowing the slope of the line, b, and any other given point (N, S) along the line, the fatigue cycle, N, associated with any other given stress amplitude, S, can be calculated. For example, let us assume that the endurance limit, S_{en}, is taken to be 10^7 for alternating stress equal to one third of the ultimate of the material ($0.33S_{ul}$). By knowing the slope of the line, b, any other number of cycles to failure associated with a given stress amplitude, S, can be written as:

$$N = (S/0.3S_{ul})^{1/b} \times 10^7$$

Or, in general, one can write:

$$N = N_A (S_a/S_A)^{1/b} \quad (2.6)$$

where (N_A, S_A) is the available point and S_a is the given stress amplitude associated with the number of cycles, N, to be determined.

The number of cycles to failure, N, obtained by the S–N curve (under load or stress control condition) is related to the total life of the part up to failure. In reality, fatigue cycles throughout the life of the structural part consist of crack initiation and propagation. Here, the term "propagation" refers to stable crack growth up to the crack instability. The S–N curve approach does not separate the crack initiation phase from the propagation. However, in some industries, it is required to assume that the crack has already initiated in the hardware and only the total number of cycles associated with the propagation are of interest to the analyst. Existing or initiated cracks assumed in the structural part are the result of load cycles induced from manufacturing, machining, or improper handling prior to its actual usage. The size of the preexisting crack can be assumed based on the capability of the inspection methods used to detect it. The analyst may assume an initial surface crack based on the inspection method applied to the part. Using the available initial flaw size, the total life of the part can be evaluated by using a fracture mechanics approach.

Inspection methods, such as penetrant inspection, magnetic particle, eddy current, ultrasonic, or x-ray (mainly for embedded cracks due to welds) are currently in practice throughout the aerospace industry. For example, for the case of a standard dye penetrant inspection, NASA has recommended that the analyst assume an initial circular surface crack of 0.075 in. in depth (with depth-to-length ratio $a/c = 1$) to use for evaluating the life of the part that are considered as the main load carrying structural component. The size of the final crack at the instability depends on the fracture toughness of the material, the crack geometry, and the applied stress. Section 5.3 of Chapter 5 will cover the inspection methods and the corresponding flaw size for different crack geometries and Chapter 4 will discuss the application of the fracture mechanics on fatigue crack growth rate concept that is currently used in aerospace and other industries.

2.4.3 Parameters Affecting the S–N Curve

For a given alloy, the shape of the S–N curve will vary according to the conditions represented. Some factors influencing the shape of the S–N curve are:

- *Material conditions*
  - Heat treatment
  - Cold working
- *Types of load*
  - Tension
  - Compression
  - Torsion
  - Combined load
- *Stress ratio*
- *Rate of load application*
- *Environment*
  - Corrosive
  - Inert
  - Temparature

Example 2.3

A component of a space structure is made of 4130 alloy steel. The space structure is orbiting around the earth every 90 minutes (see also Fig. 2.4). The maximum thermal stresses induced in the part due to exposure to the sun (once every 90 minutes) is 14 ksi. Determine if the part can survive the environment during its 30 years in space ($R = -1$).

The equation describing the S–N diagram for 4130 alloy steel is extracted from Reference [12] and is expressed as:

$$\log N = 9.65 - 2.85 \log [S_{\max}(1-R)^{0.41} - 6.13]$$

The total number of cycles during the 30 years in space is:

$$N_{\text{total}} = (30 \times 365 \times 24 \times 60)/90$$
$$= 175{,}200 \text{ cycles}$$

The number of cycles to failure based on 14 ksi stress ($R = -1$) can be calculated as:

$$S_{\max}(1-R)^{0.41} = 14 \times (2)^{0.41}$$
$$= 18.6 \text{ ksi}$$
$$\log N = 9.65 - 2.85 \log(18.60 - 6.13)$$
$$N_F = 10.0^{6.52} \text{ cycles}$$

It can be seen that the number of cycles to failure ($N_f = 10.0^{6.52}$ cycles) is larger than the total number of cycles ($N_{\text{total}} = 1.75 \times 10^5$ cycles) required for the structure to orbit around the sun during its 30-year mission in space. Therefore, the structure can survive 30 years in the space environment.

2.5 Constant-Life Diagrams

The purpose of constant-life diagrams is twofold: (1) Most of the available fatigue data are obtained through testing by applying the fully reversed loading case ($R = -1$), where mean stress, S_M, is zero. The cases of completely reversed cycles ($R = -1$) and zero to maximum stress ($R = 0$) are most commonly used in developing the S–N curve. However, in the majority of cases, the analyst would like to design the structural components for the loading case where the mean stress, S_M, is different from zero. It is, therefore, important to be able to utilize the test data pertaining to the case of $R = -1$, collected in the laboratory, for the life assessment of other loading cases. Constant-life fatigue diagrams are a family of curves generated for different fatigue life, N, each representing the variation of either $S_{\max}$ versus $S_{\min}$, $S_{\max}$ versus S_M, or S_a versus S_M. The most common type of constant life fatigue diagram (that can be interpreted easily) is the plot of S_a versus S_M. This diagram was presented by Goodman as a straight line relationship shown by equation 2.8. The Goodman equation relating S_a and S_M can easily convert the available fatigue data of a given material from the case of $R = -1$ to another format (where the mean $S_M \neq 0$) useful to the analyst. (2) Constant-life diagrams are also useful when designing structural components for an infinite life where $N > 10^7$ cycles or when designing for finite life, N.

All methods of plotting fatigue diagrams provide the same type of information. Because the maximum and minimum stress values of each load cycle are recorded directly by the testing machine, it would be easier to use these stresses for plotting, for example, the S–N diagram, rather than to convert them to alternating stress, $(S_{max} - S_{min})/2$, versus mean stress $(S_{max} + S_{min})/2$ or maximum versus mean stress. Since the generation of a constant-life diagram is required to have several experimental data, which are difficult and expensive to generate, it is much easier to develop an empirical relationship that can relate the alternating stress, S_a, to the mean stress, S_M. In Sections 2.5.1 and 2.5.2, the empirical and experimental methods of developing the constant-life diagram are discussed.

2.5.1 Development of a Constant-Life Diagram

Constant-life diagrams can be generated either by empirical relationship or by the test data. From a series of fatigue test data developed for S–N curves, the variation of S_a versus S_M with various combination of stress ratio, R, was plotted [14] and the general trend is illustrated in Fig. 2.16. The test data illustrated in Fig. 2.16 were collected for a fixed number of cycles to failure, N. In Fig. 2.16, the compressive mean stress (the region associated with $-S_M$) is not shown because the data is more sensitive to positive mean stress, S_M. Note that the alternating stress, S_{a0}, represents the fully reversible case ($R = -1$) where $S_M = 0$. Other stress amplitudes along the dotted line have nonzero mean stress and they are referred to by S_a. The experimental data indicates that as the alternating stress, S_a, decreases, the mean stress increases. When stress amplitude, S_a, approaches zero, the data points tend to approach the ultimate tensile strength of the material, as shown in Fig. 2.16. Based on this observation, Gerber from Germany (1874) and Goodman

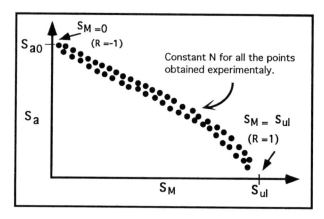

Figure 2.16 Illustration of the variation of alternating stress, S_a as a function of mean stress, S_m

2.5 Constant-Life Diagrams

(England, 1899) [15, 16] proposed an empirical relationship that would approximate the experimental data describing the constant-life stress relationship illustrated in Fig. 2.16.

$$(S_a/S_{a0}) + (S_M/S_{ul})^2 - 1 = 0 \quad \text{Gerber [10]} \quad (2.7)$$

$$(S_a/S_{a0}) + (S_M/S_{ul}) - 1 = 0 \quad \text{Goodman [11]} \quad (2.8)$$

Where experimental data are not available, the proposed Gerber and Goodman constant-life stress relationship can be used. Goodman's original law pertaining to infinite life design assumed the alternating endurance limit (where $S_M = 0$) to be one third of the ultimate tensile strength of the material. This assumption has since been modified to the relationship shown by Eq. (2.8). Note that the above constant-life approximation equations described by Eq. (2.7) and (2.8) do not apply to brittle material.

For an infinite life design, the quantity S_{a0} is set equal to S_{en} (endurance limit when $S_M = 0$). Figure 2.17 shows the Gerber and Goodman diagrams describing the constant-life stress relationship.

It is customary to plot the Goodman and Gerber constant-life diagram in terms of nondimensional quantities, for example S_a/S_{a0} versus S_M/S_{ul}, as shown in Fig. 2.18. Experimental data for aluminum alloys and steels pertaining to infinite life design had shown that [17] for constant-life analysis, the Goodman line may be considered as the lower bound and the Gerber line as an upper bound (see Fig. 2.18 where experimental data are falling between the two lines).

2.5.2 Development of a Constant-Life Diagram by Experiment

The procedure for experimentally obtaining a constant-life diagram for ductile material is to extract data from several S–N curves having different stress ratios. Two extreme values of $R = -1$ (where $S_M = 0$) for fully reversed cyclic loading

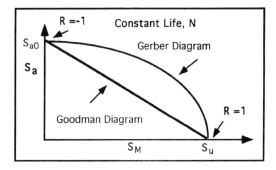

Figure 2.17 Illustration of the Goodman and Gerber diagrams

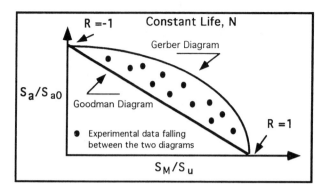

Figure 2.18 Illustration of the Goodman and Gerber diagrams in terms of nondimensional quantities S_a/S_{a0} vs. S_M/S_u

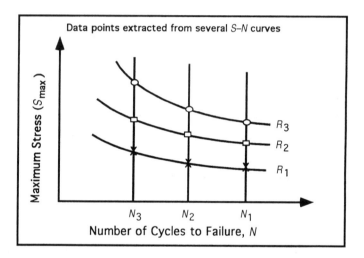

Figure 2.19 Illustration of typical S–N curves for three stress ratios of R_1, R_2, and R_3

and $R = 1$ (where $S_a = 0$) for steady load with no cyclic loading are essential to construct the constant-life diagram. Figures 2.19 and 2.20 illustrate the method of plotting constant-life lines from a family of S–N curves. In Fig. 2.19, three S–N curves, corresponding to three stress ratios R_1, R_2, and R_3, were converted to constant-life diagrams with lives of N_1, N_2, and N_3, respectively, as illustrated in Fig. 2.20.

Example 2.4

A fracture critical rod (i.e., the failure of the rod will cause a catastrophic failure to the main structure and therefore, it is considered a high risk part) is made of

2.5 Constant-Life Diagrams

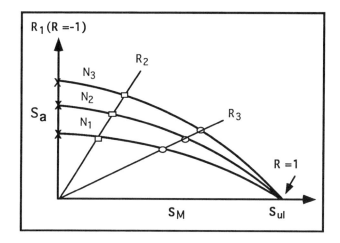

Figure 2.20 Illustration of the method of obtaining the constant-life diagram through experimental data

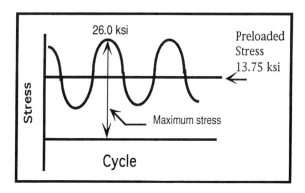

Figure 2.21 Load environment for the High Risk Preloaded Bolt in Example 2.4

2219-T851 aluminum alloy and is subjected to maximum cyclic stress, $S_{max} = 26$ ksi, as shown in Fig. 2.21. The rod is preloaded to 25% of its ultimate ($S_{ul} = 55$ ksi). The endurance limit for this material under fully reversed loading ($R = -1$) is 8 ksi (where $N = 10^7$ cycles). Fully reversed stress at 3000 cycles is provided by the S–N curve as 30 ksi. Use the Goodman constant-life relationship to calculate the life of this component.

Solution

The following quantities can be calculated:

$$S_M = 0.25 \times S_{ul}$$
$$= 13.75 \text{ ksi}$$
$$S_{min} = 2 \times S_M - S_{max}$$
$$= 1.5 \text{ ksi}$$

And the stress amplitude, S_a, is given by:

$$S_a = (S_{max} - S_{min})/2$$
$$= 12.25 \text{ ksi}$$

From the Goodman relationship [Eq. (2.8)] the value of S_{a0}, when $S_M = 0$, can be calculated by using a second point on the Goodman line where $S_M = 13.75$ and $S_a = 12.25$ as:

$$(S_a/S_{a0}) + (S_M/S_{ul}) - 1 = 0$$
$$(12.25/S_{a0}) + (13.75/55) - 1 = 0$$
$$S_{a0} = 16.33 \text{ ksi}$$

The constant-life diagram based on the Goodman relationship is shown in Fig. 2.22 for $S_{a0} = 30.0$, 16.33, and 8.0 ksi.

Using Eq. (2.6), the number of cycles to failure for the preloaded rod can be calculated. By using any two points (m and n) along the straight line, the slope associated with the S–N curve can be evaluated. Utilizing Eq. (2.5):

$$\log S_m - \log S_n = -b (\log N_n - \log N_m)$$

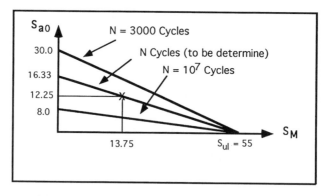

Figure 2.22 Constant Life and Goodman Diagram

Substituting $S_m = 30$ ksi and $S_n = 8$ ksi, with $N_m = 3000$ cycles and $N_n = 10^7$ cycles, respectively;

$$\log 30 - \log 8 = -b(\log 10^7 - \log 3000)$$

$$b = -0.16$$

For $S_{a0} = 16.33$ ksi, the number of cycles to failure can be calculated from Eq. (2.6) as:

$$N = (16.33/8)^{-1/0.16} \times 10^7$$

$$N = 115{,}650 \text{ cycles}$$

All the points on the curve corresponding to $N = 115{,}650$ cycles have the same fatigue life, including the preloaded bolt with mean stress $S_M = 13.75$ ksi.

2.6 Linear Cumulative Damage

The S–N diagram is useful to determine the number of cycles to failure associated with a given constant amplitude applied cyclic stress. When the total damage induced on a given part is a result of several load environments, with different fluctuating stresses, the contributing damage caused by each environment should be evaluated. Consider the case of a satellite that will experience fluctuating cycles of different stress level, for example, launch, on-orbit, and thermal cycles, during its total life. Similarly, the frames of aircraft are subjected to many different cyclic loads, depending on the altitude, speed, takeoff and landing, and air turbulence. The ratio of the partial damage, d_i, due to each individual environment having stress level, S_i, over the total damage, D_i, must be additive. The sum of these partial damage fractions should be unity:

$$d_1/D_1 + d_2/D_2 + d_3/D_3 + \cdots = 1 \tag{2.9}$$

Based on this assumption, the Palmgren–Miner rule [2, 3] (first proposed by Palmgren and later developed by Miner) states that the fatigue damage contribution by each individual load spectrum at a given stress level is proportional to the number of cycles applied at that stress level, n_i, divided by the total number of cycles required to fail the part at the same stress level (N_{fi}). It is obvious that each ratio can be equal to unity if the fatigue cycles at the same stress level would continue until failure occurs. The total failure, in terms of partial cycle ratios, can be written

$$n_1/N_{f1} + n_2/N_{f2} + n_3/N_{f3} + \cdots = 1 \quad (2.10)$$

Equation (2.10) is a useful tool to determine the life of a given structure subjected to several cyclic load cases of different stress magnitude.

Example 2.5

A component of a space structure made of 2219-T851 aluminum alloy is subjected to fluctuating loads with different stress magnitudes as shown in Table 2.9.

Table 2.9 Launch and on-orbit load spectrum for Example 2.5

Launch		On-orbit	
# cycles N	Limit stress (ksi)	# cycles N	Limit stress (ksi)
3	20.5	100	18.29
18	18.5	300	15.97
13	16.4	1300	14.50
27	14.4	2700	13.29
95	12.3	9500	12.00
159	10.3	1.5E + 4	11.68
343	8.20	3.4E + 4	9.36
1134	6.15	11.3E + 4	8.04
5349	4.10	53.4E + 4	6.72
5537	2.05	55.3E + 4	5.40
7939	1.40	79.3E + 4	4.70
		on-orbit Thermal spectrum	
		175000 cycles	2.0 ksi

Including a life factor of 4 (4 × number of cycles occurring during the service life of the structure), determine if the part will survive the load environment during 15 years of its mission on orbit (one mission is equivalent to one service life). The equation describing the S–N curve described by Eq. (2.2) is:

2.6 Linear Cumulative Damage

$$\log N = 9.65 - 2.85 \log (S_{\max}(1-R)^{0.41} - 61.3) \quad (2.2)$$

The load spectrum for launch and on-orbit (including on-orbit thermal cycles when the structure is exposed to cold and hot temperature; see Fig. 2.4) in terms of the limit load (maximum operating stress when no factor of safety is included) are given in Table 2.9:

Solution

The number of cycles to failure, N_f, associated with each step corresponding to the launch and on-orbit environment can be computed by using Eq. (2.2) (see Table 2.10). The ratio of N_i/N_{fi} is related to partial damage described by Eq. (2.10). The total failure, in terms of the partial cycle ratios, can be calculated by employing the Palmgren–Miner rule, see Eq. (2.10). With a life factor of 4 (the safety factor of 4 is included to account for the uncertainties in the material properties, load environment and analysis approach), the total summation is less than 1 (see Table 2.11). Therefore, the structure can survive the environment.

Table 2.10 Number of cycles to failure N_{fi}, for each stress magnitude corresponding to Example 2.5

Launch		On-orbit	
# cycles N_{fi}	Limit stress (ksi)	# cycles N_{fi}	Limit stress (ksi)
3.6E + 5	20.5	5E + 5	18.29
4.8E + 5	18.5	7.3E + 5	15.97
6.8E + 5	16.4	9.7E + 5	14.50
10E + 5	14.4	12E + 5	13.29
15E + 5	12.3	16E + 5	12.00
26E + 5	10.3	18E + 5	11.68
49E + 5	8.2	34E + 5	9.36
112E + 5	6.15	52E + 5	8.04
356E + 5	4.1	87E + 5	6.72
257E + 6	2.05	16E + 6	5.40
710E + 6	1.40	24E + 6	4.70
		Thermal	
		276E + 6	2.0

Table 2.11 Number of cycles to failure N_{fi}, and the sum of n_i/N_{fi} used in minor's rule for Example 2.5

Launch			On-orbit		
N_{fi}	n_i	n_i/N_{fi}	N_{fi}	n_i	n_i/N_{fi}
3.6E + 5	3	8.3E − 6	5E + 5	100	2.0E − 4
4.8E + 5	13	2.7E − 5	7.3E + 5	300	4.1E − 5
6.8E + 5	18	2.6E − 5	9.7E + 5	1300	1.3E − 4
10E + 5	27	2.7E − 5	12E + 5	2700	2.3E − 3
15E + 5	95	6.3E − 5	16E + 5	9500	5.9E − 3
26E + 5	159	6.1E − 5	18E + 5	1.5E + 4	8.3E − 3
49E + 5	343	7.0E − 5	34E + 5	3.4E + 4	.01
112E + 5	1134	1.0E − 4	52E + 5	11.3E + 4	.02
356E + 5	5349	1.5E − 4	87E + 5	53.4E + 4	.06
257E + 6	5537	2.1E − 5	16E + 6	55.3E + 4	.03
710E + 6	7939	1.1E − 5	24E + 6	79.3E = 4	.03
	TOTAL	5.3E − 4		TOTAL	0.166
			Thermal		
			276E + 6	175200	6.3E − 4

Sum n_i/N_{fi} = 0.166 + 5.3E − 4 + 6.3E − 4
       = 0.167 < 1

with a safe-life factor of 4, the summation is:

4 × 0.167 = 0.668 < 1

2.6.1 Comments Regarding the Palmgren–Miner Rule

The Palmgren–Miner rule, described by Eq. (2.10), has been used by engineers in the past and is still considered a simplified and versatile tool for determining the total life of a given structure under study. However, the oversimplified assumptions that the damage summation described by Eq. (2.10) is linear and that no account is made of the sequence in which the crack tip is experiencing cyclic stresses have drawn criticism by investigators in this field. For example, the total life analysis should yield shorter life if the crack tip experiences high stress cycles during its early stage of life followed by the smaller amplitude cycles rather than the other way around. In this case, the smaller stress cycles, following the larger amplitude cycles, are more effective in damaging the structure (damage becomes more sensitive to load cycles as it advances and grows in size). It is possible that using the high-amplitude cycles first and low cycles next may result in a Palmgren–Miner sum less than 1, while using the low-amplitude cycles first and high-amplitude stress next may result in a damage summation greater than 1. Moreover, load interactions between high and low amplitude load cycles causing a retardation effect (i.e. a delay in damage growth) is not considered in Miner's rule.

2.7 Crack Initiation (Stage I) and Stable Crack Growth (Stage II)

2.7.1 Fracture Surface Examination

The fatigue failure mechanism of metals occurs in three steps: (1) a microscopic crack is initiated at some localized spot after N number of cycles, (2) the initiated crack will grow in a stable manner (stable crack growth), and (3) the growing crack will reach its critical length and create instability. In brittle materials, the crack instability means fast fracture where separation occurs by the cleavage mechanism along low-index crystallographic planes, known as cleavage planes, by direct breakage of atomic bonds. Brittle failure also occurs in relatively ductile metals under the plane strain condition where a very small amount of energy is consumed for the separation mechanism with little or practically no plastic deformation. In contrast, in ductile materials under plane stress condition, shear type rupture occurs on shear planes inclined at approximately 45° to the applied tensile load. In studying the two fractured surfaces, two distinct zones, related to stable crack growth and instantaneous fracture, are evident and they are called the fatigue and rupture zones.

The fractured surfaces contain information related to the nature of the applied load (load history), severity of the environment (e.g., corrosion), and the material quality (type of heat treatment). The two fractured surfaces associated with the fatigue zone tend to be macroscopically flat, shiny, and smooth and show a "beach marking" or "clam shell" appearance at uniform intervals due to the variation in the load amplitude during each cycle. The beach markings (also called striation marks) can be absent on the cracked surfaces when uniform cyclic loading with little variation in the load amplitude is applied in the laboratory. Striation marks represent the position of the advancing crack when it is subjected to fluctuating load amplitude and, in most cases, each striation mark represents the amount of crack growth per load cycle. These markings are perpendicular to the direction of the advancing crack and are helpful in locating the exact origin of crack initiation on the structural part by backward tracking of the crack growth. A schematic representation of fatigue and rupture zone and the surface marks produced on smooth specimens (circular cross sections) under tension–tension and reverse bending subjected to high and low loading conditions are shown in Figs. 2.23 and 2.24a, respectively. An electron microscope picture of striation marks for 6061-T6 aluminum alloy is also shown in Fig. 2.24b which clearly illustrates the position of the advancing crack when it is subjected to fluctuating load cycle. Comprehensive representations of beach markings for other loading conditions are available in Reference [18].

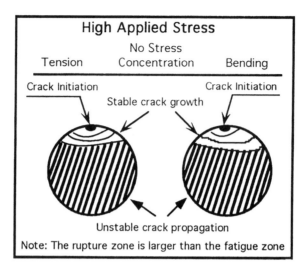

Figure 2.23 Generation of beach marks due to tension–tension and reversed bending high load for unnotched specimen

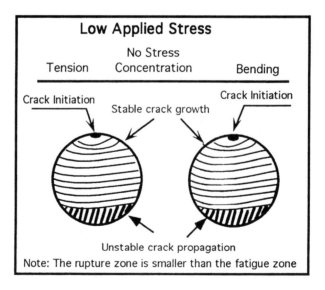

Figure 2.24a Generation of beach marks due to tension–tension and reversed bending low load for unnotched specimen

2.7 Crack Initiation (Stage I) and Stable Crack Growth (Stage II) 53

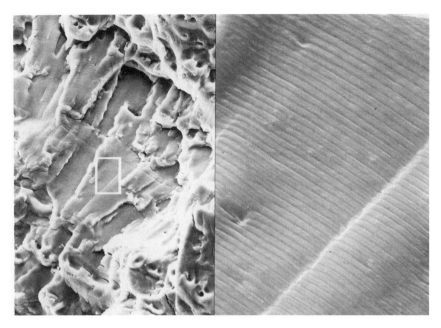

Figure 2.24b The striations marks provided by the electon microscopic technique for 6061-T6 aluminum alloy which represents the position of the advancing crack

Figs. 2.25a and 2.25b show the fracture surface appearance of a bicycle spoke made of 7075-T6 aluminum alloy that is fractured in two pieces due to in-service fluctuating load, using a light microscope with $25 \times$ and $100 \times$ magnification, respectively (see also Fig. 2.25c which was taken from the same specimen surface by an electron microscope with $2000 \times$ magnification). Fatigue crack initiation usually starts at the free surface or subsurface and macroscopically it can be traced to either surface irregularity (such as a notch, a scratch, or an abrupt change in section) or to corrosion pitting. Note that the presence of residual compressive stresses can cause the crack initiation not to occur on the surface of the material. Investigation of the fatigue surface shown in Fig. 2.25 indicated that the crack initiation was due to surface scratches together with a corrosive environment.

Careful examination of the fatigue and rupture zones on the fracture surface can give information about the kind of load, its magnitude, and the material's fracture toughness. A short fatigue zone with a large final fracture surface indicates that the applied stress intensity factor exceeded the fracture toughness (material resistance to fracture) of the material at a relatively short crack length (see Chapter 3 for an in-depth study of the stress intensity factor and fracture toughness concepts) or the material fracture toughness was low. Looking at the fracture surface of the bicycle spoke shown in Fig. 2.25 indicates that the area of the final fracture (rupture zone) is small as compared with the fatigue zone. It can be deduced that,

54   Chap. 2 Conventional Fatigue (High- and Low-Cycle Fatigue)

Figure 2.25a Fracture surface appearance of a bicycle spoke made of 7075-T6 aluminum alloy (light microscope with 25 × magnification)

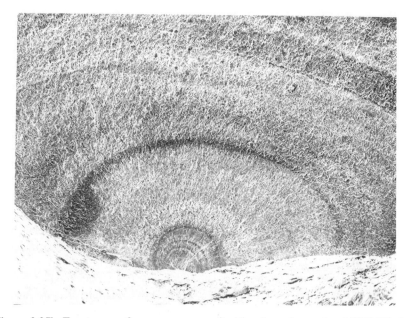

Figure 2.25b Fracture surface appearance of a bicycle spoke made of 7075-T6 aluminum alloy (light microscope with 100 × magnification)

2.7 Crack Initiation (Stage I) and Stable Crack Growth (Stage II)   55

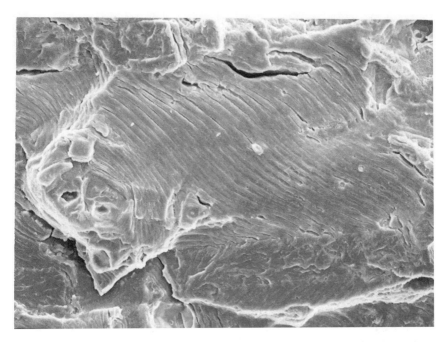

Figure 2.25c The Fracture surface of the bicycle spoke failure viewed by the electron microscope technique with a magnification factor of 2000× (7075-T6 aluminum alloy)

in the case of the bicycle spoke failure shown in Fig. 2.25, the applied load was relatively low and the material fracture toughness was high.

Microscopic examination of fatigue fracture surfaces began with a study by Zapffe and Wooden using a light microscope [19] with magnification capability of up to 500×. The most distinguishing microscopic feature of a fracture surface is its beach mark or striated surface appearance. Under low-cycle fatigue, where the applied cyclic load is high, the striations are coarse enough to be observed by the light microscope [18]. At lower stress levels, however, the striation marks are difficult to observe even with the help of the light microscope. Higher magnifications are required to resolve these fine marks. The restricted depth of field of the light microscope limits such surface examination, and thus electron fractography is the principal technique used to analyze the striated surfaces. Figure 2.25c shows the fracture surface of the bicycle spoke failure viewed by the electron microscope technique with a magnification factor of 2000×.

Another example of a fractured surface that has beach marking appearance is shown in Fig. 2.26 using a light microscope with 16× magnification. The fractured part is a 4130 steel bolt (with 125 ksi ultimate) that failed during the truck transportation of a missile to Cape Canaveral, in Florida. The bolt attaches the

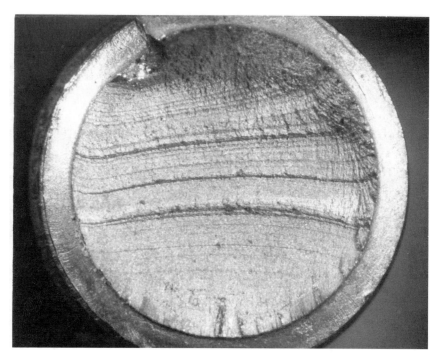

Figure 2.26 Fracture surface of a 4130 steel bolt viewed by a light microscope with a magnification factor of 16 × (striation marks on the fatigue zone can be seen on the surface of the part.)

rocket engine exhaust nozzle to a struts support that was under cyclic loading during the transportation. Looking at the surface appearance of the 4130 steel bolt (see Fig. 2.26), it can be said that the fatigue zone (stable crack growth region) is larger than the rupture zone from which one may conclude that the material had good fracture toughness and, moreover, the magnitude of the cyclic load was relatively low. A careful examination of Fig. 2.26 reveals that the bolt was exposed to a few tensile overload cycles creating a few coarse striation marks.

From a macroscopic point of view, the fracture surfaces due to fluctuating load, exhibit a flat, shiny surface appearance, as if the failure was associated with a brittle material, that is, the lack of necking or shear lip formation on the two fatigue surfaces. This phenomenon occurs even in metals that would be considered quite ductile when tested under monotonically increasing tensile load. The flat and shiny appearance of the fracture surface led metallurgist to believe that the metal had undergone recrystallization and had thereby become brittle.

2.7 Crack Initiation (Stage I) and Stable Crack Growth (Stage II)

2.7.2 Introduction to Crack Initiation

Although there are large amounts of laboratory test data and publications on this subject, there has not yet been established any fundamental theory that can fully explain the cause of the crack initiation phenomenon. Most of the development in this area is based on theoretical equations containing parameters that can be obtainable through laboratory experiments. In all these cases, the results of the theoretical work were checked against the experimental results to verify the validity of the model. In general, the formation of stage I (crack initiation) is influenced by the magnitude of the applied cyclic load, the severity of the corrosive environment, and temperature [20]. There are several proposed models that can describe the initiation of fatigue cracks in the load varying environment, all of which rely on the localization of plastic deformation by a slip movement mechanism and the formation of intrusion and extrusion on the surface of the material [20–22]. Stage I crack initiation by the slip band mechanism is presented in Section 2.7.3. In Section 2.7.2.1 the fracture mechanism when applied monotonic load is the main cause of failure is briefly discussed.

2.7.2.1 Fracture Due to Monotonic Applied Load

When the applied monotonic tensile load is the principal cause of fracture (a single load application to failure versus cyclic load), the structural failure can be described by a process known as microvoid coalescence. Second phase particles, inclusions, and dislocation pileups are the source of strain discontinuities that are suitable regions for initiation of microvoids [23, 24]. As the applied load increases, the microvoid density increases; they grow and coalesce to a critical dimension to cause fracture. The two fracture surfaces contain numerous dimples having a cuplike conical shape and for that reason, this type of fracture is referred to as dimple rupture (illustrated in Fig. 2.27a). The size of the dimples observed on the two fracture surfaces is governed by the density of nucleated microvoids in the volume of the material. A large number of microvoids result in the formation of small dimples and, conversely, when the nucleation sites for microvoids are few, the individuals grow to a larger size before the coalescing mechanism occurs and results in the formation of large dimples. Figures 2.27b, c, and d show the fracture surfaces of 4340 and A286 steels and 2014-T6 aluminum alloy subjected to monotonic applied load which clearly show the density of nucleated microvoids and the corresponding size of the formed dimples after failure. Figures 2.27b, c, and d were taken with electron microscope techniques using $2000 \times$ magnification.

The fracture surfaces due to shear overload exhibit dimples of different shape as compared with tension overload. For mode II fracture, the elongated dimples of

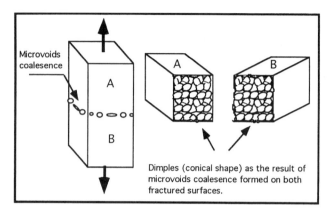

Figure 2.27a Microvoids coalescence and the formation of equiaxed dimples for tension overload.

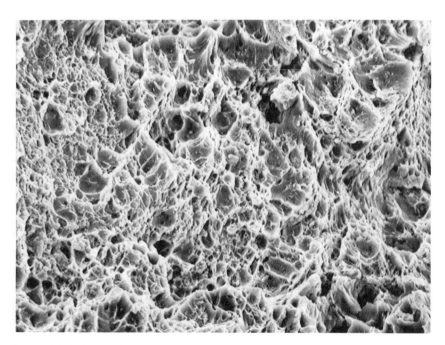

Figure 2.27b Formation of dimples for tension overload provided by the electron microscope (4340 steel with 180/200 ksi)

2.7 Crack Initiation (Stage I) and Stable Crack Growth (Stage II) 59

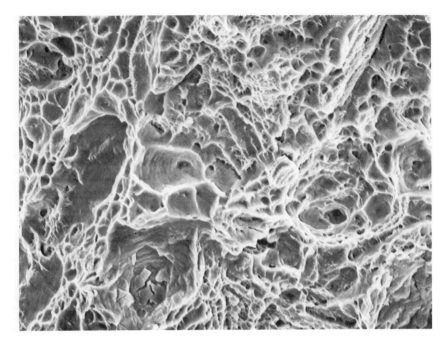

Figure 2.27c Formation of dimples for tension overload provided by the electron microscope (A286 steel with 160/180 ksi)

each mating surface are formed in such a way that their heads are oriented in opposite directions (see Fig. 2.28). In contrast, in mode I fracture (tension overloading), the dimples on the two surfaces are almost of equal size and are referred to as having equiaxed dimples. Equiaxed dimples that have equal rim area in general are not equal in the depth direction.

Prior to explaining the crack initiation in the load varying environment, where the initiated crack length is smaller than or on the order of the material grain size, it is important for the reader to have some basic knowledge of material crystal structure and the concept of slip mechanism.

2.7.2.2 Crystal Structures of Metals and Plastic Deformation

Metals generally consist of regions with a regular geometrical arrangement of atoms, called crystals or grains. The grain is merely a crystal that does not have smooth faces (Fig. 2.29). However, within a grain, the arrangement of atoms are as perfect as a crystal. The average diameter of a grain is in the range of 0.01 to 0.001 in. and varies greatly with the mechanical and heat treatment received. Just

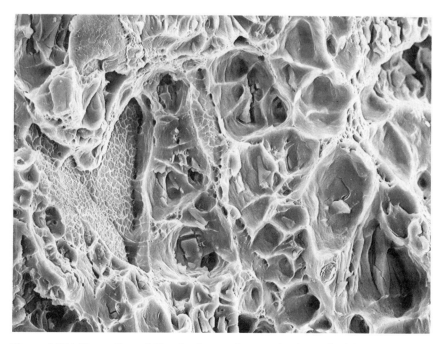

Figure 2.27d Formation of dimples for tension overload provided by the electron microscope (2014-T6 aluminum alloy)

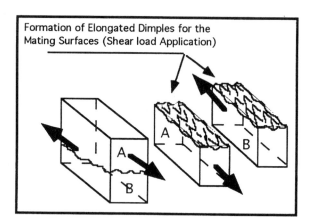

Figure 2.28 Microvoid coalescence and the formation of elongated dimples during shear overload

2.7 Crack Initiation (Stage I) and Stable Crack Growth (Stage II)

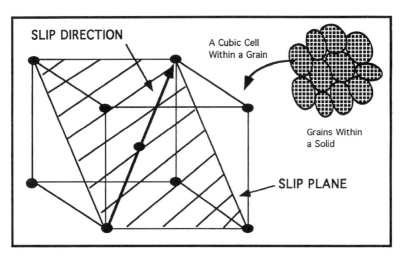

Figure 2.29 Illustration of slip plane and slip direction of a body centered cubic cell

like a crystal, a grain is made of repeated groups of atoms, called unit cells [25, 26]. A body-centered cubic unit cell of pure alpha iron is shown in Fig. 2.29.

Plastic deformation of metals can occur by a process called slip, where the adjacent planes of atoms move within the crystal. In simple terms, a slip mechanism can be visualized as a shearing forces that slides one card in a deck of cards over another. This slip takes place only in certain crystallographic planes, usually those with the most dense atomic packing. A slip plane and slip direction for the body-centered cubic unit cell structure are shown in Fig. 2.29. Slip planes for face-centered cubic structures (such as Cu, Ag, Au, Al, and Ni) are {111} and {110} for body-centered cubic metals (Li, Cr, K, and Fe). To illustrate plastic deformation, let us assume we have a number of single crystals of copper in the form of a rod. The rod can be utilized as a tensile specimen by fixing it at one end and pulling at the opposite end. The force transmitted across a particular plane of the rod can be resolved into its components normal and tangent to the plane, as illustrated in Fig. 2.30. Experiments have shown that when the tangential components of the force exceed a certain value on the area associated with the slip plane, sliding will occur. This critical force value, when acting on the area of the slip plane, is called the yield stress. Its value is independent of the normal components of the force on the slip plane. In perfect metals (where imperfection is minimized), the magnitude of the maximum shear stress in the plane (yield stress), using laws describing the interatomic forces, is about 10^7 psi. However, the actual yield value associated with pure iron, obtained by a standard tensile test, is several orders of magnitude lower than the theoretical value. The difference between the theoretical and laboratory values mentioned above is based on the imperfections in the crystal, such as voids, inclusions, and dislocations. The concept of crack initiation by

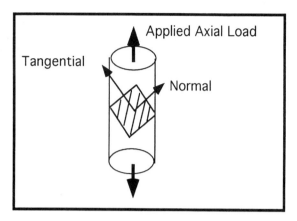

Figure 2.30 Resultant of an axial force into normal and tangential components

slip movement within a crystalline lattice structure of a grain is viewed as the displacement of dislocations in the lattice under the action of shear. Prior to fatigue failure, dislocations or mismatched atoms tend to pile up along the grain boundaries and/or slip bands and cause crack initiation and growth in the structural part. Microscopic cracks of neighboring grains (initiated along the slip band with the similar mechanism) will eventually combine to form a sizable crack able to propagate perpendicular to the applied load. Figure 2.31 shows a single dislocation that created a regional lattice imperfection that is surrounded by a perfect cubic lattice. In this figure, the upper portion of the crystal is displaced by one atomic spacing, b, creating a mismatch and internal stress between atoms. In a metal, there may be 10^7 to 10^9 dislocations in a square inch of area of the crystal throughout the metal.

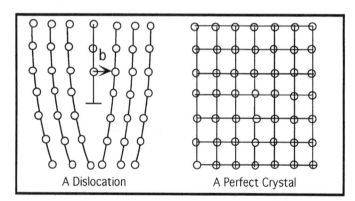

Figure 2.31 Single dislocation in otherwise perfect crystal: introducing mismatch and internal stresses among atoms

2.7 Crack Initiation (Stage I) and Stable Crack Growth (Stage II)

The sliding mechanism under low-amplitude long life cyclic loading is somewhat different from the monotonic tensile loading case shown in Fig. 2.32. Section 2.7.3 describes the slip plane cracking mechanism and creation of intrusion and extrusion on the surface of the part, which can cause crack initiation under cyclic loading.

2.7.3 Crack Initiation Concept (Intrusion and Extrusion)

When the applied tensile load is monotonic, the slip steps that appear on the surface of the specimen have simple geometry, as shown in Fig. 2.32. Coarse slip is normally associated with monotonic tensile loading; however, under low-amplitude fatigue loading, the surface of the material tends to deform by cyclic slips which are finer and more localized in nature. Slip lines (illustrated in Fig. 2.32) tend to concentrate and group into a bundle called slip bands. The surface topology corresponding to a cyclic fine slip looks like irregular notches at the specimen edge consisting of extrusions and intrusions, as shown in Fig. 2.33 [23, 24, 27–29]. An extrusion is a small piece of material that is squeezed out from the surface of a slip band due to the presence of voids and dislocations pileup. The inverse of extrusion is intrusion and is also illustrated in Fig. 2.33. Each of these effects can lead to the localization of plastic strain by the formation of stress risers on the surface of the material. Continuation of the cyclic load leads to deepening of the intrusion and, eventually, the formation of a microcrack along the slip band (stage I).

In ideal case of a single crystal of copper, the crack initiation based on slip bands formation can be described as follow [30–33]: Upon application of low-amplitude cyclic loading, in those crystals that are oriented favorably with respect to the ap-

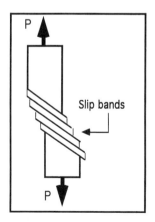

Figure 2.32 Illustration of slip by a monotonic tensile load

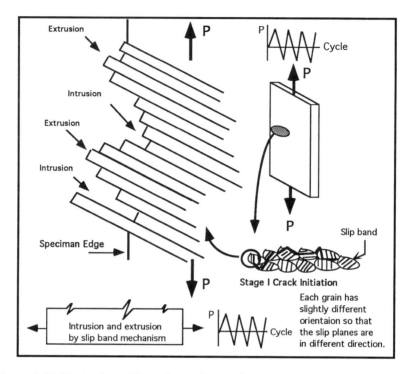

Figure 2.33 Illustration of intrusion and extrusion due to slip as a result of cyclic loading

plied fluctuating load, after N number of cycles fine localized slip lines will appear on the surface of material (see Fig. 2.34, case A). At the early stages of cycling the slip lines appear on the surface for the tensile portion of the cycle and disappear when the unloading portion of the cycle takes over. This reversibility continues for a few cycles until permanent slip bands are formed (see case B). The concentration of slip lines that are grouped in bundles and tend to increase in width (slip bands 10 to 50 μm wide) are referred to as persistent slip bands (PSBs) which later become the soft spot for the formation of microcracks. The appearance of slip bands on the surface occurs if we assume the material is free from any residual compressive stresses, otherwise the crack initiation may not occur on the surface of the material. When cycling continues, any one of the PSB can result in the formation of crevices (intrusion and extrusion; see case C) which are localized stress concentration sites for the nucleation of a fatal crack. R. E. Peterson under "Fatigue Cracks and Fracture Surfaces" indicated that when penetrating oil was placed on the surface of a fatigue specimen with heavy slip bands (case C of Fig. 2.34) bubbles are formed under repeated fluctuating load, thus indicating that separation between sliding plates had occurred. Additional subsequent cycling created crack initiation along

2.7 Crack Initiation (Stage I) and Stable Crack Growth (Stage II)

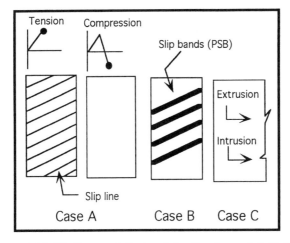

Figure 2.34 Slip lines and slip bands leading to crack initiation

the slip band and crack growth perpendicular to the applied load. The crack initiation and growth mechanism at the early stage of fatigue growth is strongly influenced by microstructural parameters such as grain size and slip band orientations. Further continuation of the cyclic load will result in the formation of stage II (stable crack growth), until finally, when the critical length is reached, the crack will become unstable. In the early stages of stable crack growth, the direction of propagation is not perpendicular to the applied load, but depends on the orientation of the primary slip band within the crystals of a grain (see Figs. 2.35 and 2.36). When the

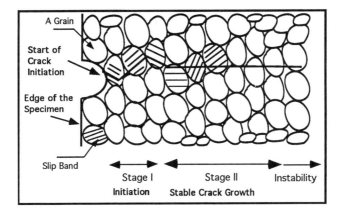

Figure 2.35 Illustration of stage I and stage II crack initiation and propagation

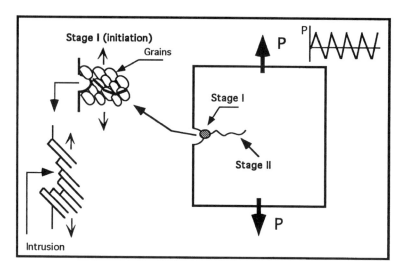

Figure 2.36 Detail description of stage I and stage II

crack proceeds to intersect the grain boundary, the growth direction of propagation alters (and may be arrested by the grain boundary) and the crack will tend to orient itself with the next grain having specific crystallographic planes suitable for crack growth. As the length of the crack increases, the crack will eventually tend to orient itself normal to the direction of the applied load, as illustrated in Figs. 2.35 and 2.36. Fatigue crack growth of stage II occurs by transgranular fracture (where crack propagates along the grain boundary) and is influenced by the magnitude of the alternating stress, the mean stress, and the severity of the environment. The extent of both stage I and stage II is governed by the material type and the magnitude of the tensile cyclic stress; the lower the magnitude of this stress, the longer the time associated with stage I will be. On the other hand, if the magnitude of the applied tensile cyclic stress is high, the number of cycles in stage I is short and the remaining life is consumed for stable crack growth (stage II). Note that there is no simple or clear outline of the boundary between stage I and stage II.

Both intrusion and extrusion can be described as the localized region of high stress concentration, where the damage initiation is most likely to occur. These critical locations can eventually cause localized plastic deformation (the material surrounding the plastic zone remains elastic) that ultimately under cyclic loading lead to the complete failure of the part. When the applied cyclic load is above the tensile yield of the material and the bulk of the structure is plastically deformed, the number of cycles to failure is small. The failure prediction must be assessed by the strain-controlled approach. The next section discusses low cycle fatigue and the strain-controlled approach.

2.8 Low-Cycle Fatigue and the Strain-Controlled Approach

A complete fatigue failure curve can be divided into both low-cycle and high-cycle regimes. In the low-cycle fatigue region, the plastic deformation size is macroscopic and the number of cycles to failure is below 10^4 cycles. High-cycle fatigue is characterized by microscopic localized plastic deformation[2] and for many materials the number of cycles to failure is above 10^4 and up to 10^7 or higher. As mentioned in Section 2.7, a fatigue crack initiates at some surface discontinuity (sharp corners, hole, or notch) where material is locally overstressed and plastically deformed. The number of cycles associated with the crack initiation and stable crack growth (stages I and II) is a function of nominal stress magnitude. For cyclic induced stress amplitude above the yield stress, the bulk of the material is plastically deformed. In this situation, the extent of stage I is short and most of the cycles are spent for stable crack growth.

In high-cycle fatigue, the cyclic induced stress is in the elastic range, and plastic deformation is highly localized. The extent of stage I is large and the bulk of the structure is elastic, with plastic deformation in the locality of stress concentrations. In both cases of high-cycle and low-cycle fatigue, the material response to cyclic loading in the critical locations (such as sharp corners, holes, or notches) is strain-controlled [34–36]. In high-cycle fatigue, where the strains are predominately elastic, it is convenient to measure the strain near the highly strained region directly from linear elasticity by knowing the applied load, since the plastic strain is small and its magnitude is difficult to calculate or measure adequately. The important point to remember is that the presence of plastic deformation in the critical region is necessary for the fatigue crack to propagate and failure to occur.

When a new hardware or a machinery part is subjected to load varying environment and is initially free from cracks, it is important for the analyst to determine the total number of cycles associated with the two stages of growth, crack initiation and crack growth (stages I and II). Crack initiation usually is defined as the number of cycles consumed to initiate a crack of a given length. In the aircraft industry, the crack initiation under low cycle fatigue is selected as a crack size of approximately 0.01 in. This corresponds to the growth of a crack by the amount of 1/10 of the notch radius (a typical hole size of 0.188 to 0.25 in. hole diameter is

---

[2] Microscopic strain or simply microstrain is defined as the strain over a gage length comparable to interatomic distances. Microstrain is not measurable by existing techniques (see ASTM E-6). On the other, hand macrostrain can be measured by several methods, including strain gages and mechanical or optical extensiometers.

used in aircraft). The remaining number of cycles associated with crack growth to failure (for the initial crack of length 0.01 in. to grow to failure) can then be determined by using fracture mechanics methodology. Thus, the total life would be the sum of crack initiation life and crack growth life. The reader should note that the establishment of crack initiation size is an arbitrary value when defining failure in the low cycle fatigue environment. Therefore, it is expected that the number of cycles to initiation will vary considerably.

2.8.1 Crack Initiation Life

As discussed previously, the testing carried out to establish an S–N curve for calculating the total life of a given part was conducted under a load-controlled environment. Under load-controlled laboratory tests, the two stages of crack initiation and stable crack growth are not distinguishable because the induced cyclic stress magnitude in the bulk of the structure is a constant under constant amplitude loading; the specimen will continue to cycle until it breaks into two pieces. However, with the strain-controlled approach, where the specimen is fluctuating under a constant amplitude strain, ε, the number of cycles required to initiate the crack, at the root of a notch, can be determined through Coffin-Manson relationship (see Section 2.12). It is recommended that the use of the S–N curve for total life evaluation be avoided when the plastic deformation at the discontinuity is considerable or the magnitude of the cyclic stress is above the yield stress and the bulk of the material is plastic. The reason is that under strain-controlled conditions, in the region of highly plastic deformation, the stress and strain are not linearly proportional.

To compute the number of cycles for a crack to initiate, the material response to cyclic loading at the geometrical discontinuity must be assessed. It is assumed that the material's cyclic stress–strain behavior for a smooth specimen, tested under strain-controlled conditions, can provide useful information to assess the local material response at the geometrical discontinuity [35, 37]. In other words, one can relate the fatigue life of a notched part to the life of a smooth specimen that is cycled to the same strain level at the geometrical discontinuity contained in the part (the similarity or similitude concept) (Fig. 2.37). The local stress and strain at the geometrical discontinuity can be related through cyclic stress–strain curve exhibited as a series of hysteresis loops generated under strain-controlled conditions. The local stresses and strains can then be related to remotely measured loads, strains, and stresses by the Neuber relationship.

In Sections 2.9 and 2.10, the engineering and true stress–strain curves are briefly reviewed for the reader to understand the concept and derivation of the cyclic stress–strain curve from the hysteresis loops method. In Section 2.11, we discuss how the cyclic stress–strain curve can be utilized by the analyst in dealing with the strain-life evaluation of notched parts subjected to fluctuating load through the Neuber relationship.

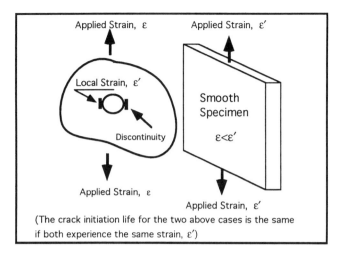

Figure 2.37 Material cyclic response for two specimens, with and without discontinuity (illustrating the concept of simulitude)

2.9 The Conventional or Engineering Stress-Strain Curve

The conventional or engineering stress–strain curve can be obtained by conducting a simple uniaxial test on standard specimens (ASTM E8, E8M). The tension test provides useful information on the strength of materials. In a uniaxial tension test, the specimen is subjected to monotonic increasing load, while its elongation and corresponding load are simultaneously plotted. All the parameters extracted from a typical engineering stress–strain curve are based on the original unloaded gage length and the original cross-sectional area of the test specimen. The engineering stress, σ, is defined as the applied load, P, divided by the original cross sectional area, A_0. The engineering strain, ε, is defined as the amount of elongation, $\Delta l = \ell_f - \ell_0$, divided by the original length, ℓ_0. A typical engineering stress–train curve is shown in Fig. 2.38.

The yield stress is defined as the stress associated with the 0.2% offset strain level and is the limit of elastic behavior. The maximum stress on the curve is called the ultimate stress and is designated by σ_{ul}. The ultimate stress can be calculated by dividing the maximum load (recorded in the chart) by the original cross-sectional area. The region to the left of the yield stress is called the elastic region, where stress and strain are linearly related.

The region to the right of the yield stress, up to the ultimate stress, has uniform straining, but the stresses and strains are not linearly related. The region from the ultimate stress, where the necking starts, up to the fracture stress, where the material breaks into two parts, displays nonuniform straining. Another important point on the engineering stress–strain curve is the fracture stress, σ_f. It can be obtained

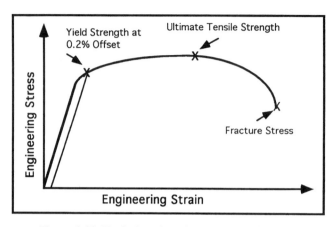

Figure 2.38 Typical engineering stress–strain curve

by dividing the final load recorded on the chart by the original cross sectional area. The strain associated with the fracture stress, σ_f, is related to the reduction of the area (RA) and it can be obtained by measuring the total elongation (by putting the two fractured pieces together) divided by the original crack length, see Appendix B for more information on this topic.

2.10 The Natural or True Stress–Strain Curve and Hysteresis Loop

2.10.1 The Natural or True Stress–Strain Curve

The natural or true stress, σ_t, is found by dividing the load, P, by the actual cross-sectional area, A, at a particular point in time (instead of the original cross–sectional area, A_0). Generally speaking, the analyst likes to compute the stresses using the original cross-sectional area, which is a measurable quantity. The use of the engineering stress is acceptable because, in engineering applications, the computed stress is required to be in the elastic range to avoid permanent deformation and to maintain structural integrity. In the elastic range, little difference can be seen between the changes in the cross-sectional area.

In the region of uniform straining, in which the specimen has undergone a considerable amount of elongation and reduction in area, the computed true stress is always larger than the engineering stress. In the region of necking, the cross-sectional area decreases rapidly, while the computed true stress increases as shown in Fig. 2.39.

The true fracture stress corresponds to the highest point on the true stress–strain curve and it can be obtained by dividing the final load by the final cross-sectional area. The true strain is based on the instantaneous gage length, ℓ,

2.10 The Natural or True Stress–Strain Curve and Hysteresis Loop

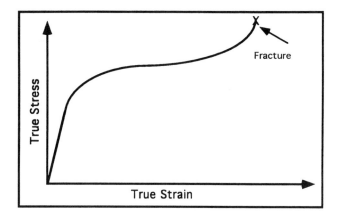

Figure 2.39 Typical true stress–strain curve

and is defined as $\varepsilon_t = \ln(\ell/\ell_0)$ or $\varepsilon_t = \ln(1+\varepsilon)$ for strain up to necking (where ε is the engineering strain). For strain above the necking up to fracture, where the gage length reading is not accurate, $\varepsilon_t = \ln(A_0/A)$.

In the elastic region, the strain and stress are related by the proportionality constant called the modulus of elasticity, E:

$$(\varepsilon_t)_e = \sigma_t/E \qquad (2.11)$$

and the true plastic straining $(\varepsilon_t)_p$ can be written in terms of the power law as:

$$(\varepsilon_t)_p = (\sigma_t/K)^{1/n} \qquad (2.12)$$

where the quantities K and n are called the strength and work hardening coefficients, respectively. The total true strain, $(\varepsilon_t)_T$, is the sum of the elastic and plastic strain:

$$(\varepsilon_t)_T = \sigma_t/E + (\sigma_t/K)^{1/n} \qquad (2.13)$$

To maintain simplicity and to avoid confusion, the subscript t shown in Eq. (2.13) that represents the true stress, σ_t, and true strain, ε_t, will be removed and replaced by σ and ε throughout the remaining part of this chapter when dealing with the cyclic stress–strain curve and the strain-life prediction method. True stresses and strains in terms of engineering stress–strain curve are discussed in Appendix B.

2.10.2 Hysteresis Loop

If we decide to load and unload a standard tensile specimen ($R = 0$) below its elastic limit stress where the bulk of the specimen is purely elastic, the generated stress-strain loop becomes a straight line with slope equal to the modules of elasticity of the material, E. By loading and unloading beyond its elastic limit for a

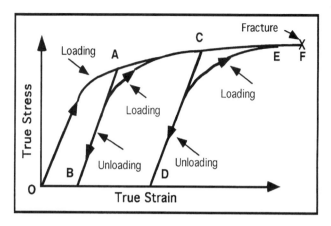

Figure 2.40 Loading and unloading a standard specimen along the true stress–strain path

few times, the true stress–strain curve will follow the path shown in Fig. 2.40. From this figure, it can be seen that, upon loading and unloading to zero stress level, the path *OABCDEF* will always follow the original curve that is associated with the monotonic true stress–strain curve. Repeated loading beyond the elastic limit will cause elevation of the yield strength of the material, as illustrated in Fig. 2.40. The work hardening exponent, n, used in Eq. 2.12 to represent the original stress–strain curve (shown by points *OAF*) is no longer valid to describe the stress–strain curve shown by *BCEF*. A new hardening exponent is, therefore, required to represent the new stress–strain curve *BCEF*. The elevation of the yield strength, when unloading along the line *BC*, can be described as the result of residual compressive stresses left behind upon loading beyond the elastic limit and unloading due to the previous loading step (as shown in Fig. 2.40 by *OAB*).

Bauschinger [38] observed that, upon unloading to zero stress level and on into compression to a stress equal to $-\sigma_{min} = -\sigma_{max}$ corresponding to point *C*, the yielding will occur before $-\sigma_{yield}$ is reached, as shown in Fig. 2.41. The same behavior was observed by reversing the direction of applied cyclic load. That is, by loading first into compression $-\sigma_{min}$ and then reloading from $-\sigma_{min}$ to stress level σ_{max}, the yield would take place before σ_{yield} is reached. This behavior was first noticed by Bauschinger and is known as the Bauschinger effect. This behavior is expected because the residual compressive stresses left behind causes yielding in compression to occur earlier than it would otherwise. The Bauschinger effect does not indicate that the yield strength in compression or tension occurs at different stress levels when cyclic loading is applied. It was later realized that the reverse yielding due to the Bauschinger effect takes place at $2\sigma_{yield}$ from σ_{max} or $-\sigma_{min}$, as shown in Figure 2.42 (G. Masing, 1923 [39]).

2.10 The Natural or True Stress–Strain Curve and Hysteresis Loop

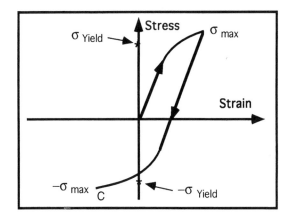

Figure 2.41 Loading and unloading into compression

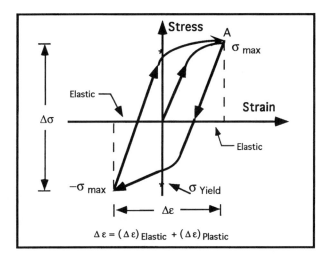

Figure 2.42 Hysteresis loop generated as a result of a complete cycle

If we now decide to apply one load cycle by unloading the specimen from point A, corresponding to stress σ_{max}, to the zero stress level and on into compression (σ_{min}) and load back up to the point A (beyond the elastic limit), a complete hysteresis loop is created (see Fig. 2.42). The total width and height of the hysteresis loop are described by the total strain range, $\Delta\varepsilon$, and the total stress range, $\Delta\sigma$, respectively. The total strain range, in terms of its components, can be written as:

$$\Delta\varepsilon = (\Delta\varepsilon)_e + (\Delta\varepsilon)_p \tag{2.14}$$

where $(\Delta\varepsilon)_e = \Delta\sigma/E$. The shape of the hysteresis loop depends on the thermomechanical process given to the material. For ductile material (for example, steel in the fully annealed condition), the width of hysteresis loop, representing the total strain, $\Delta\varepsilon$, is much wider than for high strength steel with little ductility. Moreover, the area within a hysteresis loop is related to the energy dissipated per unit volume in one cycle.

The procedure for predicting the fatigue life consists of two parts: First, we need to have the cyclic stress–strain curve that is obtainable from a set of stable hysteresis loops, and second, a mathematical relationship that represents the total strain to number of cycles to failure must be developed [40]. The cyclic stress–strain curve and the power law representation of the cyclic stress–strain curve is fully presented in Section 2.11 and in Section 2.12 the strain-life prediction models are discussed.

2.11 Cyclic Stress–Strain Curve

To obtain the cyclic stress–strain curve for a given material, several fatigue tests on a series of smooth highly polished specimens are conducted. These tests are performed under strain-controlled conditions, with stress ratio of $R = -1$ and are subjected to different strain rate amplitude. The cyclic stress–strain curve is the line drawn through tensile tips of these stabilized hysteresis loops of different strain amplitudes, plotted in the stress–strain coordinates shown in Fig. 2.43 [34]. For illustration purposes only, three points (each represents the tip of a stable hysteresis loop) are shown in Fig. 2.43 to demonstrate the method of obtaining the cyclic stress–strain curve. Such a curve can be obtained from several specimens or from one by using the incremental step test procedure [35].

Under strain-controlled conditions, a material's response to cyclic loading varies depending on the nature of the heat treatment and its initial mechanical condition (called thermomechanical history). In general, annealed materials will harden when subjected to strain-controlled cyclic conditions. The hardening effect can be observed in the hysteresis loop as an increase in the stress range, until the constant amplitude range is reached (Fig. 2.44). On the other hand, a material with prior cold-working will have a tendency to become soft when undergoing cyclic straining. This is indicated by the decrease in the stress range (Fig. 2.45), until it reaches its minimum value. Changes in stress response occur rapidly, as shown in Figs. 2.44 and 2.45 and become reasonably stable (steady-state or stabilized condition) after cycling about 10% to 20% of the total life [41–44]. In some material, the hardening or softening effects are not observed, and the steady-state (stable loop) condition is maintained when subjected to cyclic loading (Fig. 2.46). It should be noted that the cyclic stress–strain curve is generated when the hysteresis loop achieves its steady state condition. By comparing the cyclic stress–strain

2.11 Cyclic Stress–Strain Curve

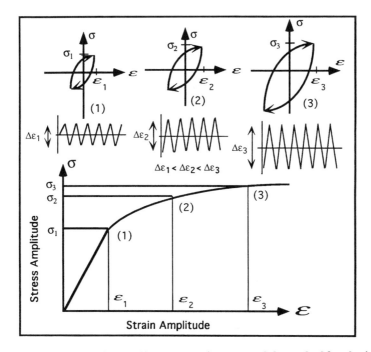

Figure 2.43 Illustration of the cyclic stress–strain curve and the method for obtaining it

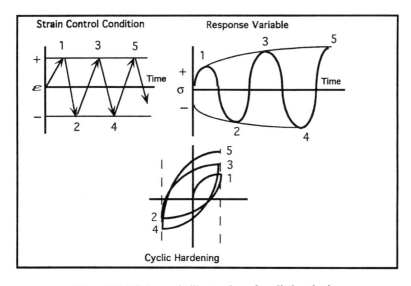

Figure 2.44 Schematic illustration of cyclic hardening

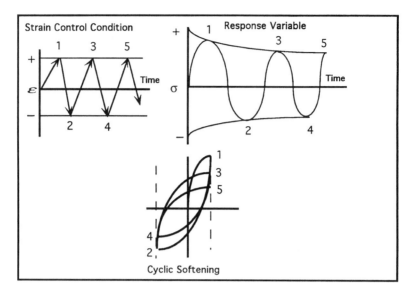

Figure 2.45 Schematic illustration of cyclic softening

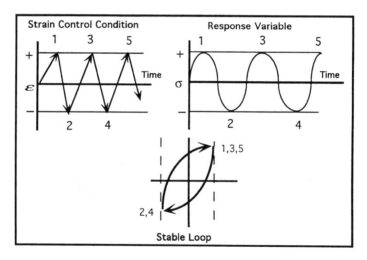

Figure 2.46 Schematic illustration of stable cyclic loop

curve with the monotonic stress–strain curve it can be concluded that material will soften or harden upon the application of cyclic load if they fall below or above the monotonic curve respectively (Fig. 2.47). For example, the cyclic stress–strain curve behavior for the SAE 4340 steel exhibits the softening effect when compared with the monotonic curve [45] as shown in Fig. 2.48.

2.11 Cyclic Stress–Strain Curve

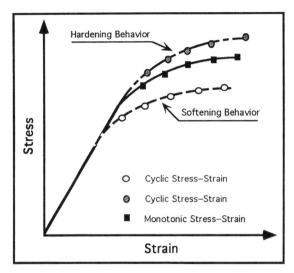

Figure 2.47 Hardening and softening behavior due to cyclic loading compared with the monotonic stress–strain curve

For materials that exhibit cyclic softening, it is important to realize that upon softening, the strength properties of the material are less than what was anticipated by the analyst. Therefore, it may be necessary for the designer to consider the softening effect of the material if a low-cycle fatigue environment is expected. In general, soft materials, such as aluminum alloys, have a tendency to undergo cyclic hardening. On the other hand, hard and strong materials, such as high strength steel, become soft. This general rule can be described in terms of the S_{ul}/S_{yield} ratio, where S_{ul} and S_{yield} are the tensile and yield strength of the material, respectively [45].

$$\text{For } S_{ul}/S_{yield} > 1.4 \quad \text{(material will harden)} \quad (2.15)$$

and

$$\text{For } S_{ul}/S_{yield} < 1.2 \quad \text{(material will soften)} \quad (2.16)$$

For ratios $1.2 < S_{ul}/S_{yield} < 1.4$, hardening or softening may occur.

The softening and hardening phenomena due to cyclic loading may be explained in terms of dislocation density. In soft materials usually after annealing, the dislocation density is low and it increases due to cycling. The interaction of dislocations as the result of density increase will cause the hardening effect. On the other hand, for an initially hardened material (usually after cold working), the dislocation density is high and it decreases due to cycling.

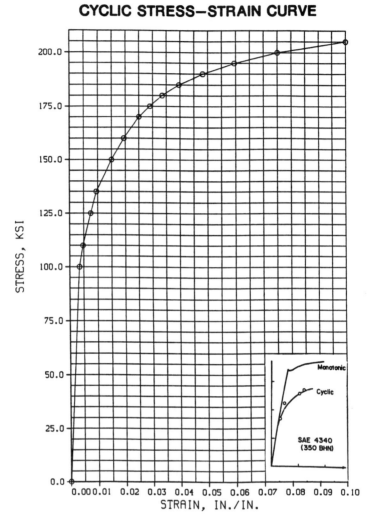

Figure 2.48 Cyclic stress–strain curve (courtesy of Douglas Aircraft Company) and monotonic stress–strain curve for 4340 steel (softening effect), taken from [45].

2.11.1 The Power Law Representation of the Cyclic Stress–Strain Curve

Just like the monotonic stress–strain curve that was described by the Ramberg–Osgood relationship, the cyclic stress–strain curve can be expressed mathematically by a power law [43]:

2.11 Cyclic Stress–Strain Curve

$$\varepsilon_T = \sigma/E + (\sigma/K')^{1/n'} \quad (2.17)$$

where the total true strain, ε_T, is given in terms of its elastic and plastic components, $\varepsilon_T = \varepsilon_e + \varepsilon_p$. The quantities K' and n' are the cyclic strength coefficient and strain hardening exponent, respectively. For most materials, the hardening exponent, n', varies between 0.1 and 0.2 [41]. If the metal is initially hard with low n value, the cyclic behavior will lead the metal to soften and cause n' to increase. On the other hand, for initially soft material, where n is high, the cyclic behavior cause n' to decrease.

From Eq. (2.17), the plastic component of the cyclic stress–strain, in terms of the power law, can be described as [43]:

$$\varepsilon_p = (\sigma/K')^{1/n'} \quad (2.18)$$

In terms of stress:

$$\sigma = K'(\varepsilon_p)^{n'} \quad (2.19)$$

The total width $\Delta\varepsilon$ and length $\Delta\sigma$ of a hysteresis loop always corresponds to a point on the cyclic stress–strain curve [Eq. (2.17)] but differ by a factor of 2. That is:

$$\Delta\varepsilon = 2\varepsilon_T \quad (2.20)$$

$$\Delta\sigma = 2\sigma \quad (2.21)$$

Therefore, an expression for the tips of hysteresis loops, in terms of the power law, can be written by utilizing Eqs. (2.20) and (2.21) as:

$$\Delta\varepsilon = \Delta\sigma/E + 2(\Delta\sigma/2K')^{1/n'} \quad (2.22)$$

Figure 2.49 is the experimentally obtained cyclic stress–strain curve for 4130 steel alloy.

Example 2.6

The cyclic stress–strain curve for a given aluminum alloy, in terms of the power law, is given as:

$$\varepsilon_T = \sigma/10^7 + (\sigma/10^4)^{1/0.2}$$

Generate the hysteresis loop for the fully reversed case and under strain–controlled conditions, with strain magnitude of 0.04. Assume that stabilization already has occurred and the loop is stable.

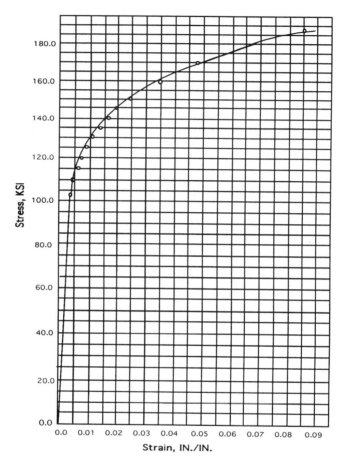

Figure 2.49 Cyclic stress–strain curve for 4130 steel (courtesy of Douglas Aircraft Company)

Solution

From Eqs. 2.20 and 2.21, any point $(\varepsilon_1, \sigma_1)$ on the cyclic stress–strain curve corresponds to $(\Delta\varepsilon_1, \Delta\sigma_1)$ on the hysteresis loop. The portion of the strain-time cycles under strain-controlled conditions, where $\Delta\varepsilon = 0.04$, is illustrated in Fig. 2.50.

The value of the stress belonging to the first point (1) can be calculated from the cyclic stress–strain equation as:

$$0.02 = \sigma/10^7 + (\sigma/10^4)^{1/0.2}$$

where the value of σ can be solved algebraically yielding:

2.11 Cyclic Stress–Strain Curve

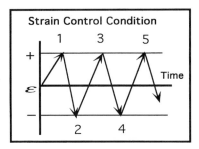

Figure 2.50 Strain–time cycles for Example 2.6

$$\sigma_1 = 4600 \text{ psi}$$

For point (2) shown in the figure, the stress can be calculated using the equation that describes the tips of the hysteresis loop [Eq. (2.22)], by considering the total strain range $\Delta\varepsilon = 0.04$, as:

$$0.04 = \Delta\sigma/10^7 + 2(\Delta\sigma/2 \times 10^4)^{1/0.2}$$

Solving the above equation, the value of the stress can be calculated as:

$$\Delta\sigma = 9200 \text{ psi}$$

and the coordinate of the second point is:

$$\varepsilon_2 = \varepsilon_1 - \Delta\varepsilon = (0.02 - 0.04) = -0.02$$

while the corresponding stress becomes:

$$\sigma_2 = \sigma_1 - \Delta\sigma = (4600 - 9200) = -4600 \text{ psi}$$

The coordinates of other points (3, 4, and 5) can be obtained through the same procedure. The resulting stress–strain hysteresis loop, corresponding to the above strain-controlled cycling, is shown in Fig. 2.51.

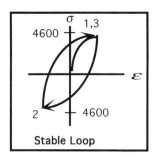

Figure 2.51 The resulting stress–strain hysteresis loop for Example 2.6

2.12 Strain-Life Prediction Models

For many materials, the log–log plot of alternating stress versus the number of cycles to failure, N, can provide the analyst with a straight line relationship. It was shown in Section 2.4 that the alternating stress, in terms of the number of cycles to failure, can be expressed in power law form. In dealing with high-cycle fatigue, the true fatigue strength, σ_a, in terms of half cycle reversal to failure ($2N_f$), can be expressed as:

$$\sigma_a = \sigma_f' (2N_f)^b \qquad (2.23)$$

where σ_f' is the fatigue strength coefficient. Its value is equal to the monotonic true fracture stress, σ_f, corresponding to $2N_f = 1$. The value of the true fracture strength, σ_f, when $N = 1$ is larger than the engineering fracture strength of the material (final load divided by the final cross-sectional area). When the fatigue strength coefficient value is not available to the analyst, an estimate of $\sigma_f = \sigma_{Ul} + 50$ ksi can be used (applicable to ferrous material only) [34], see also Appendix B.

The use of $2N_f$ reversals to failure, instead of N_f cycles to failure, is common practice throughout the literature. Imagine a test specimen that is subjected to a monotonic load up to failure. The test specimen can be regarded as having the shortest possible life. The failure process as a result of a single load application to failure is considered as a half cycle, N. Thus, based on this notation a complete cycle or full reversal can be designated as $2N$.

Equation (2.23) was derived by Basquin [13], who in 1910 first proposed the straight line relationship between the true cyclic stress amplitude and the number of cycles to failure. The quantity b is called the fatigue strength exponent (Basquin's exponent) and, for most materials, it varies between -0.05 and -0.12, with typical values between -0.085 and -0.1 [46, 47]. In the elastic range, the quantity σ_a, described by Eq. (2.23), can be replaced by the elastic strain amplitude, ε_e, as:

$$\varepsilon_e = \sigma_f'/E(2N_f)^b \qquad (2.24)$$

The plastic strain amplitude, ε_p, is also related to the number of half-cycle reversals to failure, $2N_f$ (or number of cycles to failure N_f) by a simple power law equation:

$$\varepsilon_p = \varepsilon_f' (2N_f)^c \qquad (2.25)$$

Equation (2.25) is called the Coffin and Manson law [42, 43]. The quantity ε_f' is called the fatigue ductility coefficient and its value is equal to the true monotonic fracture strain, ε_f. The fatigue ductility coefficient, in terms of reduction in area (RA), can be expressed as:

2.12 Strain-Life Prediction Models

$$\varepsilon'_f = \ln\left[\frac{1}{1 - RA}\right] \qquad (2.26)$$

The exponent C in Eq. (2.25) varies between -0.5 and -0.7, with a typical value of -0.6 [42, 43].

In Eqs. (2.24) and (2.25), the constants $\sigma'_f, b, \varepsilon'_f$, and c are all considered to be material properties. Examples of elastic and plastic strain-life curves for 4340 steel, described by Eqs. (2.24) and (2.25), are shown in Figs. 2.52 and 2.53,

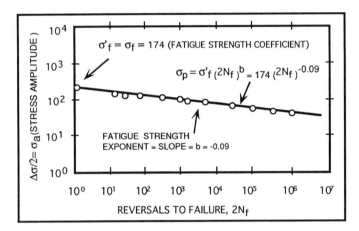

Figure 2.52 Fatigue strength-life plot for annealed SAE 4340 steel (reproduced from [45])

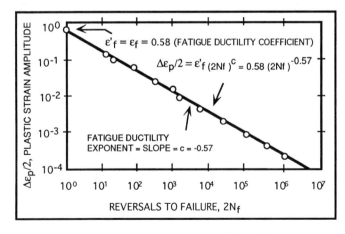

Figure 2.53 Fatigue ductility-life plot for annealed SAE 4340 steel (reproduced from [45])

respectively [45]. Combining the elastic and plastic components of the total strain-life amplitude described by the Basqin and Coffin–Manson relationship [Eqs. (2.24) and (2.25)], the total strain-life curve can be expressed as [45]:

$$\varepsilon = \sigma_f'/E(2N_f)^b + \varepsilon_f'(2N_f)^c \quad (2.27)$$

Equations (2.24) and (2.25), together with Eq. (2.27), are each plotted in log–log form and are shown schematically in Fig. (2.54). In this figure the reversal to failure cycles, $2N_f$, is the dependent variable and the strain amplitude, ε, is selected as the independent or controlled variable. The term controlled variable is used here to indicate that the total strain is controlled throughout the test duration between its maximum and minimum assigned control limits. Figure 2.55 is the strain-life curve developed for 4340 steel by plotting the elastic and plastic components of Eq. (2.27), as shown in Fig. (2.52) and (2.53), respectively.

For most metals, the strain-life curve described by Eq. (2.27) can be developed by testing at different strain amplitudes between ±2.0% and ±0.2% percent [48, 49]. Usually between 9 and 13 specimens are required to obtain the low-cycle fatigue strain-life curve of a metal [49]. It is preferable to have low cycle fatigue tests under strain-controlled condition and under axial reversed loading in order to

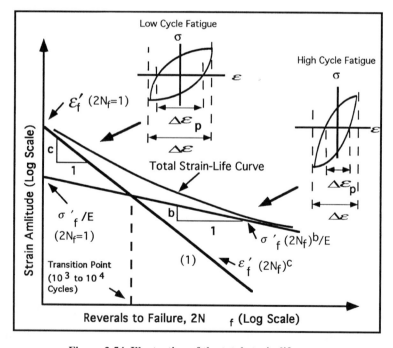

Figure 2.54 Illustration of the total strain-life curve

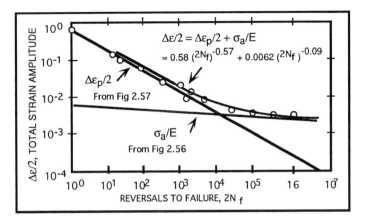

Figure 2.55 Total strain-life plot for annealed SAE 4340 steel (reproduced from [45])

avoid complexities due to the presence of non-uniform stress distribution through the thickness (low cycle fatigue test by reverse bending, see "Engineering Materials Evaluation by Reverse Bending" by M. R. Gross). Moreover, the results of axial reversed loading can easily be compared with the static properties of the metal generated under monotonic tension load. The number of cycles to failure for specimens under strain-controlled tests correspond to initiation of a crack usually between 0.01 to 0.05 in. in length. The crack initiation is detected from the test machine when the load drops after stabilization cycles by 5% to 10%.

Based on the ASTM E-606 Standard, the definition of failure under strain-controlled conditions (depending on the use of the fatigue life test data) can be defined as follows:

1. Total separation or fracture of the test specimen into two pieces.
2. Failure occurs when the ratio of modulus for unloading, E_{UN}, to loading, E_L, reaches half of the value corresponding to the first cycle in the beginning of the test. The number of cycles when $1/2(E_{UN}/E_L)$ is reached is designated as the number of cycles to failure, N_f.
3. Failure is defined as occurring when the maximum applied load is decreased by 10% because of the presence of a crack or several cracks.
4. The number of cycles associated with the presence of a surface crack detected visually which is equal to or larger than some predetermined size specified by the user of the data.

Differences in reversal to failure, N_f, in low-cycle fatigue test are expected when different failure conditions are employed (see definitions above provided by the ASTM Standards). For this reason the supplier of the ε-N data must indicate the failure definition used when the final results are submitted to the user.

In real life when damage is the result of different strain amplitudes, the sum of the all the damage accumulated by the environment on the part based on Miner's rule can be written as:

$$D = \Sigma \, n_i/N_{fi} = 1.0 \tag{2.10a}$$

where N_{fi} is defined as the number of cycles to failure (the initiation of a detectable crack or total failure) at a given strain amplitude, ε_i. In aircraft applications, when the structural part is subjected to a high fatigue load environment, the failure is associated with the accumulation of all the cycles for the initiation of a detectable crack around 0.01 to 0.05 in. in size. If the structure under consideratin is now subjected to additional load cycles with load amplitude much smaller than previous cycles, the remaining fatigue life may be determined through linear elastic fracture mechanics.

2.12.1 Transition Point, N_T

The intercept of the two straight lines at $2N_f = 1$ are ε_f' and σ_f'/E for the plastic and elastic components of Eq. (2.27), shown in Fig. 2.54. The slopes of the two lines, b and C, are shown separately in the same figure. It can be seen that the two curves, corresponding to the elastic and plastic components of Eq. (2.27), are intersecting at a point called the "transition point." For a number of cycles less than the transition point, the plastic component dominates; this represents the low-cycle regime (short lives). For a number of cycles longer than the transition point, the elastic component dominates the plastic one; this represents the high-cycle regime (long lives). At the intercept of the two lines, the plastic and elastic strain amplitudes are the same. The plastic component is governed by the ductility, defined in terms of the reduction of the area, described by Eq. (2.25), and the elastic component described by the term σ_f'/E.

The number of cycles to failure associated with the transition point, N_T, can be obtained by equating Eqs. (2.24) and (2.25). At the transition point, the quantity N_f is replaced by N_T:

$$\sigma_f'/E(2N_T)^b = \varepsilon_f'(2N_T)^c$$

Solving for $2N_T$ (reversals):

$$2N_T = [(\varepsilon_f' E)/\sigma_f']^{-1/(c-b)} \tag{2.28}$$

For most materials, the number of cycles to failure associated with the transition point, $2N_T$, falls between 10^3 and 10^4 cycles. The experimental data shows that most metals have almost the same reversal to failure at strain amplitude of ± 0.01 ($\Delta\varepsilon = 0.2$) [17].

2.12 Strain-Life Prediction Models

The selection of material to assess fatigue problems is strongly related to the magnitude of the fluctuating load. In high-cycle fatigue problems, where the nominal stress is in the elastic range, the selection of materials with high tensile strength and lower ductility, or ductile material with surface treatment (such as shot peening, carburizing, or nitriding) is appropriate. Thus, for a high-cycle fatigue design life requirement, where reversal to failure is above 10^3 to 10^4 cycles ($N_f > N_T$), it would be appropriate to select a high strength material. On the other hand, for fatigue life requirements below the transition point, where reversal to failure is lower than 10^3 to 10^4, one would pick a ductile material [50, 51].

The following example problem will assess the type of surface treatment needed when low or high cyclic loads are applied to a given material selected for use in a structural component.

Example 2.7

A part is made of 2014-T6 aluminum alloy with the following fatigue properties:

Modulus of elasticity:	$10E + 7$ psi
Tensile strength:	55,000 psi
Fatigue ductility coefficient:	0.40
Fatigue strength exponent:	-0.10
Fatigue ductility exponent:	-0.65

It is recommended by the customer to surface treat the part for better fatigue properties. Determine whether this treatment is necessary for a load environment where the part is cycling under a fully reversed constant strain amplitude of 0.018 in./in.

Solution

We would like to know if the applied fully reversed constant strain amplitude of 0.018 is associated with low- or high-cycle fatigue. By employing Eq. (2.28), the total number of cycles at the transition point can be computed for the 2014-T6 aluminum alloy:

$$2N_T = [(\varepsilon_f' E)/\sigma_f']^{-1/(c-b)}$$
$$[(10^7 \times 0.4)/55{,}000]^{-1/(-0.65+0.1)}$$
$$2N_T = 2358 \text{ reversals}$$

The total strain at the transition (where $2N_T = 2358$ reversals) can be computed by using Eq. (2.25) or (2.24), since, at the transition point the two equations are equal:

$$\varepsilon_p = \varepsilon_f' (2N_f)^c$$
$$= 0.4(2358)^{-0.65}$$
$$\varepsilon_p = 0.00256$$

The total strain at the transition point is smaller than the fully reversed applied constant strain amplitude of 0.018; thus it falls under the low-cycle fatigue classification. For this reason, the surface treatment is not recommended. It should be noted that increasing the strength of materials through cold working or heat treatment may improve fatigue properties in the high-cycle fatigue regime (make them worse in the low-cycle fatigue environment).

2.12.2 The Mean Stress Effect

Most of the available fatigue data are collected in the laboratory through testing by applying the fully reversed loading case ($R = -1$), where the mean stress, S_M, is zero. In real situations, where S_M is not zero, the influence of the mean stress on the fatigue life of the part cannot be addressed through Eq. (2.27), where $R = -1$. However, Eq. (2.27) and the strain-life curve shown in Fig. 2.55 can be corrected to account for the effect of the mean stress, S_M. A correction for the mean stress can be accomplished by simply adding the mean stress to (if $S_M > 0$) or subtracting it from (if $S_M < 0$) the strength coefficient, σ_f', of the elastic part of Eq. (2.27) [48]:

$$\varepsilon_e = (\sigma_f' - S_M)/E(2N_f)^b$$

$$\varepsilon = (\sigma_f' - S_M)/E(2N_f)^b + \varepsilon_f'(2N_f)^c \tag{2.29}$$

Another equation that is available to correct the effect of the mean stress on the strain-life prediction method was suggested by Smith–Watson–Topper [52]. This approach is based on the strain life test data obtained for various mean stress values. They considered the effect of the mean stress on the predicted life through the maximum stress, where $S_{max} = S_M + S_a$. The S_{max} of a cycle for the case of $R = -1$ (equation 2.33) was multiplied by the strain-life equation, ε, described by equation 2.27.

$$S_{max}\varepsilon = \sigma_f'^2/E(2N_f)^{2b} + \sigma_f'\varepsilon_f'(2N_f)^{c+b} \tag{2.30}$$

2.12.3 Fatigue Notch Factor, K_f

Fatigue failure of structural parts during their service life always has its origin at some surface discontinuity, sharp corner, hole, or the root of a notch, where material has deformed locally. For a given nominal stress, the presence of a notch will create stress concentration at the discontinuity and reduce the service life of the structure. The ratio of the fatigue strength of an unnotched part (specimen with no stress concentration) to the fatigue strength of a notched part at N number of cycles (under the same environment) is designated by K_f and is called the "Fatigue Notch Factor or Fatigue Stress Concentration Factor." It is also called the Fatigue Strength Reduction Factor [53]:

$$K_f = (\text{Fatigue Strength})_{unnotched}/(\text{Fatigue Strength})_{notched} \tag{2.31}$$

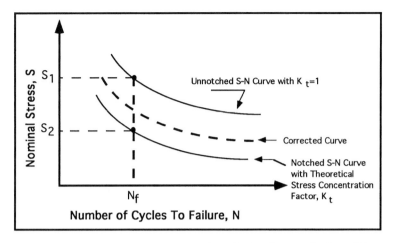

Figure 2.56 S–N curves for two cases of notched and unnotched conditions

The quantity K_f varies with material, type of notch, and loading. Figure 2.56 illustrates the S–N curves for a given material developed for two cases of notched and unnotched conditions where $K_f = S_1/S_2$ corresponding to N_f number of cycles (note that in figure 2.56 the nominal stress, S_a, is plotted against fatigue life cycle, N, on a Logarithmic scale). For example, the value of fatigue notch factor, K_f, at 5×10^6 cycles (for specimen tested in rotating bending test machine where $K_t = 1.6$) are given as 1.0 and 1.6 for Gray Iron and hard steel respectivelly [54]. In terms of the theoretical stress concentration factor of a monotonic tension load in the elastic range, K_t, the fatigue notch factor, K_f, can be approximated [53] as:

$$K_f = 1 + \frac{K_t - 1}{1 + a/r} \tag{2.32}$$

where the quantity r represents the notch root radius and a is a material constant given in units of length and is dependent on the material's strength and ductility. In Figure 2.56, the notched curve is calculated from the unnotched S–N curve by dividing each stress point on the curve by the theoretical stress concentration factor, K_t. A less conservative curve can be obtained by using the value of K_f, from Eq. (2.32), in place of K_t, for life estimation when long life is expected on the structural part, see the corrected dotted curve shown in Figure 2.56. In Eq. (2.32), the inverse of the quantity in the denominator is called the notch sensitivity factor, q:

$$q = \frac{1}{1 + a/r}$$

The estimated value of a for heat-treated steel is given by Peterson [33] as:

Table 2.12 Material constant for the Quantity a (average values)

Material	a, inches
Aluminum alloy	
2024-T4 Sheet	0.05
bar	0.008
7075-T6 Sheet	0.02
bar	0.003
Steel	
Annealed and Normalized	0.01
Quenched and Tempered	0.0025

$$a = \left(\frac{300{,}000}{S_{Ul}(\text{psi})}\right)^{1.8} \times 10^{-3} \quad \text{(in inches)} \tag{2.33}$$

where S_{Ul} is the ultimate of the material. Some average values of a for 2024-T4 and 7075-T6 aluminum alloys and steel are shown in Table 2.12 [53].

From Eq. (2.32) it can be seen that the value of K_f varies between $K_f = 1$ (no notch effect when $r = 0$) and the theoretical stress concentration factor K_t (for large notch with large r). When sufficient data are not available, the value of $K_f = K_t$ can be used where a large number of cycles are expected. On the other hand, the K_f value is lower, compared to the theoretical stress concentration ($K_f < K_t$), when short fatigue lives are expected. In general, designing for fatigue using theoretical stress concentration factor, K_t, is safe but conservative when the experimental value of K_f is not available to the analyst.

Example 2.8

A plate that is made of steel (in quenched and tempered condition) is subjected to cyclic load, as shown in Fig. 2.57. The plate is subjected to fully reversed cyclic loading with amplitude of 18.0 kips for 2×10^5 cycles. Determine the factor-of-safety with respect to applied stress (S_a) and life ($2N_f$). The stress-life relationship in high cycle fatigue ($N_f > 10^3$), is given as:

$$S_a = (\sigma'_f/E)(2N_f)^b = 295(2N_f)^{-0.88}$$

Solution

The S–N curve, from the stress-life (S–N) equation described above, is shown in Fig. 2.57. The theoretical stress concentration factor, $K_t = 3$, for a hole in a plate

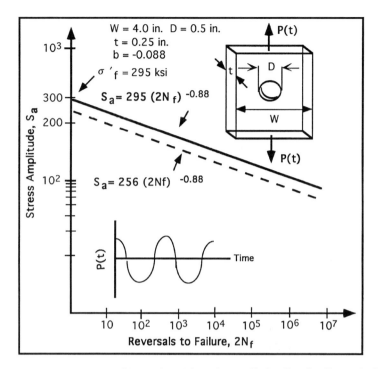

Figure 2.57 Steel plate with a hole subjected to cyclic loading for Example 2.8

and the fatigue notch factor, K_f, in terms of the theoretical stress concentration factor (K_t) from Eq. (2.32), is:

$$K_f = 1 + \frac{K_t - 1}{1 + a/r}$$

where $a = 0.0025$ (for quenched and tempered steel) and $r = 0.25$ in. Using $K_t = 3$, the value of $K_f = 2.980$. The stress amplitude, σ_a, at the stress concentration can be calculated as:

$$\sigma_a = K_f[P(t)/A]$$

where

$$A = (W - D)t = (4 - 0.5) \times 0.25 = 0.875 \text{ in.}^2$$

$$\sigma_a = 2.980 \times (18.0/0.875) = 61.30 \text{ ksi}$$

For $2N_f = 2 \times 10^5$ cycles, the fatigue strength from the S–N diagram shown in the Fig. 2.57 is:

$$S_a = 100.77 \text{ ksi}$$

The computed factor of safety, FS = 100.77/61.30 = 1.64 (with respect to applied stress). The total number of cycles to failure associated with $\sigma_a = 61.30$ ksi can be computed as:

$$\sigma_a = 61.30 \text{ ksi} = 295(2N_f)^{-0.88}$$
$$(2N_f) = 6 \times 10^7 \text{ cycles}$$
$$N_f = 3 \times 10^7$$

The computed factor-of-safety, FS = $6 \times 10^7 / 2 \times 10^5$ = 300 (with respect to life)

Example 2.9

Repeat Example 2.8 for the case of cyclic loading with amplitude of 18.0 kips and a mean stress value of 11.5 kips.

Solution

The same procedure should be applied. The mean stress, σ_M, can be computed the same way as σ_a, and is given by:

$$\sigma_M = K_f[P(t)/A]$$
$$2.980 \times (11.5/0.875)$$
$$\sigma_M = 39.16 \text{ ksi}$$

The S–N curve, corrected for the mean stress, is described by Eq. (2.29) to include the effect of the mean stress:

$$S_a = (\sigma_f'/E - \sigma_M)(2N_f)^b$$
$$(295 - 39.16)(2N_f)^{-0.088}$$
$$255.84(2N_f)^{-0.088}$$

The S–N curve for the mean correction is shown in Fig. 2.57 as the dotted line. The fatigue strength for $2N_f = 2 \times 10^5$ cycles is found from the above equation:

$$S_a = 255.84(2 \times 10^5)^{-0.088}$$
$$S_a = 87.39 \text{ ksi}$$

The computed factor of safety, FS = 87.39/61.3 = 1.42 (with respect to applied stress). The total number of cycles to failure associated with σ_a = 61.30 ksi can be computed by:

$$61.3 = 255.84(2N_f)^{-0.088}$$

$$2N_f = 1.2 \times 10^7 \text{ cycles}$$

The computed factor of safety, FS $\approx 7 \times 10^6 / 2 \times 10^5$ = 35 cycles (with respect to life)

2.12.4 Stress and Strain at Notch (Neuber Relationship)

In evaluating the fatigue life of a part that contains stress concentration (as shown in Fig. 2.58), the actual stress and strain at the notch must be computed. Finite element methods (FEM) can enable the analyst to transfer the applied load to the structure and distribute it in the form of stress across the body (the values for the stress concentrations can also be obtained experimentally through the use of photoelasticity). As a result, the stresses at every point in the body, specifically in localized areas, such as sharp corners, holes, or the roots of a notch, can be obtained by the FEM. In the elastic range, when the nominal, S, and local stresses, σ, are below the yield, the theoretical stress concentration factor (K_t) is:

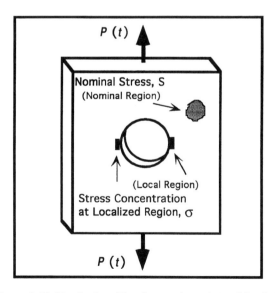

Figure 2.58 Nominal and local stress in a plate with a hole

$$K_t = \frac{\sigma}{S} \tag{2.34}$$

Under this condition, where the plasticity at the locality of the notch is absent, there is no fatigue failure accumulated in the part. In the presence of local yielding at the notch, the stress concentration can no longer be defined by Eq. (2.34). Both the stress and strain concentrations in that locality must be addressed. The strain and stress concentration factors (K_ε, K_σ) for the case of local yielding (when applied stress is in the elastic range) can be expressed as:

$$K_\sigma = \frac{\sigma}{S}, \quad K_\varepsilon = \frac{\varepsilon}{(S/E)} \tag{2.35}$$

Based on Nueber's rule [4], the product of the two quantities K_σ and K_ε during the plastic deformation is related to the theoretical stress concentration (in the elastic range), K_t, by:

$$K_\sigma K_\varepsilon = K_t^2 \tag{2.36}$$

Rearranging Eq. (2.36) in terms of the two quantities (σ, ε) described by Eqs. (2.34) and (2.35), it follows that:

$$\sigma\varepsilon = \frac{(K_t S)^2}{E} \tag{2.37}$$

For a given nominal stress, S, the localized stress, σ, and strain, ε, induced at the notch can be calculated, provided that the stress–strain equation defined by Eq. (2.13) is available for solving the two unknowns quantities (local stress and strain) shown by Eq. (2.37). When the applied load is monotonic, the stress–strain relationship defined by Eq. (2.13) can be expressed by:

$$\varepsilon = \sigma/E + (\sigma/K)^{1/n}$$

For a cyclic stress–strain curve:

$$\varepsilon_T = \sigma/E + (\sigma/K')^{1/n'} \tag{2.17}$$

Note that, in the case of cyclic loading, the cyclic stress–strain curve defined by Eq. (2.17) can be replaced by hysteresis loops or by multiplying the cyclic stress–strain curve by 2 (cyclic $\sigma - \varepsilon \times 2$). Moreover, the nominal and local stresses and strains, shown in Eq. (2.37), can be replaced by stress and strain ranges:

2.12 Strain-Life Prediction Models

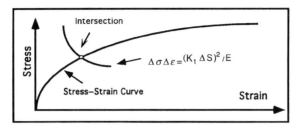

Figure 2.59 Illustration of the graphical method of obtaining the stress and strain at the notch

$$\Delta\sigma\Delta\varepsilon = (K_t \Delta S)^2 / E \qquad (2.38)$$

The graphical solution for obtaining the local stress and strain (σ, ε) at the notch is also illustrated in Fig. 2.59. Here, the intersection of two curves, described by Eqs. (2.13) and (2.37) for the case of a monotonic loading and (2.17) and (2.37) for a cyclic loading, respectively, will define the desired values of and σ and ε.

Example 2.10 (Monotonic Tension Load Case)

Determine the stress and strain at the stress concentration point for the applied nominal stress of 50 ksi, shown in Fig. 2.60. The values of n and K describing the monotonic stress–strain curve for SAE 4340 steel (annealed condition) are 0.1 and 110.0 ksi, respectively. The value of E for SAE 4340 steel is 3×10^4 ksi.

Solution

Using Eqs. (2.13) and (2.37), the two unknowns, σ and ε, can be computed as:

$$\varepsilon = \sigma/E + (\sigma/K)^{1/n}, \qquad \sigma\varepsilon = \frac{(K_t S)^2}{E}$$

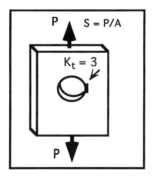

Figure 2.60 Determination of stress and strain at the stress concentration site for Example 2.10

From (2.37)

$$\sigma\varepsilon = (3 \times 50)^2/3 \times 10^4$$

And from (2.13)

$$\varepsilon = \sigma/3 \times 10^4 + (\sigma/110.0)^{1/0.1}$$

Solving for σ and ε:

$$\sigma = 70 \text{ ksi and } \varepsilon = 0.011$$

Employing Eq. (2.35), the stress and strain concentration can be obtained by:

$$K_\sigma = 70/50$$
$$K_\sigma = 1.4$$
$$K_\varepsilon = \frac{\varepsilon}{(S/E)} = 0.011(50/3 \times 10^4)^{-1}$$
$$K_\varepsilon = 6.43$$

where

$$K_t^2 = K_\sigma K_\varepsilon = 1.4 \times 6.43 = 9.0$$

Example 2.11 (Cyclic Load Case)

For the same problem, the cyclic load, $P(t)$, is available and is shown in Fig. 2.61. Properties related to the material are given in Table 2.13.
Determine the number of cycles to failure.

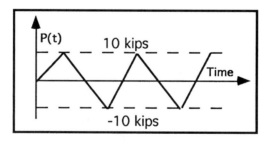

Figure 2.61 The applied cyclic load, $P(t)$, for Example 2.11

2.12 Strain-Life Prediction Models

Table 2.13 Strength and fatigue properties

Modulus of elasticity	3E + 4 ksi
Tensile strength	180 ksi
Fatigue ductility coefficient	0.626
Fatigue strength coefficient	−0.080
Fatigue ductility coefficient	−0.73
n'	0.16
K'	238 ksi
Cross-sectional area:	0.75 in.^{2}

Solution

Using Eqs. (2.17) and (2.38), the two unknowns ε and σ can be computed as follows:

$$\varepsilon = \sigma/E + (\sigma/K')^{1/n'} \quad \text{and} \quad \Delta\sigma\Delta\varepsilon = (K_t \Delta S)^2/E$$

ΔS can be computed by:

$$\Delta S = P(t)_{max}/A$$
$$= 20 \text{ kips}/0.75 \text{ in.}^2$$
$$\Delta S = 26.66 \text{ ksi}$$

The local stress and strain associated with applied $\Delta S = 26.66$ ksi can be found by solving the two Eqs. (2.17) and (2.38):

$$\Delta\sigma\Delta\varepsilon = (K_t \Delta S)^2/E$$
$$\Delta\sigma\Delta\varepsilon = (3 \times 26.22)^2/3 \times 10^4$$
$$\Delta\sigma\Delta\varepsilon = 0.213$$

Using Eq. (2.17) and applying the trial and error approach to obtain $\Delta\sigma$ and:

$$\Delta\varepsilon = 90/3 \times 10^4 + (90/238)^{1/0.16}$$
$$\Delta\varepsilon = 0.0023 \text{ in./in.}$$
$$\Delta\sigma = 90.0 \text{ ksi}$$

The number of cycles to failure can be obtained by using Eq. (2.27):

$$\varepsilon = \sigma'_f/E(2N_f)^b + \varepsilon'_f(2N_f)^c$$

At the transition point the number of cycles (reversals to failure) is:

$$2N_T = [(\varepsilon_f' E)/\sigma_f']^{-1/(c-b)}$$

$$2N_T = [(3 \times 10^4 \times 0.625)/180/]^{-1/(-0.73+0.08)}$$

$$2N_T = 1222 \text{ cycles}$$

The strain associated with $2N_T = 1222$ cycles is:

$$\varepsilon = 180/3 \times 10^4 (1222)^{-0.08} + 0.625(1222)^{-0.73}$$

$$\varepsilon = 0.007 \text{ in./in.}$$

This value is larger than $\Delta\varepsilon = 0.0023$ and therefore falls into the high cycle fatigue regime:

$$\varepsilon = \sigma_f'/E(2N_f)^b$$

$$0.0023 = 180/3 \times 10^4 (2N_f)^{-0.08}$$

$$2N_f = 10,000 \text{ cycles}$$

2.13 Universal Slope Method

The universal slope method is a useful tool for estimating fatigue life when access to fatigue data (mainly the exponents b and c of Equation 2.27) is difficult. For this reason, a great deal of effort has been put into finding ways to utilize the monotonic properties that are available for use in cyclic study. Given the ultimate strength, fatigue ductility coefficient, and modulus of elasticity (σ_{ul}, ε_f' and E) that are obtainable from monotonic tensile testing, the total number of cycles to failure can be calculated. Figure 2.62 illustrates the method of universal slopes. This approach is also called Manson's method of universal slope [55] and it is the simplified version of Eq. (2.27). The equation that describes the universal slope method is:

$$\Delta\varepsilon = 3.5(\sigma_{ul}/E)(N_f)^{-0.12} + \left\{\ln\left[\frac{1}{1-RA}\right]\right\}^{0.6} (N_f)^{-0.6} \qquad (2.39)$$

where the exponents b and c from Eqs. (2.24) and (2.25) were replaced by an average slope value of -0.12 and -0.6, respectively. The only variables that are effective in computing the fatigue cycles for different materials are the static properties, σ_{ul}, ε_f', and E that are obtainable through uniaxial tension test. Note that the quantity ε_f' is replaced by its equivalent quantity, the reduction of the area (RA), described by Eq. (2.26).

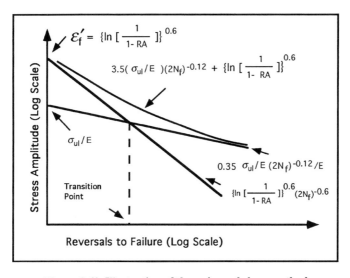

Figure 2.62 Illustration of the universal slope method

It can be seen from Eq. (2.39) that the universal slope method assumes the exponents b and c to be the same for all materials. It should be emphasized that the values assigned to exponents b and c are not always -0.12 and -0.6 and that somewhat different values have been reported for other materials. Therefore, caution must be taken in employing Eq. (2.39) for fatigue life estimation.

References

1. "Wöhler's Experiments on the Strength of Metals," Engineering, August 23, 1967, p. 160.
2. A. Palmgren, "Ball and Roller Bearing Engineering," translated by G. Palmgren and B. Ruley, SKF Industries, Inc., Philadelphia, 1945, pp. 82–83.
3. M. A. Miner, "Cumulative Damage in Fatigue," Appl. Mech., Trans. ASTM, Vol. 12, September 1945, pp. A-159-164.
4. H. Neuber, "Theory of Stress Concentration for Shear-Strained Prismatical Bodies with Arbitrary Nonlinear Stress–Strain Law," Trans., ASME, Appl. Mech., December 1961, pp 544.
5. Fatigue Technology Inc. (FTI), Extending the Fatigue Life of Metal Structures, Material Testing, Seattle, Washington.
6. N. E. Dowling, W. R. Brose, and W. K. Wilson, "Fatigue Failure Prediction for Complicated Stress–Strain Histories," T. A&M. Report No. 337, University of Illinois, Urbana, January 1971.
7. T. Indo and M. Matsuishi, "Fatigue of Metals Under Random Strain," Preprint of Japan Society of Mechanical Engineering.

8. S. J. Brodeur, and M. I. Basci, "Fracture Mechanics Loading Spectra for STS Payloads," AIAA-83-2655-CP, 1983.
9. Fatigue Crack Growth Computer Program "NASA/FLAGRO" Version 2.2, JSC-22267A, May 1994.
10. ASTM, Manual on Fatigue Testing. ASTM Special Technical Publication NO. 91. Philadelphia, ASTM Committee E-9 on Fatigue, 1949.
11. ASTM, A Tentative Guide for Fatigue Testing and the Statistical Analysis of Fatigue Data. ASTM Special Technical Publication No. 91-A. Philadelphia, ASTM, 1963, pp. 19–22.
12. Mil-Handbook 5, Department of Defense, Washington, D.C.
13. O. H. Basquin, "The Exponential Law of Endurance Tests," Proc. ASTM, Vol. 10, Part II, 1910, p. 625.
14. B. P. Haigh, "The Relative Safety of Mild and High-Tensile Alloy Steels Under Alternating and Pulsating Stresses," Proc. Inst. Automob. Engin, Vol. 24, 1929/1930, p. 320.
15. W. Gerber, "Besdtimmung der Zulossigen Spannungen in Eisen Constructionen," Z. Bayer Arch. Ing. Ver., Vol. 6, 1874, p. 101.
16. J. Goodman, Mechanics Applied to Engineering. Longmans, Green & Co., 1899.
17. H. O. Fuchs, and R. I. Stephens, "Metal Fatigue in Engineering," 1980, pp 69–76.
18. G. Jacoby, "Fractographic Methods in Fatigue Research", Exp. Mech., March 1965, pp. 65–82.
19. C. A. Zapffe and M. Clogg, Jr., "Fractography—A New Tool for Metallurgical Research", Preprint 36, American Society for Metals, 1944, later published in Trans. ASM, Vol. 34, 1945, pp. 71–107.
20. Short Fatigue Crack, edited by, K. J. Miller and E.R. de los Rios, ESIS, Publication 13, p. 31.
21. Small Fatigue Cracks, edited by R. O. Ritchie and J. Lankford, "Proceedings of the Second Engineering Foundation International Conference/Workshop, Santa Barbara, California, January 5–10, 1986." A Publication of the Metallurgical Society, Inc.
22. The Behavior of Short Fatigue Cracks, edited by K. J. Miller and E. R. de los Rios, EGF, Publication 1 (Collection of Papers and References in Crack Initiation).
23. G. F. Vander Voort, "Conducting the Failure Examination," Met. Eng. Q., Vol. 15, May 1975, pp. 31–36.
24. C. A. Zapffe et al., "Fractography: The Study of Fracture at High Magnification," Iron Age, Vol. 161, April 1948, pp. 76–82.
25. The Making, Shaping and Treating of Steel, Ninth Edition, edited by H. E. McGannon, United State Steel, Pittsburgh, December, 1970.
26. R. A. Flinn, and P. K. Trojan, Engineering Materials and Their Applications, Third Edition, Houghton Mifflin, 1986.
27. H. Mughrabi, F. Ackermann, and K. Herz, (1979) "Persistent Slip Bands in Fatigued fcc and bcc Metals," Fatigue Mechanism, ASTM STP 675, pp. 69–105.
28. W. A. Wood, "Recent Observation on Fatigue Fracture in Metals," ASTM STP 237, (1958) pp. 110–121.

References

29. P. J. E. Forsyth, The Physical Basis of Metal Fatigue, American Elsevier, 1969.
30. Achievment of High Fatigue Resistance in Metals and Alloys, edited by J. C. Grosskreutz and C. E. Feltner, ASTM, STP-467, June 1969.
31. T. Broom, and R. K. Ham, "Hardening of Copper Single Crystals by Fatigue" Proc. R. Soc., Vol. A251, 1959, pp. 186–199.
32. D.S. Kemsley, and M. S. Paterson, "Influence of Strain Amplitude on Work Hardening of Copper Crystal in Alternating Tension any Compression" Acta Metall., Vol. 8, 1960, pp. 453–467.
33. C. E. Feltner, Philos. Magazine, Vol. 12, 1965, pp. 1229–1248.
34. R. C. Juvinall, "Supplement to Engineering Considerations of Stress, Strain, and Strength," McGraw-Hill, 1967.
35. B. M. Wundt, "Effect of Notches on Low-Cycle Fatigue," ASTM, STP-490, 1972.
36. Fatigue Design Handbook, edited by J. A. Graham, prepared under the auspices of the Fatigue Design Subcommittee of Division 4 of SAE Iron and Steel Technical Committee, Vol. 4.
37. S. S. Manson and M. H. Hirschberg, "Low-Cycle Fatigue of Notched Specimens by Consideration of Crack Initiation and Propagation," NASA TN D-3146, June 1967.
38. J. Bauschinger, "On the Change of the Position of Elastic Limit of Iron and Steel Under Cyclic Variations of Stress," Mitt. Mech. Tech. Lab., Munich, Vol. 13, No. 1, 1886.
39. S. Timoshenko, "Strength of Materials", Part II, Advanced Theory and Problems, Third Edition, Robert E. Krieger Publishing Company, 1983, pp. 393-430.
40. R. W. Landgraf, JoDean Morrow, and T. Endo, "Determination of Cyclic Stress–Strain Curve," J. Mater., Vol. 4, No. 1, March 1969, pp. 176–188.
41. R. W. Landgraf, "The Resistance of Metals to Cyclic Deformation," Achievement of High Fatigue Resistance in Metals and Alloys, " ASTM STP 467, American Society for Testing and Materials, 1970, pp. 3–36.
42. T. Endo and JoDean Morrow, "Cyclic Stress–Strain and Fatigue Behavior of Representative Aircraft," J. Mater., Vol. 4, No. 1, March, 1969, pp. 159–175.
43. J. Morrow, "Cyclic Plastic Strain Energy and Fatigue of Metals," Internal Friction, Damping and Cyclic Plasticity, ASTM, STP- 378, 1965, pp. 45–87.
44. L. F. Coffin, Jr. and J. F. Tavernelli, "The Cyclic Straining and Fatigue of Metals," Trans. Metall. Soc. of AIME, Vol. 215, October 1959, pp. 794–807.
45. R. W. Smith, M. H. Hirschberg, and S. S. Manson, "Fatigue Behavior of Materials Under Strain Cycling in Low and Intermediate Life Range," NASA, TN D-1574, April 1963.
46. J. F. Tavernelli and L. F. Coffin, Jr., "Experimental Support for Generalized Equation Predicting Low Cycle Fatigue," Trans. ASME, J. Basic Eng., Vol. 84, No. 4, December 1962, p. 533.
47. S. S. Manson, discussion of Reference 23, Trans. ASME J. Basic Eng., Vol. 84, No. 4, December 1962, p. 537.
48. J. A. Graham (Ed.), Fatigue Design Handbook, SAE, 1968.
49. ASTM E-606 Standards

50. JoDean Morrow and T. A. Johnson, Material Research and Standards, MTRSA, Vol. 5, No.1, January 1965, pp. 30–32.
51. R. E. Peterson, Materials Research & Standards, MTRSA, Vol. 3, No. 2, February 1963, pp. 122–139.
52. K. N. Smith, R. Watson, and T. H. Topper, "A Stress–Strain Function of the Fatigue of the Metals," J. Mater., Vol. 5, No. 4, December 1970, pp. 767–778.
53. R. E. Peterson, "Analytical Approach to Stress Concentration Effect in Fatigue of Aircraft Materials," Proceeding of Symposium on Fatigue of Aircraft Structures, pp. 273–299, Wadc Technical Report 59-507, August 1959.
54. H. J. Grover, S. A. Gordon, and L. R. Jackson, "Fatigue of Metals and Structures" Thames and Hudson, London, 1956 pp. 66–85.
55 S. S, Manson, "Fatigue: a Complex Subject-Some Simple Approximation," Exp. Mech., Vol. 5, No. 7 July 1965, p. 193.

Chapter 3

Linear Elastic Fracture Mechanics

3.1 Energy Balance Approach (the Griffith Theory of Fracture)

The quantitative statements of the Griffith theory [1] are based on elasticity theory and are applicable only to those materials that obey Hooke's law up to the instant of fracture. In confirming his theory, Griffith used glass as the test material because at room temperature glass does follow Hooke's law to the stress at fracture. Griffith's work must be considered unique, because the surface tension that appears in his theory as the energy required for fracture was determined independently of fracture tests.

Griffith's theory [1] evolved from a consideration of the potential energy of the system, that is, the equilibrium position of an elastic body deformed by specified surface forces is such that the potential energy of the whole system is minimum. Griffith took into consideration the increase of potential energy that occurs due to the formation of two new crack surfaces. Catastrophic crack propagation under fixed grip conditions begins because the decrease in internal energy of the system as the crack extends an incremental length becomes available to extend the crack and is equal to the energy consumed in the dissipative mechanisms of the growing crack. For brittle materials that do not exhibit plastic deformation at the crack tip, the critical condition for through cracks can therefore be stated as:

$$\frac{\partial}{\partial c}[U_E - U_S] = 0 \qquad (3.1)$$

U_E is the energy per unit thickness available to create the new crack surfaces resulting from a through crack of length $2c$ in a plate of infinite size under a tensile stress, σ, normal to the plane of the crack. The term U_S is the elastic surface energy and is defined as the energy consumed per unit thickness in creating the new crack surfaces. The quantity U_E can be written [1] as:

$$U_E = \frac{\pi \sigma^2 c^2}{E} \quad (3.2)$$

and the elastic surface energy (U_S) is given by:

$$U_S = 4cT \quad (3.3)$$

where T is the surface tension of the material, that is, the work done in breaking the atomic bonds. At the atomic level, one can assume that the residual strength capability of a material depends on the strength of its atomic bonds. That is, fracture takes place when atomic bonds break between atoms in which two new crack surfaces are created. Fracture at the atomic level occurs either by cleavage where the breaking bonds are perpendicular to the fracture plane, with the theoretical cohesive strength $\sigma_c \approx E/10$, or by a process called shear across the fracture plane, where $\tau_c \approx G/10$ (E and G are the elastic and shear moduli, respectively). In terms of critical stress, σ_c, the surface energy, T, can be written as [2]:

$$\sigma_c = \sqrt{\frac{ET}{\alpha_0}}$$

where:

$$T = (0.394 \times 10^{-8}) \frac{E\alpha_0}{20} \quad (3.4)$$
$$= 1.97 \times 10^{-10} E\alpha_0$$

and α_0 is the atomic spacing in Angstroms (1Å = 0.394×10^{-8} in.)

The equilibrium condition described by Eq. (3.1) can further be simplified by setting the first derivitive of U_E and U_S [using Eqs. (3.2) and (3.3)] with respect to crack length, c, equal to zero:

$$\frac{2\pi \sigma^2 c}{E} = 4T \quad (3.5a)$$

and for one half of a center crack in an infinite plate:

$$\frac{\pi \sigma^2 c}{E} = 2T \quad (3.5b)$$

The expression on the left of Eq. (3.5b) is the elastic energy release rate (or the crack extension force, G [3]) and the expression on the right side is the energy absorption rate for creation of new crack surfaces. The critical energy release rate, G_c, is a constant for truly brittle materials because T is constant.

The energy balance approach described by Eq. (3.5) simply states that, for some applied stress σ, an ideally sharp crack of length $2c$ in an infinite body will become unstable. The energy balance approach is applicable only to the crack analysis cases where the condition of crack instability is a result of monotonic increasing tensile load to failure. However, this approach is not applicable to other important cases, such as stable crack growth that occurs when a crack is subjected to cyclic loading. The stress intensity parameter, K, is a useful tool that can address the cracking problems of interest in aircraft, pressure vessels, ships, and space vehicle structures, such as stable crack growth under fluctuating load, as well as the condition of instability. The two quantities, energy release rate, G, and the stress intensity factor, K, are linked together in the energy balance equation by Irwin [3] as:

$$G = \pi\sigma^2 c/E = K^2/E \qquad (3.5c)$$

The stress intensity factor approach using linear elastic fracture mechanics is discussed extensively in the remaining part of this chapter. Later in Chapter 6 under "Fracture Mechanics of Ductile Metals," the modified Griffith fracture criterion is discussed, which considers the dissipated energy absorption rate that results in the formation of plastic deformation at the crack tip of a metallic material.

3.2. The Stress Intensity Factor Approach

3.2.1 General

As mentioned in Section 3.1, the energy balance approach (expressed as the energy release rate, G) to crack tip problems leads to the same results as the stress intensity factor approach. The stress intensity factor, K, characterizes the crack tip stress field and its applications to fatigue crack growth and life prediction problems make this a very important parameter in the field of fracture mechanics. For this reason, its derivation is discussed in Section 3.2.2. The critical value of the stress intensity factor, called "fracture toughness," is discussed in Section 3.3. In Section 3.4, methods of constructing the residual strength diagram of cracked structures are described. The resistance curve approach, known as the R-curve, is introduced (Section 3.4) for obtaining the fracture toughness value of ductile metals with tearing fracture behavior. The failure of structural metals when subjected to a fluctuating load environment, is in most cases, the result of having surfaces exposed to scratches during machining, corrosion pitting, or surface damage

106 Chap. 3 Linear Elastic Fracture Mechanics

through improper handling. The behavior of these cracks and their growth can be assessed through the stress intensity factor parameter. The derivation of crack tip stress intensity factor for surface cracks (also called part through cracks) is presented in Section 3.7. Finally, a brief discussion of the plane strain and plane stress fracture toughness testing is included in Section 3.8.

3.2.2 Crack Tip Modes of Deformation

The stress fields and displacement modes at the crack tip can be classified into three types (see Fig. 3.1 for crack surface displacement in the x, y, and z directions):

  I. The opening mode (mode I) is characterized by displacement of the two crack surfaces moving directly apart from each other. In this case, the applied load and displacement, v, are perpendicular to the crack surfaces. The stress intensity factor corresponding to this mode is denoted by K_I, [see Fig. 3.1 case (a)].
  II. The shear or sliding mode (mode II) occurs when the two crack surfaces are displaced by sliding over each other. The direction of the applied load and displacement, u, are parallel to the crack surfaces [see Fig. 3.1 case (b)].
  III. The tearing mode (mode III) occurs when the crack surfaces slide over each other with displacement w in a direction parallel to the leading edge of the crack [see Fig. 3.1 case (c)].

Note that the symbols I, II, and III are Roman numerals which refer to the modes of fracture. The opening mode is shown by mode I. The shear and tearing modes are represented by II and III, respectively.

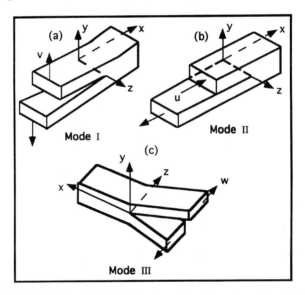

Figure 3.1 Illustration of the three basic modes of crack surface displacement

The tensile opening mode, mode I, type of failure represents the most frequent type of separation that engineers design against and that must be prevented. Failure of the structural parts by mode II and mode III, where fracture is induced by shear stresses, can also occur. However, these types of fracture seldom occur during the service life of the part. In addition, most of the available test data generated for the critical value of the stress intensity factor, K, (fracture toughness) have been established for mode Mode I type failure, see appendix A. Therefore, throughout this book, attention is given to the opening mode, mode I, and its application to fatigue crack growth when evaluating the service life of a structural part.

3.2.3 Derivation of Mode I *Stress Intensity Factor*

Linear Elastic Fracture Mechanics (LEFM) is based on the application of classical linear elasticity to cracked bodies. The elasticity assumptions made in deriving the mode I stress intensity factor are that displacements are small and that linearity between stresses and strains must be obeyed. Moreover, the assumption that the material is both homogeneous and isotropic is also included in the derivation of the stress intensity factor, K. The use of the crack tip stress intensity factor, K, for assessing small crack behavior is not recommended. For more information pertaining to the anomalous behavior of a small crack, where crack dimensions are on the order of material microstructure, the reader may refer to Section 4.3.

The following steps are used in deriving the mode I stress intensity factor. No attempt is made to include all the derivations. The reader may refer to References [4, 5] for a more in-depth treatment of this topic if he or she so chooses.

1. The equilibrium equations of stresses
2. Hooke's law and the strain–displacement relationship
3. The compatibility equations of strains
4. The Airy stress function
5. The Westergaard function
6. The use of the complex variable method
7. The use of boundary conditions to solve the constants
8. The stress intensity factor equation for different crack geometries.

In solving any two-dimensional stress problem, the equilibrium equations that describe the relationship between forces in the x and y directions can be written as [6, 7]:

$$\partial \sigma_x / \partial x + \partial \tau_{xy} / \partial y = 0 \tag{3.6}$$

$$\partial \sigma_y / \partial y + \partial \tau_{xy} / \partial x = 0$$

From the two above equilibrium equations, there are three unknowns—σ_x, σ_y, τ_{xy}—and two equations. To obtain the quantities σ_x, σ_y, and τ_{xy}, additional equations are needed.

The stresses at each point in the body are related to the strains by Hooke's law as:

$$E\varepsilon_x = \sigma_x - v\sigma_y$$
$$E\varepsilon_y = \sigma_y - v\sigma_x \quad (3.7)$$
$$\frac{E\gamma_{xy}}{2(1+v)} = \tau_{xy}$$

where E and v are the modulus of elasticity and poisson's ratio respectively.

From Eqs. (3.6) and (3.7), there are total of five equations and six unknowns (σ_x, σ_y, τ_{xy}, ε_x, ε_y, and γ_{xy}). The strain–displacement relationship can be written as:

$$\varepsilon_x = \frac{\partial u}{\partial x}$$
$$\varepsilon_y = \frac{\partial v}{\partial y} \quad (3.8)$$
$$\gamma_{xy} = \frac{\partial u}{\partial y} + \frac{\partial v}{\partial x}$$

where u and v are displacement functions. From Eq. (3.8) the strain–compatibility equation is:

$$\frac{\partial^2 \gamma_{xy}}{\partial x \partial y} = \frac{\partial^2}{\partial y^2}\left(\frac{\partial u}{\partial x}\right) + \frac{\partial^2}{\partial x^2}\left(\frac{\partial v}{\partial y}\right) \quad (3.9a)$$

In terms of strains Eq. (3.9a) becomes:

$$\frac{\partial^2 \gamma_{xy}}{\partial x \partial y} = \frac{\partial^2}{\partial y^2}(\varepsilon_x) + \frac{\partial^2}{\partial x^2}(\varepsilon_y) \quad (3.9b)$$

From Eqs. (3.6), (3.7), and (3.9b), there are six equations and six unknowns so that, together with the boundary conditions, the state of stress at each point in a body subjected to the plane stress condition can be solved.

3.2.3.1. Airy Stress Function, ϕ

Airy showed that, for any elastic equilibrium problem, there is a function that satisfies the equation of equilibrium [6]. Consider a function $\phi(x, y)$ such that:

3.2. The Stress Intensity Factor Approach

$$\sigma_x = \frac{\partial^2 \phi(x,y)}{\partial y^2}, \quad \sigma_y = \frac{\partial^2 \phi(x,y)}{\partial x^2}, \quad \tau_{xy} = \frac{\partial^2 \phi(x,y)}{\partial x \partial y} \quad (3.10)$$

Using Eq. (3.10) to replace σ_x, σ_y, and τ_{xy}, it is easy to demonstrate that the equilibrium condition described by Eq. (3.6) is satisfied. Writing Eq. (3.7) in terms of the Airy stress function and substituting the results into the compatibility equation described by Eq. (3.9b), we obtain:

$$\frac{\partial^4 \phi}{\partial y^4} + \frac{\partial^4 \phi}{\partial x^4} + 2\frac{\partial^4 \phi}{\partial y^2 \partial x^2} = 0 \quad (3.11a)$$

In more compact form:

$$\nabla^4(\phi) = \nabla^2[\nabla^2(\phi)] = 0 \quad (3.11b)$$

where the quantities $\nabla^4(\phi)$ and $\nabla^2(\phi)$ are called the biharmonic and harmonic function, respectively. Any function that satisfies Eq. (3.11b) is called an Airy stress function.

Let $Z(z)$ represent the Westergaard function [5] such that when the Airy stress function, ϕ, is written in terms of Westergaard real, $\overline{\overline{Z}}(z)$, and imaginary, $\overline{Z}(z)$, components, it can satisfy the biharmonic equation:

$$\phi = \operatorname{Re} \overline{\overline{Z}}(z) + y \operatorname{Im} \overline{Z}(z) \quad (3.12)$$

In Eq. (3.12), the Westergaard function, $Z(z)$, is a function of the complex variable $z(x + iy)$ and, furthermore, the two quantities $\overline{\overline{Z}}(z)$ and $\overline{Z}(z)$ are the second- and first-order integrals $(Z(z) = d\overline{Z}(z)/dz$ and $\overline{Z}(z) = d\overline{\overline{Z}}(z)/dz)$. Using this function can enable us to solve a through center crack in an infinite plate that is subjected to a remotely applied biaxial load, $\sigma(\infty)$, shown in Fig. 3.2. Differentiating Eq. (3.12) to obtain the stresses defined by Eq. (3.10), the stresses σ_x, σ_y, and τ_{xy} in terms of the Westergaard function, $Z(z)$, can be expressed as:

$$\sigma_x = \operatorname{Re} Z(z) - y \operatorname{Im} Z'(z)$$
$$\sigma_y = \operatorname{Re} Z(z) + y \operatorname{Im} Z'(z) \quad (3.10a)$$
$$\tau_{xy} = -y \operatorname{Re} Z'(z)$$

where $Z'(z)$ is the first-order derivative of the Westergaard function, $Z(z)$. The stresses defined by Eq. (3.10a) are in terms of $Z(z)$ and $Z'(z)$; however, their exact form can be determined by selecting a suitable Westergaard function that can satisfy the boundary conditions. The selected function Z must be an analytical function and its derivitive, dZ/dz, must be continuous (Cauchy–Riemann condition).

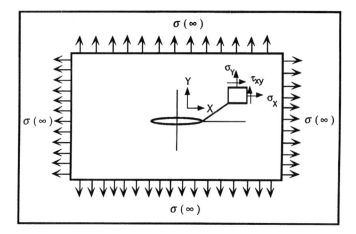

Figure 3.2 Illustration of a biaxially loaded plate containing a central crack

For a center through crack in an infinite plate subjected to a biaxial load (illustrated in Fig. 3.2), the boundary conditions are: at $x = \pm\infty$, the applied stress, $\sigma(\infty) = \sigma_x$ and accordingly at $y = \pm\infty$, the applied stress $\sigma(\infty) = \sigma_y$; moreover, the value of $\sigma_y = 0$ for $-a < x < a$ where $y = 0$. At the crack tip where $x = \pm a$, the stress $\sigma_y = \infty$ (crack tip singularity).

A function that can satisfy all of these boundary conditions can be written as:

$$Z(z) = \frac{\sigma z}{\sqrt{z^2 - a^2}} \tag{3.13a}$$

where $z = re^{i\theta}$. This function is called the Westergaard stress function. Moving the coordinate system from the center to the tip of the crack by simply replacing the quantity z with $(z + a)$, Eq. (3.13a) can be expressed as:

$$Z(z) = \frac{\sigma(z + a)}{\sqrt{(z + a)^2 - a^2}} \tag{3.13b}$$

where the derivative of this function is:

$$Z'(r, \theta) = -\frac{\sigma(2are^{-i\theta/2} - a^2 e^{-3i\theta/2})}{(2ar)^{3/2}} \tag{3.13c}$$

Obtaining the real and imaginary components of $Z(r, \theta)$ and $Z'(r, \theta)$ [from Eqs. (3.13b) and (3.13c)], the stress components described by Eq. (3.10a) can be further simplified. From (3.13b) and (3.13c), the real and imaginary components of $Z(r, \theta)$ and $Z'(r, \theta)$ in terms of (r, θ) can be written as:

3.2. The Stress Intensity Factor Approach

$$\operatorname{Re} Z(z) = \frac{\sigma a \cos \theta/2}{(zar)^{1/2}} \quad (3.13d)$$

$$\operatorname{Im} Z'(z) = \frac{\sigma a(-2ar \sin \theta/2 + \sin 3\theta/2)}{(2ar)^{3/2}} \quad (3.13e)$$

where $re^{i\theta} = \cos \theta + i \sin \theta$, $x = r \cos \theta$, and $y = r \sin \theta$. Inserting the quantities shown in Eq. (3.13e) into Eq. (3.10) and replacing $y = r \sin \theta$ with its equivalent $2r \sin(\theta/2) \cos(\theta/2)$, the crack tip stresses can be written as:

$$\sigma_x = \frac{\sigma \sqrt{\pi a}}{\sqrt{2\pi r}} \cos \theta/2 [1 - \sin \theta/2 \sin 3\theta/2] \quad (3.14a)$$

$$\sigma_y = \frac{\sigma \sqrt{\pi a}}{\sqrt{2\pi r}} \cos \theta/2 [1 + \sin \theta/2 \sin 3\theta/2] \quad (3.14b)$$

$$\tau_{xy} = \frac{\sigma \sqrt{\pi a}}{\sqrt{2\pi r}} \sin \theta/2 \cos \theta/2 \cos 3\theta/2 \quad (3.14c)$$

Note that the out of plane stress $\sigma_z = 0$ for the plane stress condition and $\sigma_z = v(\sigma_x + \sigma_y)$ for the plane strain condition.

The derivation of Eq. (3.14) is based on the assumption that the stress region at the crack tip is highly localized and the quantity r is limited to a small region at the crack tip where the term $r/a \approx 0$. From Eq. (3.14), it can be seen that there is a stress singularity at $r = 0$ and this implies that, as the distance, r, from the crack tip decreases, the magnitude of the stress increases and approaches infinity. We show in Section 3.6 that most aerospace material at the crack tip will deform plastically and yield before the stresses approach infinity.

The stresses described by Eq. (3.14) can be written in terms of their radial and angular positions with respect to the crack tip parameters $(1/\sqrt{2\pi r})f(\theta)$ and the quantity $\sigma \sqrt{\pi a}$ which defines the magnitude of the applied stress and the crack length. The second quantity, $\sigma \sqrt{\pi a}$, is called the mode I (or opening mode) stress intensity factor and it is designated by the symbol K_I having units of (ksi $\sqrt{\text{in.}}$). Thus, the stress intensity factor, K_I, can relate the crack tip stresses (σ_x, σ_y, and τ_{xy}) to parameters that are measurable quantities (σ and $\sqrt{\pi a}$). The stress intensity factor should not be confused with the stress or strain concentration factor (K_t or K_g) or with any other similar symbols, such as the strain hardening exponents (k or k').

The influence of external variables, such as the magnitude and the method of loading, and the crack geometry, on the crack tip stresses can also be described by the stress intensity factor. The general form of the stress intensity factor, including the effect of the crack geometry and loading condition, can be written as:

$$K_I = \beta_1\beta_2\sigma\sqrt{\pi a} \qquad (3.15)$$

where β_1 and β_2 are the corrections to the loading and crack geometry, respectively. A more comprehensive solution to different crack geometries and loading conditions is available in [8, 9].

The critical value of $K_I = \sigma_{cr}\sqrt{\pi a}$ is called the fracture toughness (K_c) and it is a measure of the material's resistance to unstable cracking. Irwin [10] failure criteria simply state that if the level of the crack tip stress intensity factor exceeds a critical value, K_{cr}, unstable cracking will occur. This is analogous to the case of a stress at a point and the criteria of yield or ultimate strength. The critical stress intensity factor and the material's resistance to fracture (fracture toughness) are discussed in Sections 3.2.6 and 3.3, respectively. A brief description of the ASTM practice for the determination of plane strain and stress fracture toughness is presented in Section 3.8.

The stresses defined by Eq. (3.14) are applicable to the case of an infinite plate containing a through center crack that is subjected to a remotely applied biaxial load. For the case of a uniaxial tensile load, where remote stress $\sigma_x(\infty) = 0$, this quantity must be subtracted from the crack tip stresses. However, because the crack tip stresses are much higher than the farfield or applied stresses the effect is negligible. The remote stress $\sigma_x(\infty)$ is parallel to the crack surfaces and has little or no influence on mode I crack opening or its stability.

When the applied load is perpendicular to the crack surfaces ($\sigma_x(\infty) = 0$ and $\sigma_y(\infty) \neq 0$), the variation of crack tip stresses (local stresses, σ_y and σ_x) along the x-axis where $\theta = 0$ is illustrated in Fig. 3.3. The reader should note that there is a local stress σ_x even though the applied load is in the y direction. The induced stress, σ_x, is the result of straining in the x direction due to the Poisson effect.

The stress intensity factor solutions for a few crack geometry cases important for use in aircraft and space vehicle applications, subjected to combined applied load, are discussed in Sections 3.2.4 and 3.2.5.

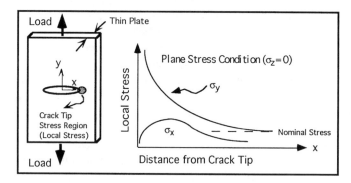

Figure 3.3 Stress variation in the region of crack tip

3.2. The Stress Intensity Factor Approach

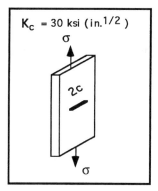

Figure 3.4 A Center through crack plate made of 2219-T87 aluminum for example 3.1

Example 3.1

An aluminum plate made of 2219-T87 has a center through crack, as shown in Fig. 3.4. The plate is subjected to an applied stress (design stress) of σ. The value of the stress intensity factor, K, at the instability, where material fracture is 30 ksi $\sqrt{in.}$. Plot the variation of the design stress, σ, as a function of half crack length, c.

Solution:

From Eq. (3.15) for a center cracked specimen subjected to axial load (shown in Fig. 3.4), the two quantities $\beta_1 = \beta_2 = 1$. The relationship between the applied stress and the critical half crack length for $K_c = 30$ ksi $\sqrt{in.}$, can be written as:

$$\sigma_{cr} = \frac{30}{\sqrt{\pi c}}$$

The plot of applied stress and the critical half crack length is shown in Fig. 3.5 for Example 3.1.

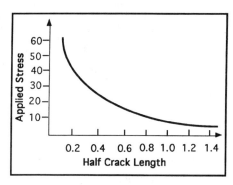

Figure 3.5 Variation of stress versus half crack length for $K_c = 30$ ksi $(in.)^{1/2}$

3.2.4 Combined Loading

When a cracked plate is subjected to more than one type of mode I induced load (combined loading), the resultant stress intensity factor is additive. For example, for the case of a center through crack in an infinite plate that is subjected to an axial tensile load, P, and bending moment, M, as shown in Fig. 3.6, the total or combined stress intensity factor characterizing the crack tip stresses can be written as:

$$K_{Total} = K_{Axial} + K_{Bending}$$

In general, the combined or superposition approach to the stress intensity factor of a crack plate subjected to different loading conditions that all produce mode I of fracture can be expressed as [8, 9, 11]:

$$K = (\sigma_0 \beta_0 + \sigma_1 \beta_1 + \sigma_2 \beta_2 + \sigma_3 \beta_3)\sqrt{\pi a} \qquad (3.16)$$

where σ_0, σ_1, σ_2, and σ_3 are the applied stresses due to tension, bending through the thickness, bending through the width, and pin bearing, respectively. Figure 3.7 illustrates different kinds of loading conditions that can all induce the mode I crack tip stress intensity factor. In Eq. (3.16) the quantities β_0, β_1, β_2, and β_3 are the corrections to the crack geometries and loading conditions. An example of combined loading is a bolted joint where the bolted plates experience most of the load types described by Eq. (3.16) (local plate bending due to bolt tension, bearing stress due to shear, and tension due to far field stress).

The principle of superposition is helpful, not only to evaluate the stress intensity factor solution of a crack plate subjected to different loading conditions (as

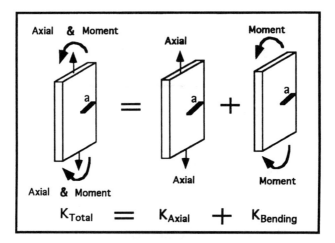

Figure 3.6 Illustration of the superposition concept for an edge crack with axial and bending loading

3.2. The Stress Intensity Factor Approach   115

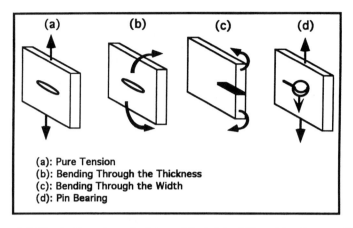

Figure 3.7 Illustration of crack plates subjected to different loading conditions

long as the crack geometry remains the same for all loading cases), but also to obtain the stress intensity factor solutions to the crack cases where the solution does not exist or is difficult to obtain. For example, consider the case of a finite width plate containing a center through crack that is wedge loaded, as shown in Fig. 3.8, case (a). The stress intensity factor solution to wedge loaded center cracked geometries alone is available [11] and it can be expressed as:

$$K_I = \frac{(P/B)}{\sqrt{\pi a}} \left(\sqrt{\frac{2\pi a}{W \sin(2\pi a/W)}} \right) \tag{3.17}$$

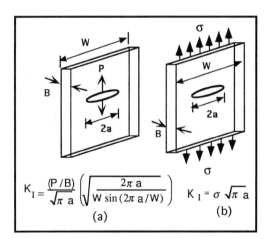

Figure 3.8 Stress intensity factor solutions for wedge loaded and uniform tension center crack

116   Chap. 3 Linear Elastic Fracture Mechanics

Consider the case of an identical plate with the same crack geometry that is now subjected to a remote uniform tension load, as shown in Fig. 3.8, case (b). The stress intensity factor solution for case (b) is also available and expressed by Eq. (3.15) where $\beta_1 = \beta_2 = 1$. The additive properties of the stress intensity factor can enable us to find the stress intensity factor solution for the case of a plate that is loaded by uniform tension and concentrated force, P, shown in Fig. 3.9. The equation that describes the combined case can be written as:

$$K_{\text{Total}} = \frac{(P/B)}{\sqrt{\pi a}} \left(\sqrt{\frac{2\pi a}{W \sin(2\pi a/W)}} \right) + \sigma \sqrt{\pi a} \qquad (3.18)$$

In Section 3.2.5, the stress intensity factor solution for a few crack geometries and loading cases that are frequently applied to the analysis of a structural part are discussed. The reader should pay attention to the fact that when dealing with combined loading, it is understood that each load contributes to the opening mode I of fracture. The effective stress intensity factor is the superposition of all the individual stress intensity factor as described by Eq. 3.16. The K solution for the case of a circular shaft subjected to combined axial and torsional loadings can no longer be described by the superposition criteria (shown by equation 3.16) because torsional applied load does not produce mode I fracture for the same crack geometry.

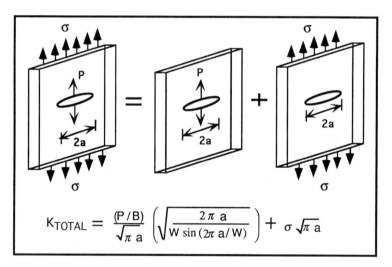

Figure 3.9 Stress intensity factor solution for combined wedge loaded and uniform tension center crack

3.2. The Stress Intensity Factor Approach

3.2.5 Stress Intensity Factor Equations for Several Through Cracks

The linear elastic fracture mechanics approach to the analysis of a cracked structure and its application to evaluating the service life of high-risk parts (also called fracture critical parts) created a widespread need for a stress intensity factor solution to other complicated crack geometries. Figures 3.10 through 3.16 are the descriptions of some of the most widely used stress intensity factor solutions for through-the-thickness crack geometry cases that are subjected to a combined loading consisting of either tension, through-the-thickness bending, through-the-width

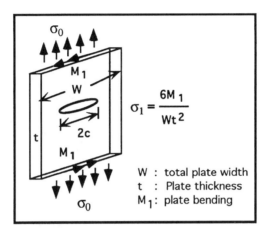

Figure 3.10 Through crack in a center of plate subjected to axial stress, σ_0, and bending through the thickness, M_1

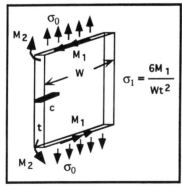

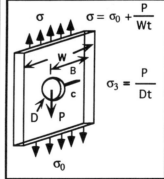

Figure 3.11 Through crack at edge of a plate

Figure 3.12 Through crack from a hole in a plate

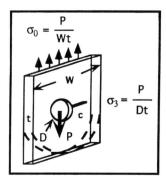

Figure 3.13 Through crack from a hole in a lug

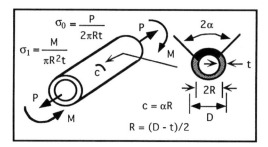

Figure 3.14 A circumferential through crack in a cylinder

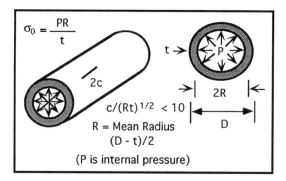

Figure 3.15 An axial through crack in a pressurized cylinder

bending, or pin bearing pressure [12]. The case of surface and corner cracks is discussed later in Section 3.7 of this chapter.

To obtain the stress intensity factor equation for the crack geometries shown in Figs. 3.10 through 3.16, one must be able to determine the geometric correction factors (β_0, β_1, β_2, β_3) applicable to each type of applied stress defined by Eq. (3.16). It does not matter how complicated the crack geometry and loading condi-

3.2. The Stress Intensity Factor Approach

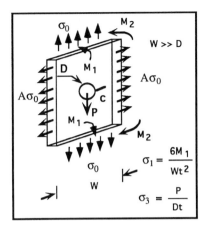

Figure 3.16 Through crack in a plate emanating from a hole subjected to combined loading

tions are, the stress intensity factor equation for mode I fracture always takes the form of $\beta\sigma(\pi a)^{1/2}$. Before attempting to write the stress intensity factor equations, it is helpful to introduce the following common symbols that are used in all crack geometries to simplify the mathematics:

$$u = c/D, w = c/W, y = D/W, \text{ and } z = (1 + 2u)^{-1}$$
$$f_0(z) = 0.7071 + 0.7548z + 0.3415z^2 + 0.642z^3 + 0.9196z^4$$
$$f_1(z) = 0.078z + 0.7588z^2 - 0.4293z^3 + 0.0644z^4 + 0.6516z^5$$

where c is the half crack length, W is the plate width, and D is the hole diameter. Stress intensity factor equation for a through center crack in a plate (Fig. 3.10) [12]:

$$K = (\sigma_0\beta_0 + \sigma_1\beta_1)\sqrt{\pi c}$$
$$\beta_0 = (\sec \pi w)^{1/2} \text{ and } \beta_1 = \beta_0/2 \tag{3.19}$$

Stress intensity factor equation for a through crack at edge of a plate (Fig. 3.11) [12]:

$$K = (\sigma_0\beta_0 + \sigma_1\beta_1 + \sigma_2\beta_2)\sqrt{\pi c}$$
$$\beta_0 = Y[0.752 + 2.02w + 0.37(1 - \sin\beta)^3], \text{ and } \beta_1 = \beta_0/2$$
$$\beta_2 = Y[0.923 + 0.199(1 - \sin\beta)^4] \tag{3.20}$$

where

$$Y = \sec \beta[(\tan \beta)/\beta]^{1/2}, \text{ and } \beta = \pi w/2$$

Stress intensity factor equation for a through crack from an offset hole (Fig. 3.12) [12]:

$$K = (\sigma_0 \beta_0 + \sigma_1 \beta_1) \sqrt{\pi c}$$
$$\beta_0 = G_0 G_w \text{ and } \beta_1 = (G_0 y/2 + G_1)/G_w$$
$$G_0 = f_0(z) \text{ and } G_1 = f_1(z) \quad (3.21)$$

where

$$G_w = [\sec \lambda (\sin \beta)/\beta]^{1/2}$$
$$\beta = D/B - 2y$$

and

$$\lambda = (\pi/2)(1 + u)/(2B/D - u)$$

Stress intensity factor equation for a through crack from a hole in a lug (Fig. 3.13) [12]:

$$K = (\sigma_0 \beta_0) \sqrt{\pi c}$$
$$\beta_0 = (G_0 y/2 + G_1) G_w G_L G_2$$

where

$$G_0 = f_0(z), \ G_1 = f_1(z), \ G_L = [\sec(\pi y/2)]^{1/2}$$

and

$$G_2 = C_1 + C_2(c/b) + C_3(c/b)^2 + C_4(c/b)^3, \ G_w = (\sec \lambda)^{1/2} \quad (3.22)$$

where

$$\lambda = (\pi/2)(1 + u)/(2B/D - u), \ b = (W - D)/2$$

and

$$C_1 = 0.688 + 0.772y + 0.613y^3$$
$$C_2 = 4.948 - 17.318y + 16.785y^3$$

3.2. The Stress Intensity Factor Approach

$$C_3 = -14.297 + 62.994y - 69.818y^3$$
$$C_4 = 12.35 - 58.644y + 66.387y^3$$

Stress intensity factor equation for a through crack in a cylinder in the circumferential direction (Fig. 3.14) [12]:

$$K = (\sigma_0 \beta_0 + \sigma_1 \beta_1)\sqrt{\pi c}$$
$$\beta_0 = (I_0/2\pi\alpha)^{1/2}$$

and

$$\beta_1 = (I_1/2\pi\alpha)^{1/2}$$

where $I_0 = [\sqrt{8}(f^2-1) + \pi\beta^2/b]\alpha^2/k$ and $I_1 = [\sqrt{8}(g^2-1) + \pi\beta^2/b)]\alpha^2/k$
The quantities f, g, h, b, and β are:

$$f = 1 + h(1 - \cot\alpha)/2\alpha$$
$$g = [1 + h(\alpha + \alpha\cot^2\alpha - \cot\alpha)/4](\sin\alpha)/\alpha$$
$$h = \sqrt{2}/(\cot[(\pi - \alpha)/\sqrt{2} + \sqrt{2}\cot\alpha])$$
$$b = \alpha/2k$$

(3.23)

where $k = (t/R)^{1/2}[12(1 - v^2)]^{-1/4}$ (see figure 3.14)

$$\beta = 1 + (\pi/16)b^2 - 0.0293\, b^3 \text{ for } b \leq 1$$

or

$$\beta = (\sqrt{8}\, b/\pi)^{0.5} + (0.179/b)^{0.885} \text{ for } b > 1$$

Stress intensity factor equation for a through crack in a cylinder in the longitudinal direction (Fig. 3.15) [12]:

$$K = (\sigma_0 \beta_0)\sqrt{\pi c}$$
$$\beta_0 = (1 + 0.52\lambda + 1.29\lambda^2 - 0.074\lambda^3)^{1/2}$$

(3.24)

where

$$\lambda = c/(Rt)^{1/2}$$

Table 3.1 Values of Coefficients A_n, B_n, C_n, D_n

n	A_n	B_n	C_n	D_n
0	−0.00074	0.70920	0.7968	0
1	0.06391	0.68902	0.5326	0.0780
2	−0.10113	0.52270	0.2767	0.7588
3	−0.29411	0.65768	0.0630	−0.4293
4	−0.79179	1.91920	−0.017	0.0644
5	—	—	1.7197	0.6510

Stress intensity factor equation for a through crack from a hole in a plate under combined loading (Fig. 3.16) [12]:

$$\beta_0 = \sum_{n=0}^{4} [A_n(1+A) + B_n]b^n, \beta_1 = [(1+u)b^n]^{3/2}(1+v)/(3+v) \quad (3.25)$$

and

$$\beta_2 = D/3 \sum_{n=0}^{5} C_n b^n, \beta_3 = \sum_{n=0}^{5} D_n b^n$$

where the quantities shown in Eq. (3.25) are given in Table 3.1.

3.2.6 Critical Stress Intensity Factor

The critical value of the stress intensity factor is an important parameter in the field of fracture mechanics when dealing with structural failure resulting from unstable cracking. It is analogous to other critical failure criteria that are based on yield or ultimate strength of the material. The critical value of the stress intensity factor at which unstable crack propagation occurs is called the fracture toughness. The fracture toughness data for many aerospace alloys are available in the "Damage Tolerant Design Handbook" [13].

For a thick section of a given material in which the plastic deformation at the crack tip is constrained and negligible (plane strain condition), the critical stress intensity factor for mode I at instability is designated by K_{Ic}. The failure criteria for unstable cracking when plane strain condition prevails can be written:

$$K \geq K_{Ic} \quad (3.26)$$

The plane strain fracture toughness, K_{Ic}, is dependent on the type of material, loading rate, and temperature. The failure criterion described by Eq. (3.26) simply

3.2. The Stress Intensity Factor Approach

states that abrupt failure occurs when the crack tip stress intensity factor, k, reaches or exceeds the material's fracture toughness, K_{Ic}. A detailed description of K_{Ic} is given in Section 3.3.

Equation (3.26) is valid when the plastic deformation at the crack tip is assumed localized and small, since the assumptions stated in the formulation of the stress intensity factor are based on linear elastic analysis (see Section 3.2.2). It is important to note that in performing a fracture mechanics analysis, both K_I and K_{Ic} are needed and must be available to the analyst. This is analogous to the strength analysis requirement that the limit stress, σ, must always be compared with both the yield, σ_{Yield}, and the ultimate, σ_{Ul}, strength of the material. That is, the analysis would be incomplete without both material allowables.

As the material's thickness decreases, the constraint to plastic flow decreases, and the state of plane stress is reached. The fracture toughness associated with the minimum thickness is called the plane stress fracture toughness and is designated by K_C. The plane stress fracture toughness is dependent on thickness as well as on initial crack size. A brief presentation of the ASTM testing method to obtain the plane strain and plane stress fracture toughness values of isotropic materials is available in Section 3.8. The reader may refer to Reference [14] for a detailed description of the ASTM testing practice.

An analytical approach called the Theory of Fracture Mechanics of Ductile Metals (FMDM) is available to use for calculating both the plane strain, K_{Ic}, and plane stress, K_C, fracture toughness values of isotropic materials. The FMDM computer program is currently being used in some aerospace companies and it is extremely valuable in obtaining both K_{Ic} and K_C fracture toughness values without going through complicated and costly ASTM test procedures. The FMDM theory is discussed in Chapter 6.

When fracture toughness values of a given material are available, it is easy to determine the critical flaw size for a given stress level where the structural instability occurs. Figure 3.17 illustrates the relationship between design stress

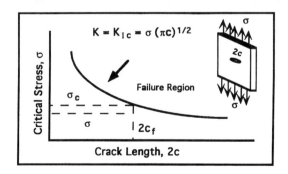

Figure 3.17 Schematic plot of critical stress as a function of crack length

and critical flaw size for a through centered crack in a plate. From Fig. 3.17, it can be seen that for a wide plate with crack length $2c_f$, unstable cracking will occur when the stress level becomes equal to or greater than σ_c. For any other stress level, σ, smaller than σ_c, unstable cracking does not occur (Fig. 3.17). When the load environment is fluctuating in nature, the fracture toughness data (critical value of stress intensity factor) are essential for a reliable crack growth analysis and should include plane strain and plane stress as well as part through fracture toughness. In Section 3.3, the concept of fracture toughness is discussed and later, in Section 3.7, surface cracks and their critical values (part through fracture toughness) are briefly reviewed. Extensive fracture toughness data for many aerospace alloys are available in Appendix A for use in crack growth study (also see Reference [13]).

3.3 Fracture Toughness

Fracture toughness is defined as the resistance of the material to unstable crack growth in a non-corrosive environment. This parameter characterizes the intensity of the stress field at the locality of the crack tip when unstable cracking takes place. The plane strain fracture toughness, K_{Ic}, is considered thickness independent; however, it can vary as a function of temperature and strain rate. The letter I in the subscript is used throughout the literature to indicate the instability as the result of mode I crack propagation with cleavage type fracture (flat surfaces) without shear lip.

For typical metals the plane strain fracture toughness, K_{Ic}, is associated with thick sections. Under the plane strain condition, the state of stress at the crack tip approaches a triaxial tensile stress state. In the three dimensional state of stress, the crack tip plasticity formation is constrained and is small compared with the crack size and other specimen dimensions. Experimental data have shown that for homogeneous material under plane strain condition the yielding (plastic deformation) occurs at a much higher stress level compared with uniaxial tension. Specifically, when the applied load is in such a way that principal stresses are almost equal in magnitude, the material will stay elastic with little plastic deformation up to failure. However, at the two extreme free edges, the state of stress is biaxial and plastic deformation occurs by a shear mechanism at a 45° plane with respect to the flat surfaces. It is shown later in this section that the material resistance to cracking is always higher at the two extreme free edges than the middle region of the crack tip due to the degree of plastic deformation resulting from biaxial or triaxiality state of stress. The effect of the triaxial constraint at the crack tip is limited to the center of the cracked section through the thickness of the material and gradually becomes biaxial at the two outer surfaces, as shown in Figure 3.18.

The cracked surfaces associated with the plane strain fracture toughness are almost flat as viewed by the naked eye. This type of failure when viewed microscopically occurs by direct separation of one atomic plane over another and is called cleavage. Cleavage is defined as low-energy fracture that propagates along well-defined low-index crystallographic planes known as cleavage planes. Theoretically, the two fractured cleavage surfaces should be completely flat. At the center of a thick plate near the crack tip where the plane strain condition exists, the state of stress is triaxial and principal stresses are almost equal. That is, the hydrostatic stress state near the crack tip, where $\sigma_x \approx \sigma_y \approx \sigma_z$ is expected and the maximum shear stress, τ_{max}, is negligible which does not cause plastic deformation [15, 16]. This is illustrated in Fig. 3.18 for the center element selected at the crack tip in the middle of the specimen. On the other hand, at the two free surfaces of the same cracked plate, where the triaxial state of stress does not exist ($\sigma_z = 0$), ductile failure (sliding of one atomic plane over another) is produced, see the edge element shown in Fig. 3.18. The material behavior in the locality of the crack tip through the part thickness (at the middle and the two ends) can be analogous to several tensile bar specimens that are situated next to each other. Those tensile bars at the two edges have the flexibility to deform and exhibit reduction of area (contraction) due to the Poisson effect. In the plane stress condition where $\sigma_z = 0$ and $\varepsilon_z \neq 0$ the material behavior due to the Poisson effect is illustrated in Fig. 3.18 where contraction in the z direction and consequently the formation of a plastic zone are permissible. In the plane strain condition where $\varepsilon_z = 0$ and $\sigma_z \neq 0$, those

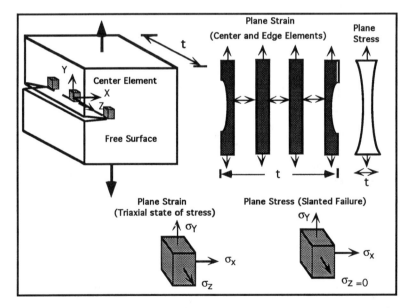

Figure 3.18 Illustration of the state of stress at the crack tip for plane strain and stress condition

bars in the middle of specimen are prevented from contraction, which results in developing stress in the z direction and therefore blocking the formation of plasticity as illustrated in Fig. 3.18.

Figure 3.19 shows the variation in the amount of flat fracture (failure by cleavage) as a function of thickness. For plane strain failure (the region of K_{Ic}), the portion of the flat surfaces is much larger than the slanted sections. However, for thin sections where the state of stress at the crack tip is not triaxial, the constraint to plastic deformation lessens and the failure is associated with plane stress. For sections with adequate thickness, in which plane strain and plane stress are combined, the state of stress is termed mixed mode.

The portion of the flat region associated with the plane stress state is much smaller than the slanted region, as indicated in Fig. 3.19. Thus, the plane strain failure is associated with flat or cleavage failure and the plane stress with slant failure (ductile failure). The fact that energy absorption at the crack tip is small for the thick section where the failure is abrupt explains why the fracture strength is generally lower for the plane strain state of stress than for the plane stress.

Figure 3.20 shows the variation in fracture toughness for three regions, called plane strain, mixed mode, and plane stress, as a function of the material's thickness, with the amount of flat and slanted surfaces corresponding to each region. The asymptotic portion of the fracture toughness curve is associated with plane strain fracture toughness and is thickness independent (Fig. 3.20). For thicknesses less than the plane strain value ($t < t_{Ic}$), mixed mode fracture toughness is obtained and the maximum fracture toughness value, K_C, corresponds to minimum thickness on the curve, as shown in Fig. 3.20.

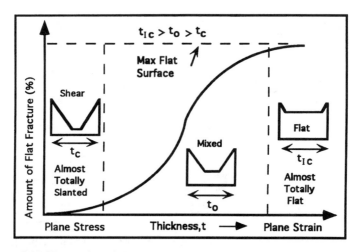

Figure 3.19 Illustration of the amount of flat fracture surface versus thickness

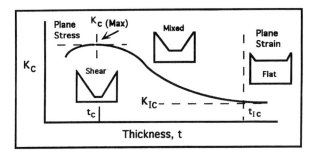

Figure 3.20 Variation of K_c as a function of thickness, t

Example 3.2

A structural part shown in Fig. 3.21 has a through crack that is emanating from a hole. The operating stress, σ_{op}, is 30 ksi. It is requested by the customer to proof test the part at room temperature, as well as at $-30°F$, to a stress level of $1.5 \times \sigma_{op}$. Nondestructive inspection indicates that the maximum flaw size in the part is $c = 0.075$ in. prior to the proof testing. Determine whether an existing crack will stay stable upon proof testing.

Solution

From Eq. (3.21), the critical crack length for a through crack emanating from a hole in terms of the critical stress can be written as:

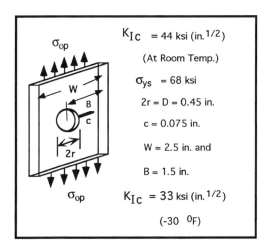

Figure 3.21 A through crack emanating from a hole in a palte (for example 3.2)

$$c = \{[(K_{Ic})/\beta_0(\sigma_p)]^2\}/\pi$$

where $K_{Ic} = 44$ ksi (in.$^{1/2}$) (at room temperature) and the proof test stress level is given as $\sigma_p = 1.5 \times \sigma_{op}$. The correction factor $\beta_0 = \beta_0(c/r)$.

$$c = \{[(44.0)/(\beta_0 \times 1.5 \times 30)]^2\}/\pi$$

where, from Eq. (3.21), the quantity $\beta_0(c/r) = G_0 G_w$. Solving for crack length, c:

$$c = 0.2 \text{ in.} > 0.075 \text{ in. (by NDE inspection)}$$

Therefore, the part will not fail upon proof test at room temperature. The second part of the problem is for the proof test conducted at $-30°F$. The related analysis is similar to the room temperature case, except that the fracture toughness value is replaced by $K_{Ic} = 33$ ksi (in.$^{1/2}$). The crack length c is:

$$c = \{[(33.0)/(\beta_0 \times 1.5 \times 30)]^2\}/\pi$$

The correction factor $\beta_0 = \beta_0(c/r) = G_0 G_w = 1.45$ is the same as the previous case:

$$c = 0.08 \text{ in.} > 0.075 \text{ in. (by NDE inspection)}$$

The part will not fail upon proof test at $-30°F$.

3.3.1 Fracture Toughness and Material Anisotropy

In most materials, the fracture toughness varies with crack orientation and loading direction. This is due to the anisotropic nature of the material that evolved during its manufacturing, cold rolling, or heat treatment (see Fig. 3.22). The ASTM E-616 coding system for manufactured material with rectangular cross-section is

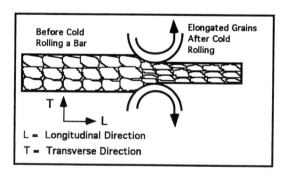

Figure 3.22 Illustration of the anisotropic nature of the material that evolved during the manufacturing

such that the first letter denotes the loading direction and the second letter represents the direction of expected crack propagation. The same system would be useful for bar and hollow cylinders. The standard nomenclature relative to directions of mechanical working (elongated grain direction) for rectangular sections is:

L = Direction of maximum deformation (maximum grain elongation). This is the longitudinal direction of the rolling, extrusion, or forging process (Fig. 3.23).
T = Direction of minimum deformation. This is the direction of long transverse (Fig. 3.23).
S = Direction perpendicular to the plane of L and T. This is the direction of short transverse (through the thickness) (Fig. 3.23).

For example, the crack orientation code L–T indicates that the loading direction is in the longitudinal direction and the direction of propagation is in the direction of long transverse. It is noteworthy that the designated orientation code for non-cracked parts contains only one letter. For example, the letter L designates material properties (not fracture properties) in the direction of maximum grain elongation.

For cylindrical sections where the direction of maximum deformation is parallel to the direction of principal deformation (for example, drawn bar, extrusions or forged parts with circular cross-section), a similar system of nomenclature for the three directions is:

L = Direction of maximum deformation (longitudinal direction)
R = Radial direction
C = Circumferential or tangential direction.

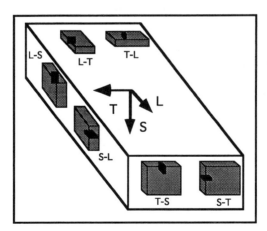

Figure 3.23 ASTM crack plane orientation code designation for rectangular cross-section

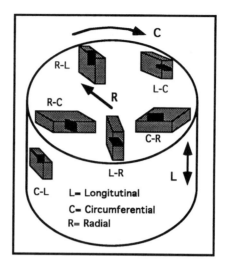

Figure 3.24 ASTM crack plane orientation code designation for cylindrical cross-section

Figure 3.24 schematically represents the loading and crack propagation directions for cylindrical sections. For example, the two-letter code *L–R* indicates that the loading is in the longitudinal direction (*L*) and the expected crack propagation is in the radial direction (*R*). It is important for the analyst to specify the crack and loading directions corresponding to a given fracture toughness for the part under consideration.

3.3.2 Factors Affecting Fracture Toughness

The change of fracture toughness with respect to material thickness was discussed in the previous section and its variation depicted in Fig. 3.20. From Fig. 3.20 it can be seen that several specimens with various thicknesses are needed to generate a complete curve that can fully describe the variation of K_c versus thickness, t. When data points are not available for analysis purposes, the following relationship may give somewhat reasonable results, as long as the plane strain fracture toughness, K_{Ic}, is available for the material under study. Irwin's equation describes plane stress fracture toughness, K_c, written in terms of K_{Ic}; plate thickness, t; and yield strength of the material, σ_{Yield}, as [17]:

$$K_c = K_{Ic}\left[1 + \frac{1.4}{t}\left(\frac{K_{Ic}}{\sigma_{\text{Yield}}}\right)^2\right] \tag{3.27}$$

Another useful relationship describing K_c in terms of K_{Ic} is available in NASA/FLAGRO material library and it is shown by equation 3.58 of Section 3.8.2.

3.3 Fracture Toughness   131

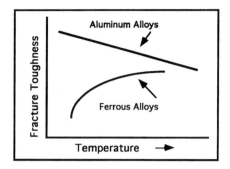

Figure 3.25 Illustration of fracture toughness variation with temperature

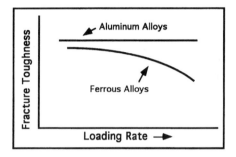

Figure 3.26 Illustration of fracture toughness variation with loading rate

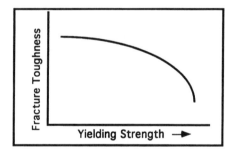

Figure 3.27 Illustration of fracture toughness variation

Other factors affecting the fracture toughness and fatigue crack growth values are loading rate, temperature, temperature rate, and yield strength (see Figs. 3.25, 3.26, and 3.27, respectively). From Fig. 3.25, it can be seen that for aluminum alloys the response of fracture toughness to temperature change increases as temperature decreases, whereas an opposite trend was observed to be true for most ferrous alloys [18, 19]. Experimental test data obtained in the laboratory indicate that most aluminum alloys possess a higher fracture toughness value at the liquid nitrogen temperature ($-320°F$) than at room temperature (see Table 3.2) [12]. This positive trend in fracture toughness value is desirable in aerospace, aircraft, and pressure vessel structures, when proof tests at both room and liquid nitrogen temperature are required. A higher fracture toughness at liquid nitrogen temperature indicates that if a proof test at room temperature is successfully completed, there is no need to conduct an additional proof test at liquid nitrogen. In the case of steel, where the material possesses sufficient ductility and good fracture toughness at room temperature, cleavage or brittle fracture can occur at a service temperature below the transition temperature. This change in material properties is known as a ductile–brittle transition (see Fig. 3.28) and was the cause of many cleavage failures occurring in ships, pressure vessels, bridges, and tanks [20–22]. Figure 3.28 shows the variation of Charpy Impact absorbed energy (representing the material's notch ductility) versus the temperature for most ferritic steels. Originally the Charpy Impact Test was developed to relate the amount of energy absorbed by a material when loaded dynamically in the presence of a notch to the notch ductility at room temperature. Later it was realized that the absorbed energy is a function of temperature for several ferritic steels. If several notched bar impact test specimens (shown in Fig. 3.28) made of ferritic low carbon steel are impacted by the pendulum of a Charpy Impact Test machine, the energy absorption value recorded on the machine shows a decrease in value when the temperature falls below room temperature (see Fig. 3.28). This can be an indication that

Table 3.2 Fracture properties at room temperature and $-320°F$

YS (ksi)	UTS (ksi)	2219-T87 (L–T) K_{Ie} ksi(in.$^{1/2}$)	K_{Ic} ksi(in.$^{1/2}$)
		Room temperature	
57	68	42	30
		Liq. nitrogen temperature	
68	83	57	41
		2219-T62 (T–L)	
		Room temperature	
43	61	41	29
		Liq. nitrogen temperature	
51	76	42	30

3.3 Fracture Toughness

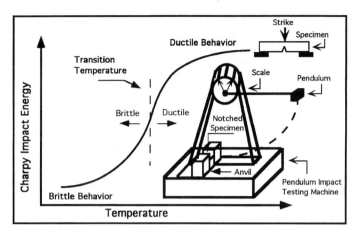

Figure 3.28 Illustration of charpy impact energy versus temperature and the pendulum impact testing machine

the material has gone brittle (plane strain mode of failure). The low-energy absorption value recorded by the Charpy Impact Test machine, when compared with room temperature, indicates brittle behavior of the material due to a low temperature environment. As a final remark, for the Charpy V-notched test to be meaningful, other parameters, such as thickness, the rate of loading, and specimen geometry must be kept constant. The reader may refer to ASTM E-23, "Standard Methods for Notched Bar Impact Testing of Metallic Materials," for more information related to this topic.

The fracture toughness for aluminum alloys seems not to vary with the loading rate, whereas ferrous alloys (such as ferritic steel) are shown to be sensitive to this parameter, as indicated in Fig. 3.26. In general, a material with body-centered cubic (BCC) crystal structure (ferritic steel) shows a reduction in the fracture toughness with an increase in rate of loading. The reverse trend is true for a face-centered cubic (FCC) crystal structures (see Fig. 3.29). The variation of the fracture toughness with respect to the temperature for material with FCC and BCC crystal structures is also shown in Fig. 3.29. A comprehensive review related to the effects of temperature, loading rate, and plate thickness on fracture toughness is available in reference [23].

The yield strength varies due to the heat treatment or cold working given to the material and its variation as illustrated in Fig. 3.27. In selecting high strength material with little ductility, to reduce the size of the structural part and save weight, the engineer must be aware that the fracture toughness value has been reduced considerably. That is, as the ability of the material to absorb energy and deform plastically decreases, the size of the flaws that could initiate instability becomes very small.

134   Chap. 3 Linear Elastic Fracture Mechanics

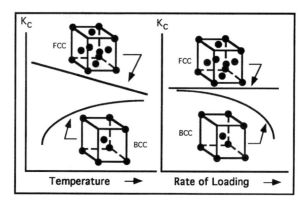

Figure 3.29 Fracture toughness variation as a function of temperature and rate of loading for material with BCC and FCC crystal structures

Additional parameters that may affect fracture toughness value and can be important when dealing with the fracture analysis of a structural part include:

- The coarse grain size may result in lowering the fracture toughness value.
- Embrittlement [segregation of phosphor (P), nitrogen (N), and possibly sulfur (S), to the grain boundary which causes an intergranular mode of fracture) due to microstructure or environmental contamination can result in lowering of the fracture toughness value.
- Work hardening lowers the fracture toughness value by lowering the hardening coefficient, n.

3.4 Residual Strength Capability of a Cracked Structure

At the atomic level, the fracture phenomenon occurs when the bonds between atoms break. For materials with perfect crystalline structure, called "whiskers," the measured fracture strength is much higher than the value obtained in the laboratory by testing a typical standard tensile specimen. For a long time, scientists believed that, for a given alloy, the actual tensile strength obtained through laboratory test should result in a lower value than the theoretical value corresponding to the situation where the alloy is free from defects, such as missing atoms, dislocations, grain boundaries, and cavities. Upon conducting tensile tests with glass fibers of fine diameter in which the probability of defects per volume is low, Griffith [24] showed a tensile strength value of 5×10^5 psi. He postulated that this result could be much lower if the diameter of the test specimen was significantly larger, due to the presence of more defects and they are known as Griffith flaws or cracks. Indeed, Weibull, in 1939, was the first to apply statistical methods to brittle material to explain the greater probability of fracture as a result of find-

3.4 Residual Strength Capability of a Cracked Structure

ing more cracks in a larger specimen than in a smaller one, leading to the possibility of the existence of a size effect on the fracture stress.

From a macroscopic point of view, the presence of a crack introduced in a structure due to manufacturing, machining, or improper handling will significantly lower the strength of the structure as compared with the uncracked condition. The amount of strength that is left in a structure after crack initiation, which is supposed to withstand the service load throughout its design life, is called the residual strength. For uncracked structures, the load carrying capacity or residual strength capability is simply the ultimate strength of the material. When the applied load in an uncracked structure exceeds the ultimate strength of the material, failure will start to occur.

The existence of a crack in a structure will result in a lowering of the residual strength of the structure below the ultimate of the material. That is, when the load on the structure exceeds a certain value called the critical stress, σ_{cr}, unstable cracking will occur. For brittle materials or materials with low fracture toughness, unstable cracking is associated with fast fracture, causing complete fracture of the structural part. Figure 3.30a illustrates the abrupt failure (brittle failure or fast fracture) of three specimens with crack lengths $a_1 > a_2 > a_3$, in which the failure stress for each individual crack length is $\sigma_1 < \sigma_2 < \sigma_3$, respectively. Figure 3.30b also illustrates the abrupt failure points for the same crack size and crack geometry but corresponding to a different material possessing a higher fracture toughness value, $(K_{Ic})_2 > (K_{Ic})_1$. For comparison, the fracture points associated with Fig. 3.30a are also plotted in the same figure. It can be seen that the critical stress values (residual strength values) for the same crack lengths are higher for the second material ($\sigma_2 > \sigma_1$), since it has a higher fracture toughness value.

For material with adequate fracture toughness, the existing crack may grow in a stable manner until it becomes unstable (tearing failure). Figure 3.31 illustrates the tearing failure for three specimens with different initial crack lengths, $a_1 > a_2 > a_3$.

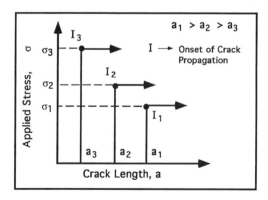

Figure 3.30a Illustration of abrupt failure for three crack sizes

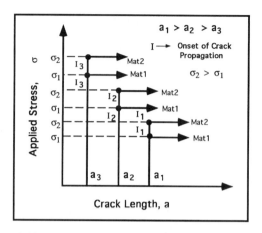

Figure 3.30b Illustration of abrupt failure for two materials

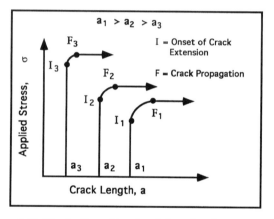

Figure 3.31 Illustration of tearing failure for three crack sizes

From Figs. 3.30 and 3.31, it can be concluded that the load carrying capacity of a cracked structure is a function of the crack size, a, and fracture toughness, K_{Ic} or K_c, and also crack geometry, β.

3.4.1 Residual Strength Diagram for Material with Abrupt Failure

The residual strength capability diagram for material with abrupt failure can be plotted by simply employing the equation of the stress intensity factor, K, that relates the critical applied stress, σ_{cr}, to the critical crack length, a_{cr}, for a given crack geometry and by replacing the critical stress intensity factor, K_{cr}, with the material's fracture toughness, K_{Ic}, obtained through testing. That is:

3.4 Residual Strength Capability of a Cracked Structure

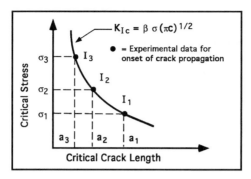

Figure 3.32 Residual strength diagram and critical stress data (I_1, I_2, I_3) extracted from Fig. 3.30

$$K \geqslant K_{cr} = K_{Ic}$$

where

$$K = \beta\sigma(\pi a)^{1/2} \quad (3.28a)$$

Figure 3.32 shows the plot of the residual strength diagram from Eq. (3.28a), in terms of the critical stress, σ_{cr}, versus the critical crack length, a_{cr}, for a center through crack structure where the correction factor $\beta = 1$. It is obvious from Fig. 3.32 that, as the crack length increases, the load carrying capacity of the cracked structure is reduced, and, if it falls below the maximum design stress level, failure can be expected.

The same diagram can also be obtained experimentally if the applied stresses from the test data shown in Fig. 3.30 are plotted as a function of crack length (see Fig. 3.33). Note that for thick sections or brittle material, the onset of crack

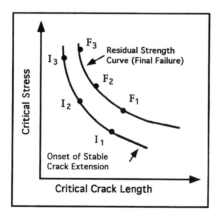

Figure 3.33 Residual strength diagram for ductile metals

growth (as indicated by the letter I in Fig. 3.30) means fast fracture and failure of the structural part. In other words, in brittle material or material with small plastic deformation at the crack tip prior to fracture (state of plane strain), the crack does not extend in a stable manner prior to abrupt fracture.

The residual strength diagram for a ductile material or metals of thin sections cannot be described in the same way as the case of a brittle material with abrupt fracture behavior. For ductile materials, the stable crack extension first occurs at some stress level below the critical stress. The residual strength diagram for material with stable crack growth prior to final failure can be constructed either by apparent fracture toughness [25] or the R-curve approach [26, 27] which will be introduced in Section 3.4.2 and 3.4.3, respectively. The apparent fracture toughness approach will give conservative results when the residual strength of the material is evaluated. The apparent residual strength curve will fall between two distinct curves, shown in Fig. 3.33. The data points with the letter I represent the onset of stable crack extension and the data points with the letter F describe the final failure.

3.4.2 The Apparent Fracture Toughness

One of the methods of constructing the residual strength diagram for a ductile material is based on the apparent fracture toughness approach [25]. In this approach, the apparent fracture toughness, K_{APP}, can be calculated by using the initial crack length (represented by the letter I) and final critical stress (as shown in Fig. 3.31 by the letter F). The apparent residual strength diagram falls between the two curves (defining the onset of stable crack extension and final crack length) represented by the dotted line shown in Fig. 3.34. The apparent residual strength diagram shown in Fig. 3.34 gives the lower bound estimate of fracture toughness for materials with tearing failure behavior since the failure criterion, $K \geq K_{cr} = K_{APP}$, is based on the

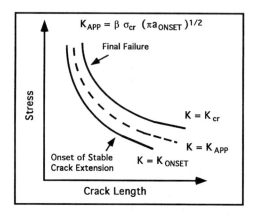

Figure 3.34 The residual strength diagram for ductile metals for $K = K_{ONSET}$, $K = K_{APP}$, and $K = K_{cr}$.

original crack length rather than the final crack length. If K_{cr} is evaluated based on the final crack length, then the R-curve approach described in the next section must be employed.

3.4.3 Development of the Resistance Curve (R-curve) and K_R

Another available method for constructing the residual strength diagram when using tough material is the crack growth Resistance curve (R-curve) [26, 27]. This method can be applied to material with a tearing fracture behavior that occurs in thin sheet metal or to ductile metals where the crack extension is slow and stable prior to final failure. The failure criterion described by the apparent fracture toughness value was considered as a single parameter failure criterion where $K \geqslant K_{cr} = K_{APP}$. The crack growth resistance approach for tearing type fracture is based on two fracture parameter criteria that simply state that fracture will occur when the stress intensity factor, K, becomes equal to or greater than the material's fracture resistance, K_R. Furthermore, fracture will also occur when the rate of change of K with respect to crack length becomes equal to or greater than the rate of change of K_R with respect to crack length, that is:

$$K \geqslant K_R,$$

and

$$\frac{\partial K}{\partial a} \geqslant \frac{\partial K_R}{\partial a} \qquad (3.28b)$$

In other words, the two failure criteria say that, at failure, when abrupt fracture occurs ($K \geqslant K_R$), the energy available to extend the crack becomes equal to or greater than the material resistance to crack growth ($\frac{\partial K}{\partial a} \geqslant \frac{\partial K_R}{\partial a}$).

To obtain the plane stress fracture toughness, K_c, for material with tearing fracture behavior, a resistance curve, known as the R-curve, must be constructed. Consider the variation of the stress intensity factor (up to the point of failure) with respect to total crack lengths for a given material's thickness, as plotted in Fig. 3.35. In this figure, the calculated stress intensity factors for each original crack length, $a_1 > a_2 > a_3$ (shown in Fig. 3.31) correspond to the onset of stable crack growth, where $K_{ONSET} = \beta \sigma_{ONSET} (\pi a_0)^{1/2}$ is shown as a dotted line. Note that a_o is the original crack length and is equal to the crack length at the onset of stable crack extension, as shown in Fig. 3.35 by the letter I. In addition, the variation of the calculated stress intensity factor corresponding to final failure (where the crack becomes unstable) as a function of the amount of stable crack extension, Δa, is plotted in Fig. 3.36. Figure 3.36 presents the crack growth resistance curve

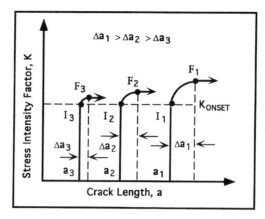

Figure 3.35 Variation of stress intensity factor versus crack length

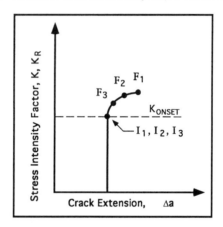

Figure 3.36 Variation of stress intensity factor versus Δa

or R-curve which covers all crack growth resistance behaviors that have been constructed for different original crack lengths (shown in Fig. 3.35). It can be concluded that the R-curve developed in Fig. 3.36 is independent of the initial crack length, but is dependent on the amount of crack extension, Δa.

In developing the R-curve for a given thickness, the K_R value is evaluated by using the measured effective crack length and the critical load obtained through testing. The equation describing K_R is [28]:

$$K_R = (P/Wt)(\pi a_{\text{eff}})^{1/2} \times f(a/W) \qquad (3.29)$$

where P is the applied load corresponding to the fracture at instability, W is the width of the specimen, $f(a/W)$ is the correction to the width, and a_{eff} is the effec-

3.4 Residual Strength Capability of a Cracked Structure

tive crack length. The effective crack length is the total crack length and is expressed as:

$$a_{\text{eff}} = a_0 + \Delta a + r_p \quad (3.30)$$

where r_p is the correction for the plastic zone (the estimation of the size and the shape of the plastic zone at the crack tip by using different yield criteria is discussed in Section 3.5). Note that the R-curve is supposed to be independent of the original crack length. However, when it is developed for a given crack length and thickness based on testing, it can be matched with the stress intensity factor curve to estimate the fracture toughness, K_C, and the load necessary to cause unstable crack propagation (Fig. 3.37). The tangent point between the developed R-curve and the stress intensity factor at $\sigma = \sigma_C$, where $K = K_C$, determines the fracture toughness, as shown in Fig. 3.37. At the tangency point, shown in Fig. 3.37, the two failure criteria described by Eq. (3.28b) are met.

In general, the construction of a residual strength diagram involves the following steps:

Step 1. The relationship between the crack length, applied stress, and the stress intensity factor for the crack geometry under consideration [Equation (3.27)] must be known or developed.

Step 2. The appropriate fracture toughness values must be available for the material under consideration. Apply the failure criteria described in Eq. (3.27) by equating the critical stress intensity factor with the fracture toughness value ($K = K_{\text{cr}} \geq K_{\text{Ic}}$ or K_c)

Step 3. Construct the residual strength diagram by plotting the variation of the fracture stress, σ_c, versus the critical crack size, a_{cr}, for the crack geometry under consideration.

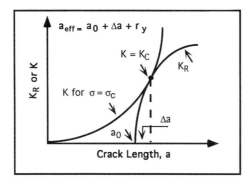

Figure 3.37 Illustration of the R-curve where K_R and K curves are tangent at the instability point

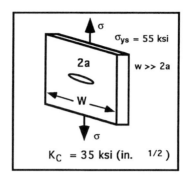

Figure 3.38 A cracked panel made of 2219-T8 aluminum shown for example 3.3

Example 3.3

Construct the residual strength diagram for the cracked panel shown in Figure 3.38. The panel is made of 2219-T8 aluminum with a fracture toughness value of 35 ksi (in.$^{1/2}$).

Solution

Applying step 1 of Section 3.4.3 by using the equation for the stress intensity factor of a wide panel ($W \gg 2a$) with a through centered crack [expressed by Eq. 3.19].

$$K = \beta\sigma(\pi a)^{1/2}$$

where

$$\beta = 1$$

The failure criteria described by Eq. (3.27) simply show the relationship that holds between the critical stress intensity factor and the applied stress. Note that for the instability to occur, the critical stress intensity factor must be equal to or greater than the fracture toughness of the material under consideration, $K \geq K_{cr} = K_c$. The residual strength diagram can now be constructed by plotting σ_{cr} versus a_{cr}, as in Fig. 3.39. For example, the residual strength or fracture stress for the cracked panel shown in the figure with crack length $2a = 1.2$ in. is 22.0 ksi. That is, the induced stress on the structural part due to the load environment must be below the residual strength of the cracked structure ($\sigma_F = 22.0$ ksi $> (\sigma)_{service}$) to preclude the failure of the part.

Example 3.4

Establish the residual strength capability of an eccentric through crack in a finite plate as shown in Fig. 3.40. The crack geometry and loading conditions are also

3.4 Residual Strength Capability of a Cracked Structure

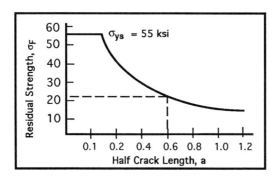

Figure 3.39 Residual strength capability diagram constructed for example 3.3

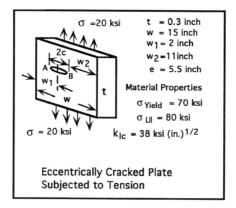

Figure 3.40 The crack geometry and loading conditions shown for example 3.4.

shown in the figure. Assume the flaw sizes obtained by two different inspection methods are a through crack of length (1) $2c = 1.5$ in. and (2) $2c = 2.4$ in.

Solution

From Fig 3.40, it is obvious that the stress intensity factor for the tip-A is more critical than for tip-B. This is due to the width correction factor, β_1, for side A versus β_2 for side B, where $w_1 < w_2$ and therefore, $\beta_1 > \beta_2$. The failure criterion based on fracture toughness, described in Section 3.4 by Eq. (3.28a), can be applied here to find the critical crack length for an applied stress of 20 ksi.

Case 1 (2c = 1.5 in.)

The critical flaw size based on tip-A can be obtained through equation 3.19 as:

$$K = \sigma \beta_w \sqrt{\pi c}$$

where

$$\beta_w = (\sec \pi c/w)^{1/2}$$

The width correction factor for this type of crack geometry is a function of distance, W_1, and the amount of eccentricity, e, as shown in Fig. 3.40. A numerical solution to the width correction factor for the above crack geometry is available in Reference [29]. Another approach, more conservative but simpler to apply, was suggested by Kaplan and Reiman [30] in which the width for the crack tip-A is taken as twice the distance from the crack center to the edge of the plate, $2W_1$. In this case, the width for tip-A is $W = 4$ in. The width correction factor based on this assumption can be obtained as:

$$\beta_w = (\sec \pi c/w)^{1/2} = 1.09$$
$$K_1 = 20(1.09)(3.14 \times 0.75)^{1/2}$$
$$K_1 = 33.454 \text{ ksi (in.}^{1/2})$$

The calculated stress intensity factor based on the initial crack length reported by inspection ($2c = 1.5$ in.) is smaller than the critical stress intensity factor $K_c = 38$ ksi (in.$^{1/2}$). The calculated critical crack length based on $K_c = 38$ ksi (in.$^{1/2}$), $\sigma = 20$ ksi, and $W = 4$ in. is:

$$38 \text{ ksi (in.}^{1/2}) = 20 \text{ ksi } (\sec \pi c/4)^{1/2} (3.14 \times c)^{1/2}$$

Solving for c:

$$c = 1.149 \text{ in. or } 2c = 2.299 \text{ in.} > 1.5 \text{ in. (by inspection)}$$

The calculated critical crack length ($2c = 2.299$ in.) is larger than the crack length found by inspection ($2c = 1.5$ in.). Therefore, the cracked plate will survive the load environment.

Case 2 ($2c = 2.4$ in.):

As shown by the analysis performed for the previous case, the calculated critical crack length $2c = 2.299$ in. is smaller than the preexisting crack reported by the second inspection ($2c = 2.4$ in.). Based on this assumption the crack at tip-A is critical and will propagate toward the edge of the plate. The new crack geometry is now a single edge crack with length $c = 2.0 + 2.4/2 = 3.2$ in., as shown in Fig. 3.41. The stress intensity factor and width correction factor for an edge crack in a finite plate are given by Eq. (3.20) that is:

3.4 Residual Strength Capability of a Cracked Structure

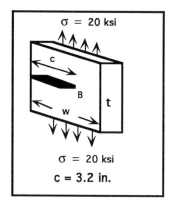

Figure 3.41 An Edge Crack in a Finite Plate (when tip-A propagate toward the edge and become an edge crack)

$$K_I = \sigma \beta_0 \sqrt{\pi c}$$
$$\beta_0 = Y[0.752 + 2.02w + 0.37(1 - \sin\beta)^3]$$

where

$$Y = \sec\beta[(\tan\beta)/\beta]^{1/2}, \text{ and } \beta = \pi c/2W$$

Calculating the quantities Y, β and the width correction β_0:

$$\beta = (3.14 \times 3.2)/(2 \times 15) = 0.334$$
$$Y = \sec(0.334)[(\tan 0.334)/0.334]^{1/2} = 1.081$$
$$\beta_0 = 1.081[0.752 + 2.02(3.2/15) + 0.37(1 - \sin 0.334)]^3 = 1.398$$
$$K_I = 20 \times 1.398 \times (3.14 \times 3.2)^{1/2} = 88.63 > 38 \text{ ksi}(\text{in.}^{1/2})$$

From the above analysis, it is clear that the calculated stress intensity factor for the new crack geometry (a single edge crack) is much higher than the fracture toughness of the material. Therefore, as soon as tip-A becomes unstable and is arrested by the left free edge, the newly formed crack geometry (a single edge crack) becomes unstable also.

In general, when analysis indicates that the residual strength capability of a given structural part is not adequate, it is recommended that either (1) the inspection method be revised to obtain smaller initial crack length or (2) the magnitude of the applied stress, σ, be reduced at the expense of increasing the part thickness. However, by doing that the fracture toughness value for the new thickness is now reduced and the material will tend to approach the plane strain condition.

3.5 Plasticity at the Crack Tip

Linear elastic fracture mechanics is based upon the assumption that the size of the plastic zone formed at the crack tip is negligible as compared to the crack length and plate thickness. That is, the crack tip plastic deformation is confined to a small region around the crack tip and the bulk of the structural components is elastic. For metals that generally go through extensive plastic deformation at the crack tip prior to failure, the use of linear elastic fracture mechanics yields conservative results when solving a given crack problem. This is true because the applied load does work on the cracked body which is stored in the form of strain energy. For brittle materials, all of the available energy will be consumed in creating two new crack surfaces. In ductile material, a large portion of the available energy will be consumed in plastically deforming the material at the crack tip (in metallic materials the energy required for plastic deformation is approximately 10^3 times larger than the surface energy). Chapter 6 presents the Fracture Mechanics of Ductile Metals (FMDM) theory which will incorporate the dissipated energy terms consumed in plastically straining the material in the region at the crack tip.

A problem that arises in applying linear elastic fracture mechanics is that the calculated crack tip stress approaches the very high value predicted by the quantity $1/\sqrt{2\pi r}$ [shown by equation (3.14)] whenever the term $r \to 0$. In real situations, there will be a finite plastic zone (r_p) ahead of a loaded crack where the material will yield prior to final failure. Irwin's first approach [10, 31] to obtain the plastic zone shape and size was based on the assumption that the plastic zone formed at the crack tip is a circle (Fig. 3.42) and that the diameter of the circle, r_p, for the plane stress condition (where $\sigma_z = 0$) can be calculated by simply equating the crack tip stress, σ_y [from Eq. (3.14)] to the tensile yield of the material. For a through center crack plate subjected to a far field stress σ:

$$\sigma_{yield} = \sigma_y = \frac{\sigma \sqrt{\pi a}}{\sqrt{2\pi r}} \cos \theta/2 [1 + \sin \theta/2 \sin 3\theta/2] \qquad (3.31)$$

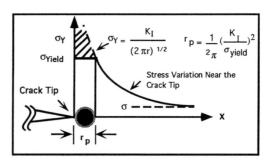

Figure 3.42 Plastic deformation at the crack tip and stress variation before and after yielding

3.5 Plasticity at the Crack Tip

At the crack tip where $\theta = 0$ and $r = r_p$:

$$\sigma_{\text{yield}} = \frac{K_I}{\sqrt{2\pi r_p}} \quad (3.32)$$

Rewriting Eq. (3.32), the plastic zone size, r_p, becomes:

$$r_p = \frac{1}{2\pi}\left(\frac{K_I}{\sigma_{\text{yield}}}\right)^2 \quad (3.33)$$

Irwin further argued that the stress distribution ahead of the crack tip (shown in Fig. 3.42 as a shaded area) cannot simply be ignored when stresses are above the tensile yield of the material. Irwin's second estimate of the plastic zone size resulted in calculating a larger plastic size than that obtained by Eq. (3.33). Furthermore, he stated that the formation of a plastic zone at the crack tip can be viewed from linear elastic fracture mechanics as if the crack length is longer than its original size, a_o, by some crack increment, Δa. This is true because the formation of plasticity at the crack tip can result in having a larger crack tip opening displacement, as if it contained a crack of larger size. That is, the new crack length (called notional crack) is now equal to $(a_o + r_p)$, with its tip a distance, r_p, ahead of the original crack tip (Fig. 3.43).

To account for the stress distribution above the yield region (as shown in Fig. 3.44), the following equilibrium equation is valid:

$$\int_0^{r_y} \frac{K_I}{\sqrt{2\pi r}} dr = (\sigma_{\text{yield}}) D_p \quad (3.34)$$

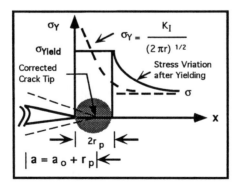

Figure 3.43 Irwin's correction for the crack tip

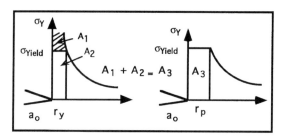

Figure 3.44 First and second estimate of plastic zone size

Equation (3.34) simply states that the total forces produced by the stresses in the shaded and unshaded regions ahead of the crack tip (designated by the letters A_1 and A_2, respectively) must be equal to the force produced by the stresses in the region A_3 (see Fig. 3.44). Replacing K_1 at $r = r_y$ with $(\sigma_{\text{yield}})\sqrt{2\pi r_y}$ and solving for D_p:

$$2r_y = D_p = 2r_p \tag{3.35}$$

Thus, the plastic zone correction based on a notional crack is to extend the original crack length by the amount of the plastic zone, $D_p/2 = r_p$, derived by Eq. (3.35) (Fig. 3.43). The stress intensity factor associated with the notional crack is $K_{\text{NEW}} = \beta\sigma\sqrt{\pi(a + r_p)}$.

3.6 Plastic Zone Shape Based on the Von Mises Yield Criterion

In the previous discussion related to the plasticity formation at the crack tip (given in Section 3.5), a simplified assumption that the shape of the plastic zone is a circle [31] was employed and the radius of the plastic zone was derived based on crack tip stresses taken to be equal to the uniaxial yield of the material [Eq. (3.31)]. That is, a given material will fail plastically when the maximum principal stress becomes equal to the uniaxial yield strength. Moreover, the dependency of the plastic zone with respect to the angle, θ, was not accounted for in Irwin's approach, described in Section 3.5.

A more appropriate yield criterion can be implemented to provide the size and shape of the plastic region for all values of θ [32, 33]. The most common acceptable yield criterion is based on the Von Mises criterion (proposed in 1913) [34], which simply states that, for yielding to occur, the maximum value of the distortion energy per unit volume in that material must reach the distortion energy per unit volume needed to yield the material in a tensile test specimen of the same material. In terms of principal stresses, the Von Mises criterion can be written as:

$$(\sigma_1 - \sigma_2)^2 + (\sigma_2 - \sigma_3)^2 + (\sigma_1 - \sigma_3)^2 = 2\sigma_e^2 \tag{3.36}$$

3.6 Plastic Zone Shape Based on the Von Mises Yield Criterion

The equivalent stress, σ_e, shown in equation 3.36, is calculated from multiaxial stress state where σ_1, σ_2, and σ_3 are principal stresses at a given point in the body. Yielding occurs when the quantity σ_e exceeds the monotonic yield value of the material.

The principal stresses σ_1, σ_2, and σ_3 are related to the crack tip stresses σ_x, σ_y, and σ_z by the following relationships:

$$\sigma_1 = \frac{\sigma_x - \sigma_y}{2} + \left[\left(\frac{\sigma_x - \sigma_y}{2}\right)^2 + \tau_{xy}\right]^{1/2} \tag{3.37}$$

$$\sigma_2 = \frac{\sigma_x + \sigma_y}{2} - \left[\left(\frac{\sigma_x - \sigma_y}{2}\right)^2 + \tau_{xy}\right]^{1/2} \tag{3.38}$$

where $\sigma_3 = 0$ for the plane stress and $\sigma_3 = v(\sigma_1 + \sigma_2)$ for the plane strain condition and v is the Poisson ratio. Substituting for the quantities σ_x, σ_y, and τ_{xy} from Eq. (3.14) of Section 3.2.3, the principal stresses, σ_1 and σ_2, in terms of stress intensity factor become:

$$\sigma_1 = \frac{K_I}{\sqrt{2\pi r}} \cos\theta/2 \,[1 + \sin\theta/2] \tag{3.39}$$

$$\sigma_2 = \frac{K_I}{\sqrt{2\pi r}} \cos\theta/2 \,[1 - \sin\theta/2] \tag{3.40}$$

Inserting the principal stresses from Eqs. (3.39) and (3.40) into the Von Mises yield criterion [shown by Eq. 3.36 where $\sigma_e = \sigma_{yield}$), an expression for the plastic zone radius, r_p, as a function of θ can be obtained. For the case of the plane stress condition, where $\sigma_3 = 0$:

$$r_p(\theta) = \frac{K_I^2}{4\pi\sigma_{Yield}^2}[1 + \cos\theta + 3/2\sin^2\theta] \tag{3.41}$$

and for plane strain where $\sigma_3 = v(\sigma_1 + \sigma_2)$:

$$r_p(\theta) = \frac{K_I^2}{4\pi\sigma_{Yield}^2}[(1 - 2v)^2(1 + \cos\theta) + 3/2\sin^2\theta] \tag{3.42}$$

The plastic zone size, r_p, for the case of $\theta = 0$ can be obtained based on the Von Mises yield criterion via Eqs. (3.41) and (3.42) for the plane stress and plane

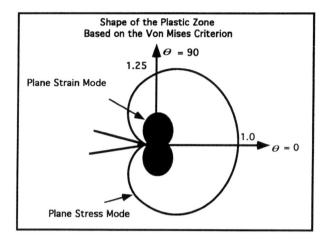

Figure 3.45 Plastic zone shape for the plane stress and plane strain condition

strain conditions, respectively. The plot of the nondimensional quantity $r_p(\theta)/r_p(0)$ versus the angle θ is plotted in Fig. 3.45. Note that the plastic zone size, r_p, for $\theta = 0$ is equal to the value of r_p that was obtained from Eq. 3.33.

The nondimensional quantity $r_p(\theta)/r_p(0)$ for the plane stress condition can be written as:

$$r_p(\theta)/r_p(0) = 1/2 + 3/4 \sin^2 \theta + 1/2 \cos \theta \tag{3.43a}$$

and for the plane strain:

$$r_p(\theta)/r_p(0) = \frac{1}{2}(1 - 2v)^2(1 + \cos \theta) + \frac{3}{4} \sin^2 \theta \tag{3.43b}$$

As discussed in Section 3.3, the fracture toughness is related to the amount of plastic deformation and varies with the thickness. For a structural part with a given thickness, the shape and the size of the plastic zone vary throughout the section. The variation of the plastic zone size and its shape through the thickness based on the Von Mises yield criterion is illustrated in Fig. 3.46. At the free edges, where $\sigma_3 = 0$, the plastic zone totally resembles the plane stress case (see Fig. 3.46). In the interior or the midsection region, the plastic zone shape and size correspond to the plane strain condition.

Example 3.4

Plot the plastic zone shape and size for the plane stress and plane strain cases by applying the Tresca yield criterion.

3.6 Plastic Zone Shape Based on the Von Mises Yield Criterion

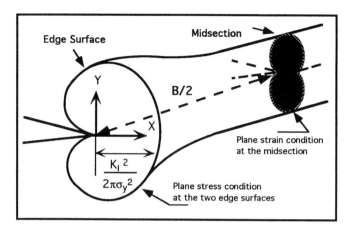

Figure 3.46 Illustration of plastic zone shape through the thickness

Solution

The Tresca yield criterion [35] is based on the maximum shear stress criterion, which simply states that a given structural component is safe under multiaxial state of stress when the maximum value of shear stress, τ_{max}, in that component is smaller than the critical value. The critical shear stress value corresponds to the value of the shearing stress in a tensile test specimen of the same material as the specimen starts to yield. In terms of principal stresses, when $\sigma_1 > \sigma_2 > \sigma_3$ and the Tresca yield criterion can be written as:

$$\tau_{max} = |\sigma_1 - \sigma_3|/2 = \frac{1}{2}\sigma_{yield} \qquad (3.44)$$

For the plane stress condition, where $\sigma_3 = 0$, Eq. (3.44) in terms of the crack tip stress intensity factor can be written [see Eq. (3.39)] as:

$$\sigma_1 = \frac{K_I}{\sqrt{2\pi r}} \cos\theta/2 [1 + \sin\theta/2] = \sigma_{Yield}$$

Solving for the plane stress plastic zone size, $r = r_p$:

$$r_p = \frac{K_I^2}{2\pi\sigma_{Yield}^2}[\cos\theta/2(1 + \sin\theta/2)]^2$$

and for the plane strain condition, where $\sigma_3 = v(\sigma_1 + \sigma_2)$, the size of the plastic zone in terms of the crack tip stress intensity factor can be expressed as:

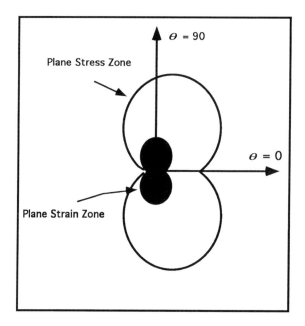

Figure 3.47 Plastic zone shape based on the Tresca yield criterion

$$r_p = \frac{K_I^2}{2\pi\sigma_{\text{Yield}}^2} \cos^2\theta/2\,[(1-2v) - \sin\theta/2]^2$$

The shape of the plastic zone, based on the Tresca yield criterion (described by the above two equations), is plotted in Fig. 3.47.

3.7 Surface or Part Through Cracks

In most structures, preexisting cracks are found in the form of surface cracks (also called part through cracks) that initiate at surface discontinuities or emanate from a hole in the form of corner cracks (see Fig. 3.48). Surface scratches are introduced into the part as the result of surface machining, grinding, forming, or may be due to improper handling during manufacturing. These surface cracks may become through cracks during the service life of the structural part before reaching their critical size. In other cases, embedded cracks found in welded parts will grow gradually to the surface and become surface cracks.

Surface cracks will grow in both length and depth directions. Therefore, in analyzing these crack geometries by linear elastic fracture mechanics, it is important to have an expression for the mode I crack tip stress intensity factor, K_I. When the load environment is fluctuating, the expression pertaining to stress

3.7 Surface or Part Through Cracks

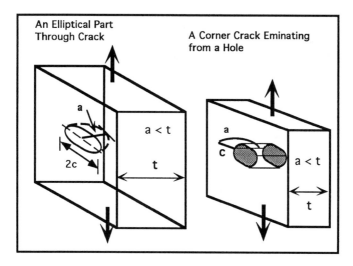

Figure 3.48 Illustration of a surface part through crack and a corner crack eminating from a hole

intensity factor, K_I, must be provided to the analyst if fatigue crack growth analysis for both the depth and length directions are needed. In Section 3.7.1, the stress intensity factor equation for a surface crack in an infinite plate is formulated. The solutions corresponding to two of the most commonly used surface crack geometries are discussed in Sections 3.7.2 and 3.7.3. Part through fracture toughness for both the depth and length directions are covered in Section 3.7.4. Finally, the concept of leak-before-burst (LBB) is introduced in Section 3.7.5, together with example problems.

3.7.1 Stress Intensity Factor Solution for Surface Cracks (Part Through Cracks)

The stress intensity factor, K_I, for a surface crack in a plate subjected to uniform tensile load (as plotted in Fig. 3.48) can be formulated by using the stress intensity factor equation corresponding to an embedded elliptical crack in an infinite body subjected to uniform load (see Fig. 3.49 for the geometry and loading [36]).

At any point, m, along the boundary of the elliptical crack, the mode I stress intensity factor can be written as:

$$K_I = \frac{\sigma\sqrt{\pi a}}{\Phi}\left(\left(\frac{a}{c}\right)^2 \cos^2\theta + \sin^2\theta\right)^{\frac{1}{4}} \qquad (3.45)$$

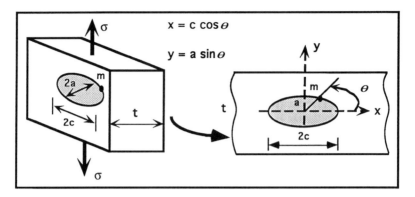

Figure 3.49 Embeded elliptical crack geometry under mode 1 loading

where θ is the angle that defines any point around the perimeter of the elliptical crack (Fig. 3.49). The quantity Φ is the complete elliptical integral of the second kind and is given by:

$$\Phi = \int_0^{\frac{\pi}{2}} \left(1 - \left(1 - \left(\frac{a}{c}\right)^2 \sin^2 \phi \right)\right)^{\frac{1}{2}} d\phi \qquad (3.46)$$

Empirical expressions that can describe the quantity Φ of Eq. (3.46) for different crack depth to crack length aspect ratios, a/c, are [28]:

$$\Phi = 1.0 + 1.464 \left(\frac{a}{c}\right)^{1.65}$$

$$\Phi = 1.0 + 1.464 \left(\frac{c}{a}\right)^{1.65} \qquad (3.47)$$

for $a/c \leq$ and $a/c > 1$, respectively.

Using the K_I solution for an embedded crack described by Eq. (3.45), the stress intensity factor equation at any point, m, around the periphery of a surface flaw in an infinite plate subjected to tensile load can be written as [36]:

$$K_I = \frac{1.1 \sigma \sqrt{\pi a}}{\Phi} \left(\left(\frac{a}{c}\right)^2 \cos^2 \theta + \sin^2 \theta \right)^{\frac{1}{4}} \qquad (3.48)$$

The maximum and minimum stress intensity factors around the crack front are associated with the angles $\theta = 90°$ and $\theta = 0°$, respectively.

When the part through crack is a circular surface crack, as shown in Fig. 3.50, where the aspect ratio $a/c = 1$, the value of the stress intensity factor around the

3.7 Surface or Part Through Cracks

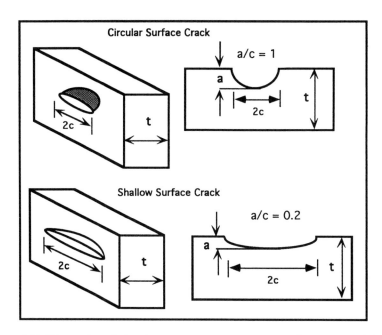

Figure 3.50 Illustration of circular and shallow cracks with $a/c = 1$ and $a/c = 0.2$, respectively

crack front is a constant. From Eq. (3.48), the stress intensity factor corresponding to a circular crack ($a/c = 1$) can be written as:

$$K_I = \frac{1.1\sigma\sqrt{\pi a}}{\Phi} \tag{3.49}$$

where the quantity $\Phi = 2.464$ [see Eq. (3.47) for $a/c = 1$]. Accordingly, for a shallow crack, where $a/c = 0.2$ and $\theta = 90°$, as shown in Fig. 3.50, the stress intensity factor can be represented by Eq. (3.49), in which Φ takes the value of 1.102.

The two previously mentioned aspect ratios of $a/c = 1$ (a circular crack) and $a/c = 0.2$ (a shallow crack) as shown in Fig. 3.50 are the limiting aspect ratio cases that are widely used in fracture mechanics analysis. For example, NASA requires all fracture critical flight hardware (such as thin-wall pressure vessel structures) to be examined for safe-life using these two limiting aspect ratios to ensure that the structural part can survive the expected environment.

The correction factor, β, employed for the equation of stress intensity factor to account for the plate width, back surface correction, and loading conditions of a part through surface crack [see Eq. (3.49)] is provided in References [37, 38] and can be written as:

156 Chap. 3 Linear Elastic Fracture Mechanics

$$K_I = \beta(a/c, a/t, c/w, \theta) \frac{\sigma\sqrt{\pi a}}{\Phi} \qquad (3.50)$$

where the correction factor β is:

$$\beta(a/c, a/t, c/w, \theta) = [M_1 + M_2(a/t)^2 + M_3(a/t)^4] * g * f_\phi * f_w \qquad (3.51)$$

The back surface correction $f_w = [\sec \pi c/w (a/t)^{1/2}]^{1/2}$. Other quantities in Eq. (3.51) are given as:

$$M_1 = 1.13 - 0.09(a/c)$$
$$M_2 = -0.54 + 0.89/(0.2 + a/c)$$
$$M_3 = 0.5 - [1/(0.65 + a/c)] + 14(1 - a/c)^{24}$$
$$g = 1 + [0.1 + 0.35(a/t)^2](1 - \sin\theta)^2$$
$$f_\phi = [(a/c)^2 \cos^2\theta + \sin^2\theta]^{1/4}$$

for the case of $a/c < 1$, and for the case of a/c > 1, the correction factor parameters are:

$$M_1 = \sqrt{c/a}\,[1 + 0.04(c/a)]$$
$$M_2 = 0.2(c/a)^4$$
$$M_3 = -0.11(c/a)^4$$
$$g = 1 + [0.1 + 0.35(a/t)^2(c/a)](1 - \sin\theta)^2$$
$$f_\phi = [(c/a)^2 \sin^2\theta + \cos^2\theta]^{1/4}$$

It should be noted that Eq. (3.51) is valid within the limits of:

$$0 \leq \frac{a}{c} \leq 2, \quad \frac{c}{w} < \frac{1}{4}, \quad 0 \leq \theta \leq \pi$$

and

$$\frac{a}{t} \leq 1.25\left(\frac{a}{c} + 0.6\right) \text{ for } 0 \leq \frac{a}{c} \leq 0.2$$
$$\frac{a}{t} < 1 \text{ for } 0.2 \leq \frac{a}{c} \leq \infty$$

The stress intensity factor solutions for other surface crack geometries are also available in Reference [39]. In this section, only the two most widely used cases in aircraft and aerospace industries for life evaluation of high-risk or fracture critical parts will be discussed. These cases are: (1) a surface crack in a pressurized pipe and (2) a corner crack emanating from the edge of a hole.

3.7.2 Longitudinal Surface Crack in a Pressurized Pipe

The stress intensity factor for a surface crack oriented in the longitudinal direction in a pressurized pipe has the form of Eq. (3.50). The U.S. Air Force and NASA require all pressurized containers to be considered as a fracture critical or high-risk part and they must demonstrate an analytical life of not less than four times the service life of the structure (see Chapter 5 for the definition of fracture critical part). A pressurized vessel is defined as a container that stores fluid with stored energy of 14,240 ft-lbs or contains gas with pressure above 14.7 psi. In a pressurized cylinder the circumferential induced stresses (hoop stress) are larger than the longitudinal stresses. It is therefore necessary to develop the stress intensity factor solution for a pressurized pipe or vessel with a longitudinal surface crack. The correction factor, β, for the stress intensity factor of a part through thickness crack in a pressurized pipe (shown in Fig. 3.51) with diameter D is [39]:

$$\beta(a/c, a/t, \theta) = 0.97 * [M_1 + M_2(a/t)^2 + M_3(a/t)^4] * g * f_\phi * f_c * f_i \quad (3.52)$$

where the quantities in Eq. 3.52 are:

$$M_1 = 1.13 - 0.09(a/c)$$
$$M_2 = -0.54 + 0.89/(0.2 + a/c)$$
$$M_3 = 0.5 - [1/(0.65 + a/c)] + 14(1 - a/c)^{24}$$
$$g = 1 + [0.1 + 0.35(a/t)^2](1 - \sin\theta)^2$$
$$f_\phi = [(a/c)^2 \cos^2\theta + \sin^2\theta]^{1/4}$$

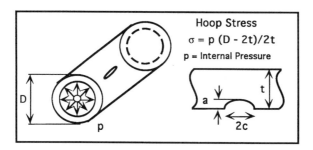

Figure 3.51 Part through crack in a pipe

The stress intensity factors for the depth and length directions are associated with the angles $\theta = 90°$ and $\theta = 0°$, respectively.

The quantities $f_c = [(1 + k^2)/(1 - k^2) + 1 - 0.5(a/t)^{1/2}] [t/(D/2 - t)]$ where $k = 1 - 2t/D$ and the value of $f_i = 1$ for internal crack and 1.1 for external crack.

3.7.3 Part Through Corner Crack Emanating from a Hole

In fracture mechanics analysis of high risk parts of a given flight hardware, NASA and the military require the analyst to assume the existence of a corner crack from the edge of a hole (when this assumption is considered as the worst crack location and orientation). Understanding corner cracks from a hole is extremly important when assessing the integrity of a bolted joint. Many structural parts, especially in aircrafts, are jointed together by fasteners which their failure can lead to loss of life and the structure. To be able to implement the above requirement, it is necessary to develop the stress intensity factor solution for corner cracks at the edge of a hole (Fig. 3.52). The solution to the general case of a quarter elliptical crack geometry is given by Eq. (3.50), with the correction factor, β, provided by Newman and Raju as [38, 39]:

$$\beta(a/c, a/t, \theta) = [M_1 + M_2(a/t)^2 + M_3(a/t)^4] * g_1 * g_2 * f_\phi \quad (3.53)$$

where

$$M_1 = 1.08 - 0.03(a/c)$$
$$M_2 = -0.44 + 1.06/(0.3 + a/c)$$
$$M_3 = -0.5 - 0.25(a/c) + 14.8(1 - a/c)^{15}$$
$$g_1 = 1 + [0.08 + 0.4(a/t)^2](1 - \sin\theta)^3$$
$$g_2 = 1 + [0.08 + 0.15(a/t)^2](1 - \cos\theta)^3$$
$$f_\phi = [(a/c)^2 \cos^2\theta + \sin^2\theta]^{1/4}$$

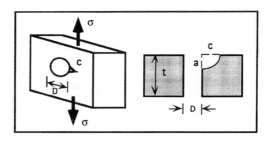

Figure 3.52 Part through corner crack emanating from a hole

3.7 Surface or Part Through Cracks

for the case of $a/c < 1$ and

$$M_1 = \sqrt{c/a}\,[1.08 - 0.03(c/a)]$$
$$M_2 = 0.375\,(c/a)^2$$
$$M_3 = -0.25(c/a)^2$$
$$g_1 = 1 + [0.08 + 0.4(c/t)^2](1 - \sin\theta)^3$$
$$g_2 = 1 + [0.08 + 0.15(c/t)^2](1 - \cos\theta)^3$$
$$f_\phi = [(c/a)^2 \sin^2\theta + \cos^2\theta]^{1/4}$$

for the case of $a/c > 1$.

3.7.4 Part Through Fracture Toughness, K_{Ie}

The critical value of the stress intensity factor for a part through crack is called the part through fracture toughness and is designated by the symbol K_{Ie}. Just like the plane strain and plane stress fracture toughness data for through cracks, the part through fracture toughness data are also an important parameter to have when the number of cycles to failure associated with a given load environment is evaluated (see Chapter 4 for an in-depth discussion of the fatigue crack growth mechanism).

The part through fracture toughness can be obtained through testing (see ASTM-E740). The ASTM test specimen for determining the part through fracture toughness value, K_{Ie}, is shown in Fig. 3.53. A dog bone shape specimen hav-

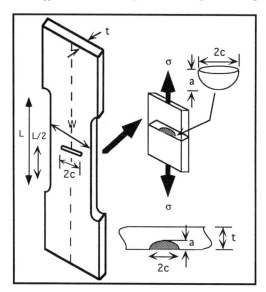

Figure 3.53 Illustration of ASTM part through crack test specimen

ing part through crack (possessing crack depth, a, and crack length, $2c$) is subjected to axial load as shown in the figure. The critical value of load at fracture, as well as final crack length and crack depth after fracture, are measured. Using Eqs. 3.50 and 3.51 [38] the critical value of stress intensity factor at fracture where $K = K_{Ie}$ can be calculated. A typical test specimen configuration can be:

> Use $2c = 0.4$ in.
> Use $a = 0.1$ in.
> Use $w = 3.5$ in.
> Use $L = 7.0$ in.
> Total length 14.0 in.

Note: The above dimensions are for $t = 0.3$ in.

In the case where the K_{Ie} value is not available for the fracture analysis, it can be approximated in terms of plane strain fracture toughness, K_{Ic}, for through cracks by the following equation [39]:

$$K_{Ie} = K_{Ic}(1 + C_k K_{Ic}/\sigma_{yield}) \tag{3.54}$$

where C_k is an empirical constant, which for most aerospace material is equal to 1.0 (in.$^{-1/2}$). Equation (3.54) gives reasonably good results for a variety of isotropic materials. The value of the part through fracture toughness, K_{Ie}, for material with high K_{Ic}/σ_{yield} ratio will be limited to $1.4K_{Ic}$ [33].

As mentioned previously, there are two stress intensity factors that are associated with a part through crack, corresponding to the angles $\theta = 90°$ and $\theta = 0°$. The critical values of K_I for the depth direction, where $\theta = 90°$ (also called a-tip), is given by Eq. (3.54) and is designated by K_{Ie}. The critical value of K_I for the length direction, where $\theta = 0°$ (c-tip), is approximated by $1.1K_{Ie}$.

One of the most important applications of K_{Ie} is in pressurized tanks used in aerospace and nuclear structures where the requirement of leak-before-burst (LBB) must be met. In the case of a pressurized tank that contains a hazardous fluid or gas, LBB is not desirable and the tank must demonstrate it has safe-life during its usage. Section 3.7.5 briefly discusses the LBB concept.

3.7.5 The Leak-Before-Burst (LBB) Concept

In most materials, it is desirable to have crack stability even though the part through crack grows and becomes a through crack [40, 41]. For this to happen, the following conditions at the onset of part through crack to through crack transition should be valid:

$$(K_I)_{\text{Part through crack}} = (K_I)_{\text{Through crack}} \tag{3.55}$$

3.7 Surface or Part Through Cracks

Moreover:

$$(K_I)_{\text{Part through crack}} < K_{Ie} \qquad (3.56a)$$

and

$$(K_I)_{\text{Through crack}} < K_c \text{ or } K_{Ic} \qquad (3.56b)$$

Under these conditions, the part through crack (now it is a through crack) will grow until the stress intensity factor, K_I, becomes equal to or greater than K_c or K_{Ic}. This condition is called leak-before-burst (LBB).

The leak-before-burst criteria described by Eqs. (3.55) and (3.56) can be applied to a pressurized tank. The assumed surface crack (based on the NDE inspection technique) will grow due to fatigue and sustain load through the thickness. From the safety point of view, for a hazardous fluid or gas, neither leak-before-burst nor the condition of instability, where K_I (at a-tip) $> K_{Ie}$ are desirable. For a nonhazardous fluid or gas, however, leak before burst is required, as illustrated in Fig. 3.54.

If the criteria for leak-before-burst described by Eqs. (3.55) and (3.56) do not hold and a leak-before-burst design is desired, the following modifications might be considered:

1. Thicken the structure to reduce the design stress.
2. Change the material.

Examples of fatigue crack growth problems related to LBB for pressurized vessels are demonstrated in Section 4 of Chapter 4.

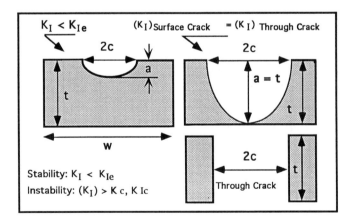

Figure 3.54 Illustration of Surface Cracks Before and After Transition to a Through Crack

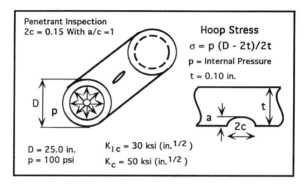

Figure 3.55 A pressurized tank subjected to internal pressure with a surface flaw (shown for example 3.5)

Example 3.5

A pressurized cylindrical tank is subjected to 200 psi pressure, as shown in Fig. 3.55. As part of the safety requirement, the tank must undergo a proof test of 1.5 × operating pressure (300 psi). The flaw size obtained by performing penetrant inspection prior to the proof test indicated that the maximum preexisting circular surface flaw that can escape the inspection has $2c = 0.15$ in. with $a/c = 1$. The pressurized tank is made of 2219-T871 aluminum alloy with $K_{Ic} = 30$ ksi (in.$^{1/2}$). Determine (1) if the tank will be leak-before-burst and (2) whether or not the through crack will be stable (assume $K_{Ie} \approx 1.1 K_{Ic}$).

Solution

The stress intensity factor solution to the crack geometry shown in Fig. 3.55 is given by Eq. (3.52). The correction factor to the stress intensity factor of the above crack geometry is:

$$\beta(a/c, a/t, \theta) = 0.97 * [M_1 + M_2(a/t)^2 + M_3(a/t)^4] * g * f_\phi * f_c * f_i$$

where a/t and the quantities M_1, M_2, M_3 for the case of $a/c = 1$ can be calculated as follows:

$$M_1 = 1.13 - 0.09(a/c) = 1.04$$
$$M_2 = -0.54 + 0.89/(0.2 + a/c) = 0.201$$
$$M_3 = 0.5 - [1/(0.65 + a/c)] + 14(1 - a/c)^{24} = -0.1$$
$$g = 1 + [0.1 + 0.35(a/t)^2](1 - \sin\theta)^2 = 1$$
$$f_\phi = [(a/c)^2 \cos^2\theta + \sin^2\theta]^{1/4} = 1$$

The stress intensity factor for the depth direction is associated with an angle of $\theta = 90°$.

3.7 Surface or Part Through Cracks

The quantities $f_c = [(1 + k^2)/(1 - k^2) + 1 - 0.5 \, (a/t)^{1/2}] \, [t/(D/2 - t)] = 1.015$ where $k = 1 - 2t/D = 0.992$. The value of $f_i = 1$ for an internal crack and 1.1 for an external crack. Substituting these values into the stress intensity correction factor equation:

$$\beta(a/c, a/t, \theta) = 0.97 * [1.04 + 0.201(0.75)^2 - 0.1(0.75)^4] * 1 * 1 * 1.015 * 1.1 = 1.21$$

Moreover, the induced hoop stress due to 200 psi proof pressure is:

$$\sigma = p(D - 2t)/2t = 37{,}200 \text{ psi} = 37.2 \text{ ksi}$$

$$K_1 = \beta(a/c, a/t, c/w, \theta) \frac{\sigma \sqrt{\pi a}}{\Phi}$$

$$= 1.21 \times 37.2 \times (3.14 \times 0.75)^{1/2}/2.464 = 8.865 \text{ ksi (in.}^{1/2})$$

where

$$\Phi = 2.464$$

Note that the calculated stress intensity factor is much smaller than the part through crack fracture toughness value, $K_{Ie} = 1.1 K_{Ic} = 33$ ksi (in.$^{1/2}$). Therefore, no catastrophic failure is expected, that is:

$$K_{Ie} = 33 \text{ ksi (in.}^{1/2}) > 8.865 \text{ ksi (in.}^{1/2})$$

Now let us assume the existing crack grows and becomes a through crack having length $2c = 2t = 0.2$ in. (the part through crack is now leaking). To check for LBB, it is necessary to calculate the stress intensity factor for a through crack of length $2c = 0.2$ in. From equation (3.56), the condition for stability can be written as:

$$(K_I)_{\text{Through crack}} < K_c$$

The equation for the stress intensity factor from Eq. (3.24) is:

$$K = (\sigma_0 \beta_0) \sqrt{\pi c}$$

$$\beta_0 = (1 + 0.52\lambda + 1.29\lambda^2 - 0.074\lambda^3)^{1/2}$$

where

$$\lambda = c/(Rt)^{1/2}$$
$$\lambda = 0.1/(12.5 \times 0.1)^{1/2} = 0.09$$

$$\beta_0 = (1 + 0.52 \times 0.09 + 1.29 \times 0.09^2 - 0.074 \times 0.09^3)^{1/2} = 1.04$$
$$K = (\sigma_0 \beta_0)\sqrt{\pi c} = 37.2 \times 1.04 \times (3.14 \times 0.1)^{1/2} = 21.68 \text{ ksi (in.}^{1/2})$$

Because the calculated value of the stress intensity factor is smaller than the fracture toughness of the material, the tank would be leak-before-burst:

$$(K_1)_{\text{Through crack}} = 21.68 \text{ ksi (in.}^{1/2}) < K_c = 50 \text{ ksi (in.}^{1/2})$$

indicating that the through crack will be stable.

3.8 A Brief Description of ASTM Fracture Toughness Determination

The present method of determining the residual strength capability of a given structure by LEFM requires the critical value of the stress intensity factor (called fracture toughness). The fracture toughness data can be produced by conducting several tests, performed in accordace with the ASTM practice. Section 3.3 has an in-depth discussion on plane strain and plane stress fracture toughness and their importance in generating the residual strength capability curve. Since fracture toughness is a function of thickness and crack length, it is therefore necessary to conduct several laboratory tests with various thicknesses that require specimen preparation, surface finish, and data collection (such as stable crack growth and load increase measurements). These preparations and measurements are costly and time consuming.

Standard test methods for plane strain fracture toughness of metallic materials are briefly discussed in Section 3.8.1 and the determination of plane stress and mixed mode fracture toughness, K_c, by the resistance curve (R-curve) method is presented in Section 3.8.2.

3.8.1 Plane Strain Fracture Toughness (K_{Ic}) Test

As mentioned in Section 3.3, the plane strain fracture toughness, K_{Ic}, is independent of thickness and crack length. To obtain a valid K_{Ic} value, the state of stress near the crack front must approach the triaxial tensile plane strain where the formation of plasticity at the crack tip is reduced. Moreover, the plastic zone size formed at the crack tip must be small compared to other dimensions (such as crack size and specimen thickness). The relationships between the crack tip plastic zone size, r_y, and specimen size requirement to ensure elastic plane strain condition throughout the body are:

$$r_y = \frac{1}{6\pi}\left(\frac{K_{Ic}}{\sigma_{YS}}\right)^2 \quad (3.57)$$

3.8 A Brief Description of ASTM Fracture Toughness Determination

$$a \geq 2.5 \left(\frac{K_{Ic}}{\sigma_{YS}}\right)^2 \quad B \geq 2.5 \left(\frac{K_{Ic}}{\sigma_{YS}}\right)^2 \quad W \geq 5.0 \left(\frac{K_{Ic}}{\sigma_{YS}}\right)^2 \quad (3.58)$$

where a, B, and W are the crack length, and specimen thickness and width, respectively. The following calculation shows that the specimen thickness should be approximately 47 times the radius of the plane-strain plastic zone, r_y, in order to meet the plane strain test specimen requirements. Dividing the crack length, a, or specimen thickness, B, from Eq. (3.57) by Eq. (3.58):

Specimen Thickness/Plastic Zone Size = $(a$ or $B)/r_y = 2.5(6\pi) \approx 47$

The plane strain plastic zone shape and size were derived in Sections 3.5 and 3.6. The Irwin and Von Mises yield criteria described by Eq. (3.33) of Section 3.5 were utilized in determining plastic zone size and shape.

3.8.1.1 Standard K_{Ic} Test Specimen and Fatigue Cracking

Two common types of standard specimen are available for K_{Ic} testing, namely the slow-bend test specimen and the compact-tension specimen (CTS or simply CT). The CT specimen is more commonly used for K_{Ic} testing than the slow-bend test specimen (see Fig. 3.56). Other specimen configurations are also available for

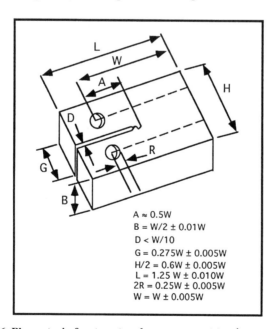

Figure 3.56 Plane strain fracture toughness compact tension specimen (CT)

K_{Ic} testing and the reader may refer to the ASTM-E399 Standards. The corresponding dimensions for the CT specimen are shown in Fig. 3.56.

Specimen dimensional measurements must be such that the thickness, B, is measured to 0.1%, and the crack length measurement after fracture is to the nearest 0.5%. The crack length measurement should be the average of three measurements at three positions along the crack front.

The crack is introduced in the specimen by a starter notch that extends by fatigue cracking the notch. The purpose of fatigue cracking is to simulate a natural crack that can provide a satisfactory plane strain fracture toughness test result. The stress ratio associated with cyclically loading the notch is $0.1 > R > -1$ and the $K_{\max}$ (for fatigue cracking) $< 60\% \, K_{Ic}$. It is common to fatigue-crack the CT test specimen by the amount $0.05W$, where W is the width of the specimen. Annex A2 of ASTM-399 outlines the procedure for fatigue cracking.

Note that before the plane strain fracture toughness specimen can be machined, some estimate of the sizing of the CT test specimen must be known. Table 3.3 provides the ASTM recommended crack length or thickness for the suggested ratio of yield strength to modulus of elasticity. For example, 2219-T851 aluminum alloys with yield strength of 53 ksi give a σ_{ys}/E ratio of 0.0053. The recommended crack length or test specimen thickness, B, is approximately 3.0 in. (see Table 3.3).

To establish that a valid K_{Ic} has been obtained, it is first necessary to calculate the conditional stress intensity factor, K_Q, which can be determined through the construction of test records consisting of an autographic plot of the applied load, P, versus displacement, Δ, (P–Δ curve). The conditional stress intensity factor, K_Q, is given [42] as:

Table 3.3 ASTM Recommended Crack Length or Thickness for K_{Ic} Testing

σ_{ys}/E	Recommended crack length or thickness
0.0050–0.0057	3.00
0.0057–0.0062	2.50
0.0062–0.0065	2.00
0.0065–0.0068	1.75
0.0068–0.0071	1.50
0.0071–0.0075	1.25
0.0075–0.0080	1.00
0.0080–0.0085	0.75
0.0085–0.0100	0.50
0.0100 and Greater	0.25

3.8 A Brief Description of ASTM Fracture Toughness Determination

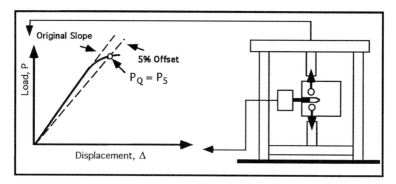

Figure 3.57 Load versus displacement for obtaining the apparent K_Q

$$K_Q = \frac{P_Q}{BW^{1/2}} \left[29.6 \left(\frac{a}{W}\right)^{1/2} - 185.5 \left(\frac{a}{W}\right)^{3/2} + 655.7 \left(\frac{a}{W}\right)^{5/2} \right. \quad (3.59)$$
$$\left. - 1017.0 \left(\frac{a}{W}\right)^{7/2} + 638.9 \left(\frac{a}{W}\right)^{9/2} \right]$$

where the applied load, P_Q, is the point on the P–Δ curve shown in Fig. 3.57. The point P_Q is obtained by drawing a secant line with slope 5% from the initial slope tangent to the linear part of the record. The obtained value of P_Q on the P–Δ curve is compared with the P_{max} value situated on the curve (see Fig. 3.58 for different types of load displacement records) to calculate the apparent plane strain fracture toughness, K_Q, to verify the conditions established by Eqs. (3.57) and (3.58). In Fig. 3.58, the value of P_5 is considered to be equal to P_Q if there is no other load

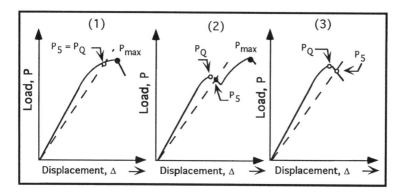

Figure 3.58 Load versus displacement for different possible types of behavior and the interpretation of K_Q

in the P–Δ record that precedes it and is higher than P_5 (see case 1 of Fig. 3.58). Otherwise, the P_Q load is the higher value on the load displacement record as shown in cases 1 and 2 of Fig. 3.58. If the ratio of P_{max}/P_Q (the load P_{max} is a load point situated above the P_5 value) is less than 1.10, then the calculated K_Q value from Eq. (3.59) should be used in Eq. (3.57) to compute the plastic zone size, r_y. This value should be compared with specimen thickness, B, and crack length, a. If both B and a are larger than r_y, then $K_Q = K_{Ic}$. In addition to the above mentioned conditions for obtaining a valid K_{Ic} value, the following two items are also important to know when conducting the plane strain fracture toughness testing:

- It is necessary to perform a minimum of three tests for each material heat treatment.
- The loading rate should be between 30 ksi (in./min.)$^{1/2}$ to 140 ksi (in./min.)$^{1/2}$

The plane strain fracture toughness, K_{Ic}, values for several selected aerospace alloys are available in Appendix A. The K_{Ic} values for aluminum alloys may range as high as 46 and as low as 16 ksi (in.$^{1/2}$) for 7075-T63 and 2020-T651 aluminums, respectively. Plane strain fracture toughness as high as 200 ksi (in.$^{1/2}$) can be found in the NASA/FLAGRO material library among the stainless steel alloys (see Appendix A). In general, a high K_{Ic} is associated with ferrous alloys that have undergone appropriate heat treatment which can introduce sufficient ductility and high resistance to fracture in the material.

Plane strain test specimen thickness as large as 12 in. is reported in the literature for alloys that possess high ductility and good resistance to fracture [34]. The specimen preparation for determining the plane strain fracture toughness value when the thickness requirement is as high as 12 in. is costly and not feasible to implement. On the other hand, the flexibility of using the FMDM theory in evaluating the fracture toughness value will eliminate the difficulties associated with specimen preparation and time-consuming laboratory testing (see Chapter 6).

Another important point that must be emphasized in conjunction with the plane strain fracture toughness value is its variation with respect to the grain orientation that evolves during the manufacturing process. The plane strain fracture toughness values for material that has undergone a forging process are usually expected to be smaller in the T–L direction than the L–T. For example, the K_{Ic} values for 7050-T7452 aluminum alloy in the forged condition are 31 and 21 ksi (in.$^{1/2}$) in the L–T and T–L directions, respectively. As another example, the K_{Ic} values for 7050-T76511 in the extrusion condition are 30 and 24 ksi (in.$^{1/2}$) in the L–T and T–L directions, respectively (see Appendix A). From these examples, the analyst must recognize that the proper K_{Ic} value must be used in assessing the service life of the structural part. Not recognizing the differences in material properties with respect to their orientation (and these may differ considerably) can lead to erroneous analysis results.

3.8 A Brief Description of ASTM Fracture Toughness Determination

3.8.2 Plane Stress Fracture Toughness (K_c) Test

In contrast to plane strain fracture toughness, K_{Ic}, the critical stress intensity factors for plane stress and mixed mode conditions, K_c, are thickness and crack length dependent. Many aircraft and space vehicle structures are made of thin sheets of aluminum alloy. The life assessment of these structures by a fracture mechanics approach requires knowledge of the plane stress fracture data that are obtainable through laboratory testing. The replacement of the K_c by K_{Ic} value yields conservative results ($K_{Ic} < K_c$) when evaluating the life of the part which in many cases leads into thickening the part to reduce the design stress. The variation of material fracture toughness, K_c, versus specimen thickness was discussed in Section 3.3 and is shown in Fig. 3.59 for 2219-T87 aluminum alloy [33]. From Fig. 3.59, it can be seen that the fracture toughness value increases as the thickness decreases. The increase in fracture toughness value is strongly dependent on the state of stress at the crack tip and the extent of plasticity formation in that region. For a thick section, the extent of the plastic zone through the thickness is at a minimum, and as the thickness decreases, the constraint to plastic flow decreases, until the state of plane stress is reached. The fracture toughness associated with minimum thickness is called the plane stress fracture toughness, K_c. The maximum K_c value for 2219-T87 aluminum alloy is 70 ksi (in.$^{1/2}$) and is associated with a minimum thickness $t \approx 0.15$ in. and half crack length $a = 1.5$ to 1.6 in.

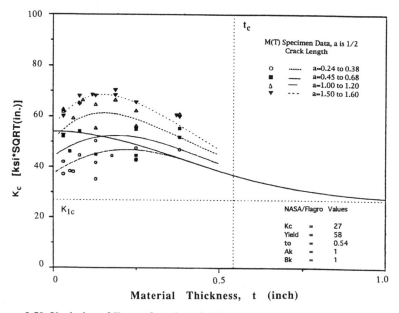

Figure 3.59 Variation of K_C as a function of thickness, t, for 2219-T87 aluminum alloy

(as shown in Fig. 3.59). The smallest K_C value (with $t = 0.15$ in.) in this figure is 44 ksi (in.$^{1/2}$) and it applies to a half crack length of 0.24 to 0.38 in.

To develop a curve similar to the one shown in Fig. 3.59 for another alloy or the same alloy with another heat treatment, it is necessary to conduct several tests with cracked specimens of various thicknesses. Each specimen will provide a distinct R-curve different from the others and represents material resistance to fracture for the thickness under consideration. The R-curve represents the material's resistance to fracture during incremental stable slow crack growth, Δa, under monotonic increasing load. As indicated in Section 3.4.3, the R-curve is constructed by plotting the material resistance to fracture, $K_C = K_R$, for different crack lengths, (see Fig. 3.60). Because the R-curve is proven to be independent of original crack length, a_o, it can be plotted as a single curve (K_R versus Δa) for a given thickness using a standard ASTM cracked specimen as recommended by ASTM E-561 (Standard Practice for R-curve Determination).

In the R-curve concept, unstable crack extension occurs when the applied K becomes equal to or exceeds the crack growth resistance, K_R, of the material. At the instability point, the crack growth resistance, K_R, is equal to K_C as shown in Fig. 3.60. That is, the plane stress fracture toughness, K_C, is determined from the tangency between the R-curve and the applied K curve (K versus physical or effective crack length, $a_o + \Delta a$) of the specimen. The specimen configuration and dimensions specified by the ASTM testing procedure for conducting a valid R-curve test are briefly discussed in Section 3.8.2.1.

The variation of fracture toughness, K_C, with material thickness for many aerospace alloys can be found in the NASA/FLAGRO material library. This variation is presented in the form of an empirical equation described in terms of plane strain fracture toughness, K_{Ic}, [39] as:

$$K_c = K_{Ic}[1 + B_k e^{-(A_k * t/B)^2}] \tag{3.60}$$

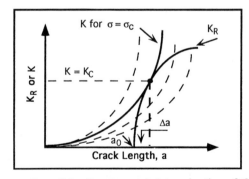

Figure 3.60 Construction of the R-curve and determination of the instability point where the applied K and K_R are tangent to each other

3.8 A Brief Description of ASTM Fracture Toughness Determination

where B is the thickness that meets the plane strain condition and t is the thickness associated with the part that could be in the plane stress or mixed mode conditions with fracture toughness, K_C, to be determined. The constants A_k and B_k are the curve fit parameters and are given in Appendix A for several aerospace alloys. Note that the curve fit relationship described by Eq. (3.60) is applicable to minimum crack length, as was shown in Fig. 3.36 for the case of 2219-T87 aluminum alloy with half crack length a = 0.24 to 0.38 in. The plane stress fracture toughness values associated with minimum crack length will yield a minimum fracture toughness value that is conservative to use when evaluating the service life of a fracture critical component.

3.8.2.1 M(T) Specimen for Testing K_c

For determination of material resistance to fracture by using the *R*-curve approach, three types of specimen are recommended by the ASTM committee. The middle center cracked wide tension panel M(T) specimen, the compact specimen C(T), or the crack-line-wedge-loaded specimen C(W) are used to conduct the plane stress fracture toughness test. Only the center cracked wide panel is discussed in this section (see Fig. 3.61). The reader may refer to E561 for the recommended specimens that can also deliver applied stress intensity factor, K, to the material. Section 3.10.2.2 briefly describes (1) the apparatus (grips and fixture) employed for uniformly transferring the applied load into the M(T) standard specimen, (2) the method of preventing buckling for the M(T) specimen, (3) specimen fatigue cracking, and (4) load and specimen dimensional measurements.

As discussed in the previous section for the plane strain fracture toughness case (per the ASTM E-399 test method), the planar dimensions of the specimens are sized to ensure that elastic conditions are met at all values of load in the net section of the specimen. In other words, the width, W, and half crack size, a, must be chosen so that the remaining ligament is below net section yielding at fracture. The ASTM guide for designing an M(T) panel is such that a specimen with $W = 27\ r_y$ width and crack length $2a = w/3$ is expected to fail at a net section

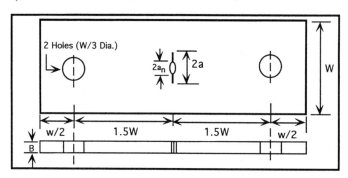

Figure 3.61 Standard center-cracked-tension M(T) specimen

Table 3.4 Specimen Width Size and Corresponding Crack Length for an M(T) Specimen.

K_{max}/σ_{ys} (in.$^{1/2}$)	Width (W) (in.)	Crack Length ($2a_0$) (in.)	Spec. Length (in.)
0.5	3.0	1.0	9.0
1.0	6.0	2.0	12.0
1.5	12.0	4.0	24.0
2.0	20.0	6.7	30.0
3.0	48.0	16.0	72.0

stress equal to the yield strength of the material. The selected width dimension must be larger than 27 r_y to avoid net section yielding. Table 3.4 is a list of minimum recommended M(T) specimen dimensions for an assumed K_{max}/σ_{yield} ratio, where K_{max} is the maximum K level obtained in the test and σ_{yield} is the 2% offset yield strength of the material. Note that for the plane stress condition, the value of r_y is given in terms of K_{max} and σ_{yield} by Eq. (3.33) as:

$$r_p = \frac{1}{2\pi}\left(\frac{K_{max}}{\sigma_{yield}}\right)^2 \quad (3.61)$$

For an M(T) specimen, the K_{max} can be calculated by:

$$K_{max} = (P/WB)[\pi a_{eff} \sec(\pi a_{eff}/W)]^{1/2} \quad (3.62)$$

3.8.2.2 Grip Fixture Apparatus, Buckling Restraint, and Fatigue Cracking

The grip fixture apparatus must distribute the applied load to the M(T) test specimen uniformly to enable one to obtain accurate readings for constructing a valid R-curve. To ensure that uniform tensile stress is transferred from the pin grip fixture (when a single pin grip is used) into the specimen, the length between the two single pins on each end shall be $3W$ (as shown in Fig. 3.62). For example, for a test specimen with width $W = 3.0$ in. and crack length $2a = 1.0$ in., the length requirement between the application of loads at the two ends is 9.0 in. In the case when the specimen width, W, is wider than 12 in. (as indicated in Table 3.4) the ASTM requires use of multiple pin grips to uniformly distribute the load, with a length requirement of only $1.5W$, as illustrated in Fig. 3.62. For example, for $W = 20$ in., the length requirement is $1.5W = 30$ in.

When the test specimen is thin, buckling can develop during loading that can affect the accuracy of the test data. To prevent the test plate from buckling, rigid face plates are required to be attached to both sides of the specimen. In the case of the M(T) specimen, the rigid plates shall be affixed to the test specimen as illustrated in Fig. 3.63. The lubricated buckling restraint plates must be attached to

3.8 A Brief Description of ASTM Fracture Toughness Determination   173

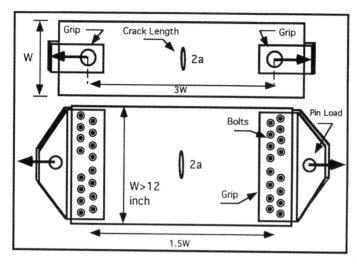

Figure 3.62 The pin grip fixture apparatus for $W = 3$ and $W > 12$ in. M(T) specimens

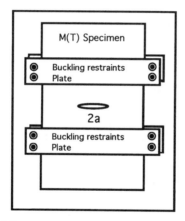

Figure 3.63 Buckling restraint plates affixed to the cracked specimen

both sides of the test specimen in such a way that the crack measurement data readings are accessible and are not blocked.

To simulate a natural crack in the M(T) specimen, the starting notch is first introduced in the specimen either by saw cutting or electrical discharge machining (EDM). The starting notch length plus fatigue precrack should be between 30% and 40% of the specimen width, W, and situated in the center of the test specimen. The extended length associated with the fatigue precrack must not be less than 0.05 in. Fatigue cracking may be eliminated if the cutting saw thickness can simulate the sharpness of a fatigue starter crack. The ASTM procedures for obtaining a valid fracture toughness value clearly indicate that extensive preparation both before and after testing is necessary.

References

1. A. A. Griffith, "The Phenomena of Rupture and Flow in Solids," Philos. Trans. R. Soc. Lond., Ser. A., Vol. 221, 1920.
2. A. S. Tetelman and A. J. McEvily, Jr., Fracture of Structural Materials, John Wiley & Sons, 1967 pp. 39–48.
3. G. R. Irwin, "Analysis of Stresses and Strains Near the End of a Crack Traversing a Plate," Trans. ASME, J. Appl. Mech., Vol. 24, 1957, p. 361.
4. C. P. Paris and G. C. Sih, "Stress Analysis of Cracks," in "Fracture Toughness Testing and Its Applications," ASTM STP No. 381, ASTM, Philadelphia, 1965.
5. H. M. Westergaard, "Bearing Pressures and Cracks," Trans., ASME, J. Appl. Mech., 1939.
6. S. P. Timoshenko and J. N. Goodier, Theory of Elasticity, 3rd Edition, McGraw-Hill, 1970.
7. N. I. Muskhelishvili, "Some Basic Problems of the Mathematical Theory of Elasticity," (1933), English Translation, Noordhoff, 1953.
8. H. Tada, P. C. Paris, and G. R. Irwin (Eds.), Stress Analysis of Cracks Handbook, Del Research Corporation, Hellertown, PA, 1973.
9. G. C. Sih, Handbook of Stress Intensity Factors for Researchers and Engineers, Institute of Fracture and Solid Mechanics, 31, Series E, No. 2, June 1964.
10. G. R. Irwin, Fracture Handbuch der Physik, Springer-Verlag, Heidelberg, VI, 1958, pp. 551–590.
11. D. P. Rooke and D. J. Cartwright, "Compendium of Stress Intensity Factors," Her Majesty's Stationary Office, London, 1976.
12. Fatigue Crack Growth Computer Program "NASA/FLAGRO," developed by R. G. Forman, V. Shivakumar, and J. C. Newman. Appendix C, Reference C1-C27, JSC-22267A, January 1993.
13. Damage Tolerant Design (Data) Handbook, Revision to MCIC-HB-01, (1975), covered under USAF Contract No. F33615-80-C-3229.
14. "1996 Annual Book of ASTM Standard," Metals Test Methods and Analytical Procedures, Vol. 03.01.
15. J. M. Craft, A. M. Sullivan, and R. W. Boyle, "Effects of Dimensions on Fast Fracture Instability of Notched Sheet," Crack Propagation Symposium, Cranfield 1961, Paper 1.
16. A. P. Parker, "The Mechanics of Fracture and Fatigue," E. & F. N. Spon Ltd., pp. 101–117.
17. M. F. Kanninen, and A. T. Hopper, "Advanced Fracture Mechanics," Oxford Engineering Science Series, 1985, pp. 150–200.
18. J. M. Barsom, "The Development of AASHTO Fracture Toughness Requirement for Bridge Steel," Engin. Fract. Mech., Vol. 7, No. 3, September 1975.

19. J. M. Barsom, "The Development of AASHTO Fracture Toughness Requirement for Bridge Steel," American Iron and Steel Institute, Washington, D.C., February 1975.
20. The Making, Shaping and Treating of Steel, Ninth Edition, edited by H. E. McGannon, United State Steel, Pittsburgh, December 1970.
21. American Association of State Highway and Transportation Officials (AASHTO) Material Toughness Requirement, Association General Offices, Washington, D.C. 1973.
22. "PVRC Recommendations on Toughness Requirements for Ferritic Materials," WRC Bulletin 175, New York, August 1972.
23. S. T. Rolfe and J. M. Barsom, "Fracture and Fatigue Control in Structures, Applications of Fracture Mechanics," Prentice-Hall, 1977.
24. R. A. Flinn and P. K. Trojan, "Engineering Materials and Their Applications," Third Edition, Houghton Mifflin, 1986, pp. 321–323.
25. W. S. Margolis and F. C. Nordquist, "Plane Stress Fracture Toughness of Aluminum Alloy 7475-1/2 in. Plate, Temperes-T7651 and T7351 and of Aluminum Alloy 2024-1/8 in. Sheet-T81 and T62 Temper," General Dynamics, Forth Worth Div., F-16 Air Combat Fighter Technical Report TIS GA2300, CDRL A031, USAF Contract F33657-75-C-0310.
26. J. E. Srawley and W. F. Brown, "Fracture Toughness Testing Method," ASTM STP 381, 1965, pp. 133–195.
27. J. M. Krafft, A. M. Sullivan, and R. W. Boyle, "Effect of Dimensions on Fast Fracture Instability of Notched Sheets," Cranfield Crack Propagation Symposium, Vol. 1, 1961, pp. 8–28.
28. J. C. Newman , Jr. "Fracture Analysis of Surface and Through Cracked Sheet and Plates," Engineering Fracture Mechanics, Vol. 5, No. 3, Sep. 1973 pp. 667–684.
29. D. P. Wilhem, "Fracture Mechanics Guidelines for Aircraft Structural Applications," AFFDL-TR-69-111, Air Force Flight Dynamics Laboratory, February 1970.
30. M. P. Kaplan and J. A. Reiman, "Use of Fracture Mechanics in Estimating Structural Life and Inspection Intervals," J. Aircraft, Vol. 13, No. 2, February 1976, pp. 99–103.
31. F. A. McClintock and G. R. Irwin, "Plasticity Aspects of Fracture Mechanics," ASTM STP 381, 1965, pp. 84–113.
32. A. R. Duffy et al., "Fracture Design Practice for Pressure Piping, Fracture I," pp. 159–232, edited by H. Liebowitz, Academic Press, 1962.
33. D. P. Rooke, "Elastic Yield Zone Around a Crack Tip," Royal Aircr. Est., Farnborough, Tech. Note CPM 29, 1963.
34. W. Johnson and P. B. Mellor, Plasticity for Mechanical Engineers, Van Nostrand, 1962.
35. G. S. Spencer, An Introduction to Plasticity, Chapman and Hall, 1968.
36. G. R. Irwin, "Crack Extension Force for a Part Through Crack in a Plate," J. of Appl. Mech., December 1962, pp. 651–654.
37. R. M. Engle, Jr., "Aspect Ratio Variability in Part-Through Crack Life Prediction," ASTM STP 687, Am. Soc. Test. and Mater., 1979, pp. 74–88.
38. J. C. Newman, Jr. and I. S. Raju, "Stress Intensity Factor Equations for Cracks in Three Dimensional Finite Bodies," NASA TM 83200, NASA Langley Research Center, August 1981.

39. Fatigue Crack Growth Computer Program "NASA/FLAGRO," Developed by R. G. Forman, V. Shivakumar, and J. C. Newman, JSC-22267A, January 1993.
40. ASTM Committee, "The Slow Growth and Rapid Propagation of Cracks," Mater. Res. Stand., 1, 1961, pp. 389–394.
41. G. R. Irwin, Fracture of Pressure Vessels: Materials for Missiles and Spacecraft, pp. 204–229, McGraw-Hill (1963).
42. J. E. Srawley, "Wide Range Stress Intensity Factor Expressions for ASTM E 399 Standard Fracture Toughness Specimens," Int. Fract. Mech., Vol. 12, June 1976, p. 475.

Chapter 4

Fatigue Crack Growth

4.1 Introduction

As discussed in Chapter 2, the traditional approach to estimate the total life of a structural part is to use the S–N diagram (alternating stress, S, versus the total number of cycles to failure, N). This diagram can be constructed by conducting several laboratory tests on standard specimens subjected to low amplitude fluctuating loads simulating the actual environment. With the S–N approach, it is assumed that the structure is initially free from defects and that the total life of the structure is defined as the sum of the crack initiation and propagation cycles. Furthermore, it was mentioned that in the localized region adjacent to the notch where material is plastically deformed, small fatigue cracks can initiate along the persistent slip band which eventually coalesce to create larger cracks (stage I growth). The number of cycles required to generate a small crack, along the active slip plane, is dependent on the load amplitude, material grain size, and temperature. Under low load amplitude, the extent of stage I is large and most of the structural life is consumed for crack initiation. On the other hand when load amplitude is raised to the low cycle regime where the bulk of the structure is plastic, the number of cycles used for stage I is small and the entire structural life is expended for stable crack growth (stage II). In the high cycle fatigue regime, the plastic deformation is limited to a small region at the tip of the crack, and estimation of the number of cycles to failure from the end of initiation stage (i.e., the start of stage II) can be determined from the linear elastic fracture mechanics approach.

The application of linear elastic fracture mechanics in estimating the life of structural hardware assumes that the crack has already initiated in the material. In reality, this is often the case, because the initial cracks may have their origin in many ways. For example, they may be introduced as cracks or as incipient cracks during the manufacture of structural parts, they may grow from defects in the parent metal, from incomplete welds, or from shrink cracks or other imperfections in weldments. The initial crack length associated with these cracks can be estimated conservatively based on the Nondestructive Inspection (NDI) method. That is, the largest crack that could escape detection by nondestructive inspection can be assumed to exist in the material and it can be used as the initial crack size for life evaluation of the structural components.

The material as received from the vendor will contain defects of a small size, such as porosities (fine holes and pores in a metal, most commonly in welds and casting); inclusions (such as oxides, sulfides, and silicates); or microcracks that can eventually lead to fracture. These inherent flaws are considerably smaller than the NDI capability to detect them and will not grow appreciably in service. From a safety viewpoint, the use of a longer initial crack length (obtained through NDI methods) is conservative, however, from a practical consideration, the longer crack size assumption should be realistic enough not to impact the weight or cause rejection of the part. It should be noted that the initial crack size to be used to evaluate the crack growth behavior must not be so small that it would violate the concepts of fracture mechanics. That is, the concept of isotropic continuum must be obeyed, such that the flaw size cannot be smaller than the grain size.

For aircraft and space vehicles, it is required to assume that cracks exist in all structures that are classified as high-risk or fracture critical parts. These cracks shall not grow to their critical size at a specified load during their usage period. Therefore, it is necessary to predict the rate of growth of the assumed flaw and the number of cycles to failure compared with the number of cycles used during their service life.

Different crack growth empirical equations capable of calculating the number of cycles to failure are discussed in Section 4.2. Section 4.2 includes the crack growth equation employed in the NASA/FLAGRO (NASGRO) computer code [1]. This program was initially developed for fracture analysis of space hardware and recently has been extended to crack growth analysis of aircraft structural problems. The NASA/FLAGRO computer program is currently the standard computer code for the NASA, the European Space Agency, the U.S. Air Force, and many aerospace companies. A brief review of the ASTM fatigue crack growth rate testing is also included in this section. Later in Section 4.3, the Elber and Newman crack closure concepts (used in crack growth equations) are presented and are followed by a few example problems. Variable amplitude loading and the retardation concept (including example problems) are discussed in Section 4.4. Structural integrity analysis of bolted joints under cyclic loading is covered in Section 4.6. This section covers an in-depth discussion on preloaded bolts subjected to cyclic loading and includes several example problems. Structural integrity analysis of a crack emanating from a hole in a bolted joint subjected to a fluctuating load environment is also included in this section. Finally, material

4.1 Introduction

anisotropy and its effect on bolted joint analysis when evaluating the life of the structure is reviewed in Section 4.7.

4.1.1 Stress Intensity Factor Range and Crack Growth Rate

It is the goal of fracture mechanics to estimate the total number of cycles for the assumed or the initial crack, a_i, to reach its final length, a_f. The crack tip stress intensity factor was extensively discussed in Chapter 3 and is an extremely useful parameter to address crack growth behavior as long as the bulk of the material is elastic and plastic deformation is limited to a small region at the crack tip. The stress intensity factor range associated with maximum and minimum stresses of each cycle ($K_{max} - K_{min}$) is used to describe the amount of crack advancement, $\Delta a = a_f - a_i$, during ΔN number of cycles. In using fracture mechanics to describe fatigue crack growth the minimum value of the stress intensity factor, K_{min}, in a cycle is taken to be zero when $R \leq 0$. The rate of crack growth, $\Delta a/\Delta N$, in terms of the crack tip stress intensity factor range, ΔK, can be written as:

$$\frac{\Delta a}{\Delta N} = f(\Delta K) \qquad (4.1)$$

Equation (4.1) simply states that the material's rate of crack growth, $\Delta a/\Delta N$, is a function of the stress intensity factor range. The function $f(\Delta K)$ can be obtained as the result of laboratory test data and can then be utilized to solve crack growth problems in which the structural part has undergone the same loading conditions ($R = \sigma_{min}/\sigma_{max}$). That is, the crack growth behavior of structural parts can be estimated by testing small specimens in laboratory that simulates the same cyclic loading conditions. It should be emphasized that when crack tip stress states at a given time are the same for two separate crack geometry and loading conditions, ΔK, the crack growth rate, $\Delta a/\Delta N$, would almost be the same for the two crack cases. For example, consider the case of a wide center crack plate with initial crack length of 1.5 in. subjected to a remote constant fluctuating load ($R = 0$) of 15 ksi. The rate of crack growth for this crack geometry will be the same as if the crack length is 0.5 in. and subjected to a remote loading of 26 ksi ($R = 0$):

$$\Delta K = K_{max} - K_{min} \quad \text{(note that } \sigma_{min} = 0 \text{ and therefore, } K_{min} = 0\text{)}$$

$$\Delta K = K_{max} = \beta \sigma_{max} (\pi a)^{1/2}$$

for $\sigma_{max} = 15$ ksi, $K_{max} = 15 \times (3.14 \times 1.5)^{1/2} = 32.6$ ksi (in.$^{1/2}$)

for $\sigma_{max} = 26$ ksi, $K_{max} = 26 \times (3.14 \times 0.5)^{1/2} = 32.6$ ksi (in.$^{1/2}$)

180   Chap. 4 Fatigue Crack Growth

In the next section, several empirical relationships will be developed that can define the crack growth behavior shown by Eq. 4.1.

4.2 Crack Growth Rate Empirical Descriptions

All of the equations that presently describe the function $f(\Delta K)$ are based on the trends developed by experimental data. In general, a centered crack specimen with original crack length, a_o, is subjected to constant cyclic loading of a given stress ratio, R (see Fig. 4.8 for other crack geometries used in generating crack growth data). The incremental crack length growth, Δa, is periodically measured and recorded together with the number of cycles (ΔN). From this information, the variation of crack advancement versus associated number of cycles is plotted as shown in Fig. 4.1 which represents the experimental fatigue crack growth data. Figure 4.2 shows the variation of crack growth, $a_o + \Delta a$, versus the number of cycles, N, for an aluminum lithium alloy tested in accordance with ASTM E-647 practice. The slope of the curve $(da/dN \approx \Delta a/\Delta N)$ at a point shown in Fig. 4.1 is computed for any crack length, $a = a_o + \Delta a$, along the curve and is called the fatigue crack growth rate or crack extension per cycles of loading. Currently there are two methods recommended by the ASTM for determination of the slope, $\Delta a/\Delta N$. In the secant method of determining the slope, two data points are needed per slope, whereas in the incremental polynomial method a minimum of five data points are required to compute the slope at a point, da/dN [2] (see Fig. 4.3). The computed slope and the corresponding stress intensity factor range at a point are to be used in generating the da/dN, ΔK curve. In the secant method of determining

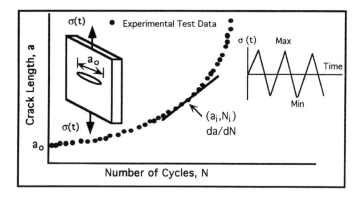

Figure 4.1 Schematic representation of fatigue crack growth curve data and slope at a given point

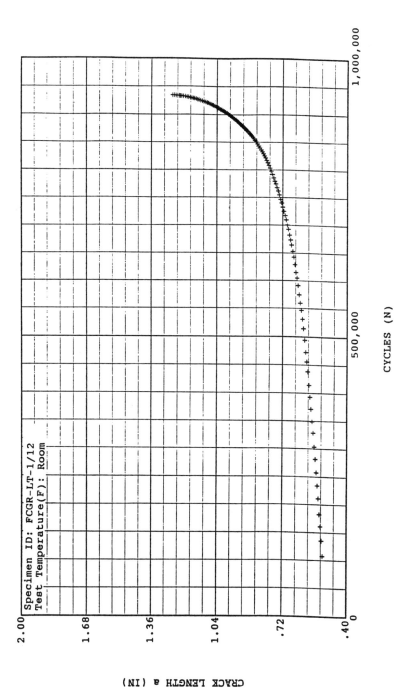

Figure 4.2 Variation of crack advancement (Δa) versus number of cycles for aluminum lithium alloy (courtesy of Westmorland Company)

181

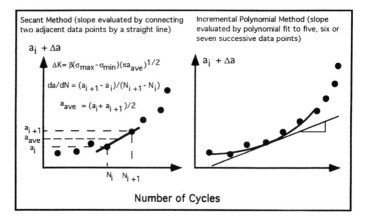

Figure 4.3 Determination of slope at a point by secant and incremental polynomial methods

slope at the midpoint of two data points, one data point has to be sacrificed. That is, from the two data points, a_i and $a_i + 1$, as shown in Fig. 4.3, only one point is generated to describe the da/dN, ΔK curve (($a_i + a_i + 1$)/2). In the incremental polynomial method, a smooth curve (a parabola) is fit through five, seven, or nine data points for the determination of the slope, da/dN. With this method, more than two data points are lost when generating the da/dN, ΔK curve. The logarithmic plot of the calculated stress intensity factor range, ΔK, versus the corresponding crack growth rate, da/dN, is shown in Fig. 4.4. The crack growth rate curve produced in Fig. 4.4 is independent of load magnitude, $\sigma(t)$, and crack geometry.

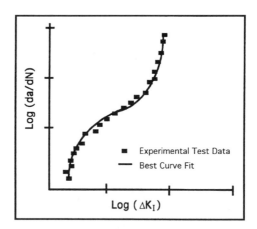

Figure 4.4 Illustration of crack growth rate versus stress intensity range

4.2 Crack Growth Rate Empirical Descriptions

Numerous fatigue crack growth rate empirical and analytical relationships have been developed and are available in [3]. The earliest relationship describing the da/dN behavior (shown in Fig. 4.4) was formulated by Paris and his colleagues at Lehigh University in 1960 [4] and expressed as:

$$\frac{da}{dN} = c(\Delta K)^m \qquad (4.2)$$

where c and m are constants that can be determined from the test data. Figure 4.5 shows typical crack growth data and the curve fit for 2024-T861 aluminum alloy (plate and sheet, T–L, room temperature) that was generated under stress ratio $R = 0.1$ (taken from the NASA/FLAGRO material library [1]). The constants c and m describing the Paris relation are 3.4 and 4.8E-8, respectively. For most metals, the Paris constants m and c are: $m = 3$ to 5 and $c = 10^{-10}$ to 10^{-6}, respectively. Fatigue crack growth data presented in Fig. 4.5 were utilized to compute the constants used in the crack growth rate curve fit relation described by Eq. (4.5) for 2024-T861 aluminum alloy [1].

The Paris relation described by Eq. 4.2 is applicable only to the middle region of the crack growth curve, where the variation of log (da/dN) with respect to log (ΔK) is linear, as indicated in region II of Fig. 4.6. In general, there are three regions associated with the crack growth curve. In region I, the crack growth rate (da/dN) is small and the corresponding stress intensity range, ΔK, approaches a minimum value called the threshold stress intensity factor, ΔK_{th}, below which the crack does not grow. The value of ΔK_{th} is not associated with $da/dN = 0$; rather, it is associated with a cutoff growth rate of 4×10^{-10} m/cycle assigned by ASTM- E647. In region III, the crack growth is rapid and accelerates until the crack tip stress intensity factor reaches its critical value. The critical value of stress intensity factor, K_c, is shown in Fig. 4.6 as an asymptotic line to the crack growth curve. The value of K_c obtained through cyclic loading (the critical value of ΔK obtained from the da/dN curve) is usually smaller than the monotonic loading case. It should be remembered that an invalid fracture toughness value (as a result of improper fracture toughness testing) incorporated into the crack growth equations can result in significant error when determining the life of the part in consideration.

The number of cycles that a cracked structural part spends in each region of the da/dN, ΔK curve (shown in Fig. 4.6) is different. More than 80% of the life of the structure is usually spent in region I, where the existing crack is short and the crack growth rate is small. Figure 4.6 illustrates the approximate percent life of a typical aerospace alloy in regions I, II, and III.

Utilizing the Paris crack growth relation described by Eq. (4.2) beyond its limit can result in life estimation error. Note that the Paris equation does not apply to

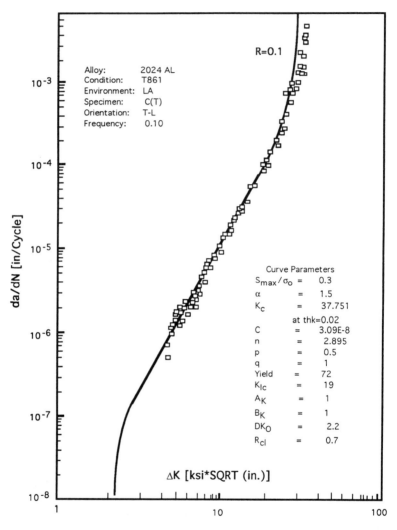

Figure 4.5 Typical crack growth data and the curve fit for 2024-T861 (from NASA/FLAGRO Material Library [1])

4.2 Crack Growth Rate Empirical Descriptions

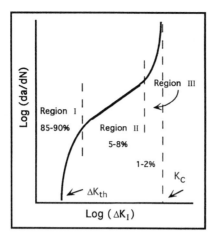

Figure 4.6 Three regions of the fatigue crack growth curve and approximate percentage of life spent

regions I or III, where crack growth is slow or rapid. In addition, the effect of the stress ratio, R, on fatigue crack growth was not considered in Eq. (4.2).

Another attempt to provide an empirical relationship for describing the crack growth data was made by Walker [5]. Walker's crack growth equation is similar to the Paris relation, but it accounts for the effect of the stress ratio, R, in the region II:

$$da/dN = C[(1-R)^m K_{max}]^p \tag{4.3}$$

where the constants C, m, and p can be obtained through experimental data.

Forman [6] formulated an empirical equation describing the crack growth behavior in regions II and III, including the effect of R:

$$da/dN = \frac{c(\Delta K)^n}{(1-R)K_c - \Delta K} \tag{4.4}$$

where c and n can be obtained through experimental data and K_c is the fracture toughness of the material and is thickness dependent (see Section 3.3.1 of Chapter 3).

Equation (4.4) was later modified by Forman–Newman–de Koning (FNK) to account for all the regions of the crack growth curve, including the stress ratio and crack closure effects [1, 7, 8]:

$$\frac{da}{dN} = \frac{C(1-f)^n \Delta K^n \left(1 - \dfrac{\Delta K_{th}}{\Delta K}\right)^p}{(1-R)^n \left(1 - \dfrac{\Delta K}{(1-R)K_c}\right)^q} \quad (4.5)$$

where C, n, p, and q are empirically derived constants; R is the stress ratio; and ΔK and ΔK_{th} are the stress intensity factor range and threshold stress intensity factor, respectively. The parameter f is called the crack opening function and it will incorporate the effect of closure behavior on crack growth rate under constant amplitude loading. The Forman–Newman–de Koning (FNK) crack growth rate relationship described by Eq. (4.5) is widely used in aerospace structures for life estimation of high risk fracture critical parts.

A detailed discussion related to constant amplitude crack closure phenomena and crack opening function, f, is given in Section 4.3.

Example 4.1

The crack growth curve for a given aluminum alloy is plotted in Fig. 4.7: (1) Evaluate the empirical constants associated with Eqs. (4.3) and (4.4) for the Paris and Forman relationships, respectively, where $K_c = 55$ ksi (in.$^{1/2}$). (2) For a centered crack plate with original crack length of $a_i = 1.0$ in. and maximum applied stress of 20 ksi, determine the number of cycles corresponding to a growth of $\Delta a = 0.7$ in. ($R = 0$).

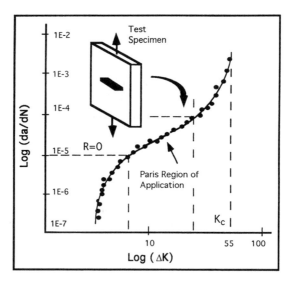

Figure 4.7 Crack growth curve fit for the aluminum alloy mentioned in the problem

4.2 Crack Growth Rate Empirical Descriptions

Solution

To use the Paris equation for the middle region of the crack growth curve, the two constants c and m must first be evaluated. Utilizing the two points on the linear portion of the curve, the constants c and m can be determined as:

For $da/dN = 1E - 4$ in./cycle, $\Delta K = 24.5$ ksi (in.$^{1/2}$)
For $da/dN = 1E - 5$ in./cycle, $\Delta K = 8.5$ ksi (in.$^{1/2}$)

Inserting the two da/dN values into Eq. (4.3):

$$0.0001 = c(24.5)^m$$
$$0.00001 = c(8.5)^m$$
$$10 = (2.88)^m$$
$$m = 2.18 \text{ and } c = 9.37 \times 10^{-8}$$

The Paris equation can be written as:

$$da/dN = 9.37 \times 10^{-8} (\Delta K)^{2.18}$$

The Forman constants c and n [from Eq. (4.4)] can be obtained through the same procedure, where $R = 0$. The same two points on the crack growth curve (Fig. 4.7) are selected:

$$0.0001 = [c(24.5)^n]/(55 - 24.5)$$
$$0.00001 = [c(8.5)^n]/(55 - 8.5)$$

Solving for c and n:

$$c = 1.03 \times 10^{-5} \text{ and } n = 1.78$$

For the second part of the problem:

$$\Delta K = K_{max} - K_{min} = \beta(S_{max} - S_{min})(\pi a_i)^{1/2}$$

where

$$S_{max} = 20 \text{ ksi}, S_{min} = 4 \text{ ksi}, \beta = 1, \text{ and } a_i = 1.0 \text{ in.}$$

Assuming the quantity β remains a constant throughout the growth, the total number of cycles, ΔN, for the crack to grow 0.7 in. can be computed. Using the Paris equation shown by Eq. (4.2):

$$dN = \int_{a_i}^{a_f} \left[\frac{1}{9.37 \times 10^{-8} \times 16^{2.18}(\pi a)^{2.18/2}} \right] da$$

where the total number of cycles $\Delta N = 3759$. Using the Forman equation [Eq. 4.4)]:

$$dN = \int \frac{(1-R)K_c - \Delta K}{1.03 \times 10^{-5}(\Delta K)^{1.78}} da$$

$$= \int \frac{(1-R)K_c}{1.03 \times 10^{-5}(\Delta K)^{1.78}} da - \int \frac{\Delta K}{1.03 \times 10^{-5}(\Delta K)^{1.78}} da$$

$$dN = \int_{a_i}^{a_f} \left[\frac{55}{1.03 \times 10^{-5} \times 16^{1.78}(\pi a)^{1.78/2}} \right.$$

$$\left. - \frac{16(\pi a)^{1/2}}{1.03 \times 10^{-5} \times 16^{1.78}(\pi a)^{1.78/2}} \right] da$$

where a_i and a_f are 1.0 and 1.7 in., respectively. Simplifying the above integration, the total number of cycles $\Delta N = 3084$.

4.2.1 Brief Review of Fatigue Crack Growth Testing

The establishment of a complete da/dN versus ΔK curve (as shown in Fig. 4.5) for the determination of the number of cycles to failure of machinery parts requires conducting steady-state fatigue crack growth rate testing in accordance with ASTM E-647, "Standard Test Method for Measurement of Fatigue Growth Rates." Two types of specimen configurations, Compact Tension, C(T), and Middle-Tension, M(T), are recommended for use in generating the da/dN versus ΔK plot (see Figs. 4.8 and 4.9). This test method involves cycling notched M(T) or C(T) specimens that have been precracked, while the incremental crack growth length, Δa, is periodically measured and recorded together with the number of cycles, ΔN. Other specimen configurations may be used, provided that stress intensity factor solutions for the crack geometries are available and moreover, the net section yielding is avoided.

In precracking the test specimen to simulate an ideal crack, the final K_{max} value associated with precracking should not exceed the initial K_{max} value that the test specimen will experience upon testing. Otherwise, the crack tip will be subjected to the retardation effect at the onset of testing. The retardation effect will slow the rate of crack growth due to overload (high to low load) and is discussed in Section 4.4. This effect is particularly important when the threshold stress intensity factor range, ΔK_{th}, is recorded. The amount of precracking should not be less than 0.05 of the width of the test specimen. Electrical discharge machining (EDM), milling,

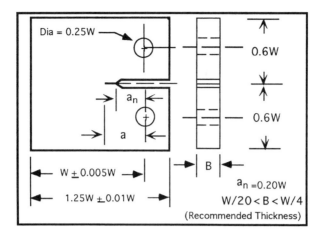

Figure 4.8 Standard compact tension C(T) specimen for crack growth rate testing (all dimensions are written in terms of *W*.)

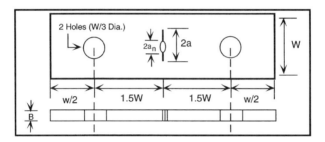

Figure 4.9 Standard M(T) specimen for crack growth rate testing (all dimensions are written in terms of *W*.)

broaching, or saw-cutting techniques can be used for notch preparation prior to precracking. Precracking dimension requirements for the ASTM fatigue crack growth test are shown in Fig. 4.10.

The incremental crack growth associated with N number of cycles is measured either visually or with some other optical technique, and the growth rate, $\Delta a/\Delta N$, as a function of the stress intensity factor range, $\Delta K = K_{max} - K_{min}$, is calculated. The fatigue crack length reading technique, irrespective of the method of measurement, must be capable of measuring as good as $0.002W$. To be capable of reading crack extension visually to $0.002W$, a low power traveling microscope with 50× magnification capability is recommended. It would be helpful, although not necessarily required, to apply reference marks at equal intervals on the test specimen prior to the start of testing to eliminate the potential for errors while reading the data. The compliance method determines an analytical relationship

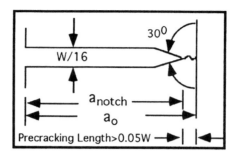

Figure 4.10 EDM notch and minimum precracking (ASTM-E647)

between the inverse of the load–displacement slope, v/p, that has been normalized to elastic modulus, E, and specimen thickness, t (Evt/p) and the normalized quantity of crack length over the specimen width, a/W, and is one nonvisual technique that can be used to monitor the crack advancement as it is subjected to cyclic loading. In the compliance method [9] a clip gage is located at the mouth of the specimen and the displacement of the gage together with the corresponding load are measured (see Fig. 4.11). Depending on the location of data measurements on the specimen mouth a theoretical relationship between the two dimensionless quantities Evt/p and a/W can be developed. For example, when the clip gage is situated at the edge of a C(T) specimen (as shown in Fig. 4.12) the polynomial equation describing a/W as a function of Evt/p can be written [9, 10] as:

$$a/W = 1.001 - 4.6695\,(x) + 18.4\,(x)^2 - 236.82\,(x)^3 + 1214.9\,(x)^4 - 2143.6\,(x)^5$$

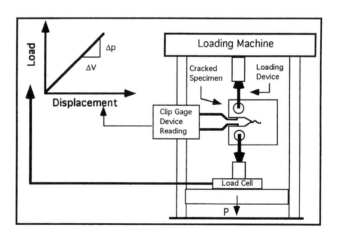

Figure 4.11 The compliance method of measuring load–displacement at the mouth of the loaded crack

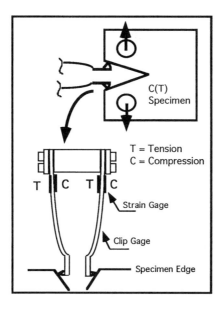

Figure 4.12 Clip gage set up for measuring crack length in terms of displacement measurement at the mouth of the specimen

where

$$x = [(Evt/p)^{1/2} + 1]^{-1}$$

When using the compliance method, it is good practice to also check the crack measurements with an optical method at several intervals during the test. Multiple reading methods are advisable because it is possible that the clip gage could become loose at the mouth of the specimen due to specimen and machine vibration which can introduce errors while data are collected.

The crack length readings recorded in Fig. 4.2 of Section 4.2 must be arranged in such a way that the da/dN data can be evenly distributed with respect to stress intensity factor range, ΔK. In region I, where the incremental crack growth, Δa, is small, the associated number of cycles, ΔN, is large. The opposite is true in the region III, where the crack growth accelerates and the elapsed cycles, ΔN, are small. For the case of the C(T) specimen type, the even distribution of da/dN with respect to ΔK can be accomplished by measuring the incremental crack growth length, Δa, to be at most equal to or smaller than $0.04W$ when $0.25 \leq a/W \leq 0.4$ (ASTM-E147). When $0.40 \leq a/W \leq 0.6$, the Δa measurement should decrease to $0.02W$. In the region III with $0.60 \leq a/W$, the crack length measurement should be more frequent ($\Delta a \leq 0.01W$). In the case of the M(T) test specimen, with

$\Delta a \leq 0.03W$ and $2a/W < 0.6$, the reading should be $\Delta a \leq 0.02W$ and $2a/W > 0.6$, respectively. Generally, the measured value of $\Delta a = 0.01$ in. is an acceptable value for both specimen types mentioned above. As an example, for a compact tension specimen having $W = 2.0$ in. and total crack length $a = 0.45$ in. (see Fig. 4.8), the incremental crack growth length measurement, Δa, visually should be equal to or less than 0.08 in. where $a/W = 0.225$. As the crack length increases and becomes equal to $a = 0.85$ in. (where $a/W = 0.425$), the crack length measurement interval will be every 0.04 in.

The results of crack growth tests are mostly thickness independent; however, materials may show thickness dependency in region III of the da/dN versus ΔK curve when crack growth accelerates and K_{max} approaches the material fracture toughness. In selecting specimen size, adequate thickness is needed to avoid buckling. The presence of buckling can introduce error in the test data measurements. For the C(T) specimen type, the recommended thickness, B, should be larger than $W/20$ and should not exceed $W/4$, as shown in Fig. 4.8.

If an M(T) specimen is used to generate fatigue crack growth data (this type of specimen is used when data for the case of $R < 0$ are of interest), the recommended thickness is $W/8$. For a thin M(T) specimen, lateral deflection may occur due to improper loading. Strain gages should be mounted on the specimen for detection of any bending-induced strain. Bending strain as high as 5% of normal strain may be acceptable in the specimen (ASTM E-647).

In both types of test specimen, the ratio of original crack length, a, to specimen width, W, must be such that net section yielding does not occur at all values of loading (LEFM limitation). That is, the specimen must be predominantly elastic, except in the localized region at the crack tip. To avoid net section yielding, an empirical relationship based on test results relating the specimen dimensions (width and crack length) to the material yielding, σ_{Yield}, and the calculated maximum stress intensity factor, K_{max}, is established [10]. For the C(T) specimen, net section yielding can be avoided when $(W - a) \geq 4/\pi (K_{max}/\sigma_{Yield})^2$ and for the M(T) specimen $(W - 2a) \geq 1.25 (P_{max}/t\, \sigma_{Yield})$, where P_{max} is the maximum load in the cycle and t is the specimen thickness. The maximum stress intensity factor, K_{max}, can be calculated for the C(T) specimen type by the following relationship [11]:

$$K_{max} = \left[\frac{P_{max}(2 + a/W)}{t\sqrt{W}(1 - a/W)}\right]\beta \tag{4.6}$$

where $\beta = [0.886 + 4.64(a/W) - 13.32(a/W)^2 + 14.72(a/W)^3 - 5.6(a/W)^4]$ and the stress intensity factor range, ΔK is given by:

4.2 Crack Growth Rate Empirical Descriptions

$$\Delta K = \left[\frac{(P_{max} - P_{min})(2 + a/W)}{t\sqrt{W}(1 - a/W)} \right] \beta \quad (4.6a)$$

Accordingly for the M(T) specimen [12]:

$$\Delta K = \frac{P_{max} - P_{min}}{tW} \left[(\pi a) \sec \frac{\pi a}{W} \right]^{1/2} \quad (4.6b)$$

The measured crack length to be incorporated into Eqs. (4.6a) and (4.6b) must be an average of the measured value from both sides of the specimen to ensure load symmetry and that the material in consideration is isotropic.

In using the M(T) specimen type, the crack measurement must be the average value of two crack tips. Including back surfaces, a total of four measurements are required [(for the C(T) specimen, only two measurements are needed]. If crack growth directions at the crack tip are not perpendicular to the applied load to within ±20°, the test must be discontinued and the data obtained are invalid.

Residual stresses can have significant influence in fatigue crack growth data, specifically in the region where the ΔK_{th} value is of interest. The residual stress is added to the applied stress, which can either lower or raise the calculated value of the crack tip stress intensity factor. For example, when the magnitude of the compressive residual stress is above or equal to the applied stress, the crack growth rate data will be close to the threshold value since the crack tip stress intensity is not effective in causing any crack growth ($\Delta a \approx 0$). This phenomenon can occur when the test specimen is machined from the weld region where post stress relief is impossible. Other parameters influencing the crack growth rate are: (1) temperature, (2) severity of corrosive environment, (3) influence of specimen thickness, (4) fatigue crack size (it is known that short cracks exhibit greater growth rate than long fatigue cracks; see Section 4.3 for further discussion related to this topic).

In summary, fatigue crack growth testing is performed to provide the analyst with the following information:

- Fatigue crack growth data can be used to determine the number of cycles to failure for an initial crack length in a given material subjected to a given cyclic load environment.
- For two or more materials under the same cyclic loading conditions, it can help to establish the selection of the material that provides the longest life.
- The plot of fracture stress versus the number of cycles to failure can be used to establish the inspection interval, an important tool for quality assurance purposes.
- The effects of heat treatment, fabrication (material anisotropic) and environment on fatigue crack growth can be determined by having laboratory crack growth data generated for different grain orientations.

4.3 Stress Ratio and Crack Closure Effect

Data from experimental testing conducted in the laboratory to generate the crack growth curve under constant amplitude loading clearly indicate that the stress ratio, R, and crack closure behavior have a significant effect on crack growth rate [1]. Figures 4.13 and 4.14 show the influence of the stress ratio on the crack

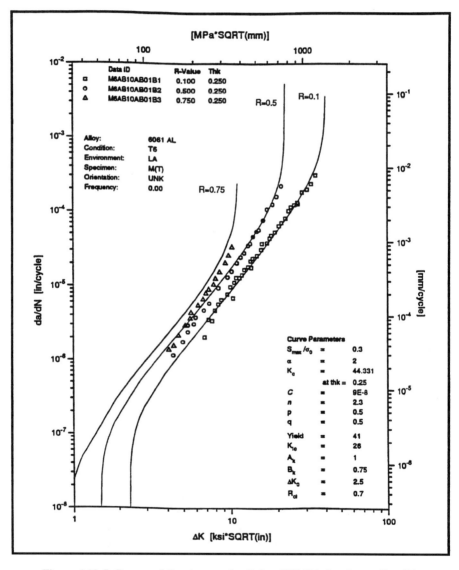

Figure 4.13 Influence of the stress ratio, R, for 6061-T6 aluminum alloy [1]

4.3 Stress Ratio and Crack Closure Effect

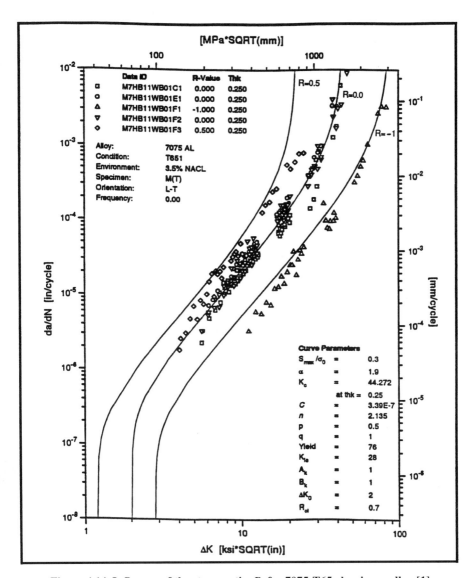

Figure 4.14 Influence of the stress ratio, R, for 7075-T65 aluminum alloy [1]

growth curve for 6061-T6 and 7075-T651 aluminum alloys [1], respectively. From Figs. 4.13 and 4.14, it can be deduced that, for a given crack growth rate, da/dN, the stress intensity range, ΔK, is generally higher as R becomes smaller and approaches a negative value. In addition to the influence of the stress ratio on fatigue crack growth rate, the crack closure behavior has significant influence on the behavior of growing fatigue crack and the rate at which it is advancing. The stress ratio and closure correction to the crack growth rate equation (da/dN versus ΔK) is usually done in one of the two ways: (1) By replacing the stress intensity factor range, ΔK, with its corrected value (called effective stress intensity range, ΔK_{eff}), and then the amount of crack growth is calculated simply by the Paris equation or (2) using the apparent stress intensity factor range, ΔK, but incorporating other effects separately, as shown by Eq. (4.5), where the stress ratio and closure effect is incorporated by the quantities $(1 - R)^n$ and $(1 - f)^n$ respectively.

In Section 4.3.1, the Elber [13–15] and Newman [8] approaches to the crack closure phenomenon will be discussed. It should be mentioned that crack closure can be caused also by other mechanisms, such as oxide, roughness, or corrosion that are not discussed in this book. In Section 4.4, the retardation effect (crack growth delay due to variable amplitude loading) is described by the two mathematical models, called the Wheeler and Wellenborg models.

4.3.1 Elber Crack Closure Phenomenon

Wolf Elber [13, 15] was the first to explain the crack closure effect of constant amplitude loading for various positive stress ratios, R, on crack growth rate behavior by using the concept of plastic yielding at the crack tip. Prior to Elber's observation on the crack closure behavior and its effect on crack growth, it was believed by the experts in the field [16] that the two crack surfaces contact each other and close the crack at zero or under compressive stresses. This assumption is valid for an ideal crack where the plastic deformation is limited to a localized region at the crack tip (an ideal crack was defined as a saw-cut crack of zero width). Elber argued that, upon application of constant amplitude cyclic loading during the increasing tensile portion of the load, a plastic zone is formed. Fatigue crack growth tests on aluminum alloys generated in the laboratory by Elber have indicated that for constant amplitude loading, the formation of plasticity occurring at the crack tip remains behind as the crack grows in a stable manner. As illustrated in Fig. 4.15, the tensile deformation will create compressive stresses which, when the specimen is unloaded, are distributed at the locality of the crack tip, together with the plastically deformed material left behind, cause the two crack surfaces to contact prior to complete unloading. The compressive residual stresses are a result of permanent deformation that stretches the material, and upon load removal it has to fit the elastic material surrounding it. The misfit causes the elastic material to induce compressive residual stress in the plastic region local to the crack tip.

In Fig. 4.15, when the crack length was at location (1), it was surrounded by a plastic zone larger than any other previous position, while its corresponding crack length was smaller. At location (2) after N number of cycles, a new plastic zone is formed that now is larger than location (1) because of its length and higher value associated with crack tip stress intensity factor. The crack tip plastic zone at all stages of advancing fatigue crack forms an envelope that contains all the previous

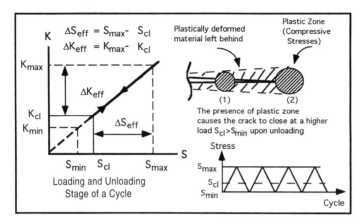

Figure 4.15 Illustration of crack closure based on plastic zone and effective stress value

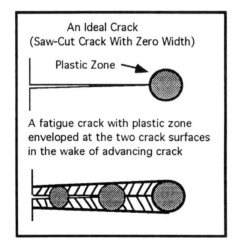

Figure 4.16 Illustration of plastic deformation formed at crack surfaces in the wake of an advancing crack (compared with an ideal crack)

plastic zone sizes and its presence is the cause of crack closure behavior. Figure 4.16 illustrates the crack tip plastic zone and the envelope of all plastic zones for an advancing crack subjected to cyclic load compared with an ideal crack with zero width.

The plastic deformation helps to partially close the crack surfaces such that the crack will close and open at a stress level higher than the $s_{\min}$. This effect results in having a smaller stress intensity factor range, ΔK, value and, consequently, a smaller crack growth rate, as illustrated in Fig. 4.15. The effective stress intensity factor range, in terms of the effective stress range $(\Delta s)_{eff}$, is written as [15]:

$$\Delta K = \beta \Delta s (\pi a)^{1/2}$$
$$\Delta K_{eff} = \beta (\Delta s)_{eff} (\pi a)^{1/2} \qquad (4.7)$$

where $(\Delta s)_{eff} = s_{max} - s_{cl}$. The quantity s_{cl} is the stress level at which the closure occurs ($s_{min} < s_{cl}$) and K_{cl} is the corresponding stress intensity factor ($K_{min} < K_{cl}$). In addition to the quantity s_{cl} (the stress level at which the two crack surfaces are in contact upon unloading), equal attention must be given to another parameter, called the crack opening stress, s_{op}, (the stress at which the two crack surfaces are not in contact with each other upon reloading). Hence, the fatigue crack will advance only during the opening portion of the cycle. In this book, it is assumed that the two quantities s_{cl} and s_{op} are the same (in general, for a given material, the closure stress, s_{cl}, may have different value than s_{op}).

The effective stress range ratio, U, that corrects the apparent stress intensity factor range to account for closure behavior, can be written [15] as:

$$U = \Delta K_{eff}/\Delta K = (\Delta s)_{eff}/(\Delta s) = (s_{max} - s_{op})/(s_{max} - s_{min}) \qquad (4.8)$$

Elber's crack growth rate equation, in terms of the apparent stress intensity range, ΔK, and quantity U, can be expressed in terms of the Paris law as:

$$da/dN = C(U \Delta K)^n \qquad (4.9)$$

Based on constant amplitude testing of 2024-T3 aluminum alloy sheet, a linear equation for U in terms of R was established [15] that can be written:

$$U = 0.5 + 0.4R \qquad -0.1 < R < 0.7 \qquad (4.10a)$$

This relationship was later modified by Schijve [17] as follows:

$$U = 0.55 + 0.35R + 0.1R^2 \qquad (4.10b)$$

The quantity U described by Eqs. (4.10a) and (4.10b) provides the correction for the closure effect (described in terms of stress ratio, R) on the crack growth relationship. From Eqs. (4.9) and (4.10) it can be seen that the two quantities, effective and apparent stress intensity factor range, become equal when the stress ratio R increases. That is, the crack closure effect becomes less effective when the R value increases.

4.3.1.1 Threshold Stress Intensity Factor, ΔK_{th}

In the limit when K_{max} reaches the K_{cl} value, the effective stress intensity factor $\Delta K_{eff} = 0$ and, therefore, no crack growth is expected. This does not necessarily imply that the stress intensity factor range, ΔK, is zero; it simply states that there is a minimum value of ΔK, called the threshold stress intensity factor (ΔK_{th}), in which the growing crack ceases to advance, see region I of crack growth curve where ΔK is a constant (asymptotic value of ΔK in region I) and da/dN approaches zero. As was mentioned previously (in Section 4.2), zero growth is approximated by a cutoff growth rate of 4×10^{-10} m/cycle assigned by ASTM-E647 that provides a practical ΔK_{th} value near the threshold region. The ΔK level at which the growth is zero (or when $da/dN \approx 4 \times 10^{-10}$ m/cycle) is a function of the stress ratio, R.

The threshold stress intensity range, ΔK_{th}, in terms of stress ratio, R, and the threshold stress intensity factor at $R = 0$, ΔK_0, in a simplified form was empirically expressed as [18]:

$$(\Delta K_{th})_R = (1 - C_0 R)^d \Delta K_0 \quad (4.11)$$

C_0 and d are the fitting constants. Note that, in Eq. (4.11) the quantity ΔK_0 must be available through laboratory testing. This quantity is independent of crack length. By allowing the constant $C_0 = 1$ the Klesnil and Lukas [19] equation is obtainable.

$$(\Delta K_{th})_R = (1 - R)^d \Delta K_0 \quad (4.12)$$

When the fitting constant $d = 1$, the Barsom [20] relationship is obtained. In cases where the two fitting constants are not available, the value of $C_0 = d = 1$ will give conservative results for the threshold stress intensity factor, ΔK_{th} for $R \geq 0$.

$$(\Delta K_{th})_R = (1 - R) \Delta K_0 \quad \text{(for } C_0 = d = 1\text{)} \quad (4.13)$$

In general, the fatigue threshold value, ΔK_{th}, decreases with increasing stress ratio, R and becomes a constant at $R = 0$ (Figure 4.13 illustrates the influence of R on ΔK_{th} value for the 6061-T6 aluminum alloy). Equation (4.11) can be modified to account for the dependency of threshold stress intensity factor on crack size [21]. Frost [22, 23] and Usami [24, 25] studied the effect of crack length on the threshold stress intensity factor and concluded that ΔK_{th} decreases with decreasing crack length. On the other hand, when crack length has sufficient dimension that its behavior can be described by the linear elastic fracture mechanics approach, the threshold stress intensity factor, ΔK_{th}, is independent of crack size but dependent on R. Generally, cracks in material can be classified as being long, short, or small in size. Long or large cracks are usually considered as being through the thickness cracks in a thick specimen (such as a crack in a notched centered crack specimen that is used to generate fracture properties of metals). The crack growth behavior of large cracks is well established in terms of the closure phenomenon through the effective stress intensity factor, ΔK_{eff} [see Eq. (4.9)]. In the threshold regime (region I), where crack growth is slow, the test measurements for long cracks can be sensitive to overloading which introduces the retardation effect, and inconsistency in data recording can occur.

Short cracks can be defined as being physically short in dimensions, such as surface cracks with short crack length smaller than 0.02 and 0.078 in. [26] in length. Cracks with short dimensions are less affected by closure phenomenon and will grow faster than corresponding long cracks subjected to the same stress intensity factor, ΔK. In general, crack closure effects increase with crack length. In addition, short cracks show a lower threshold value when compared with long cracks, as shown in Fig. 4.17. The transition from short to long crack behavior is reported by different sources [26–28] to be between 0.5 and 2 mm.

Small cracks (the authors noticed that the term "small" is sometimes used interchangeably with "short" as was evident in Reference [27]) are defined as having dimensions as small as their material characteristic microstructure dimensions, such as the grain size. Their growth is controlled by grain boundaries and cannot be described by the linear elastic fracture mechanics approach. There are several classes of small cracks that Ritchie and Lankford defined in their work "Small Crack Problems" [26]. They used microstructurally, mechanically, and physically small as useful qualifiers to classify small fatigue cracks. For example, a crack is mechanically small when the crack length, a, is smaller than the plastic zone size, r_y ($a < r_y$). When the crack length is smaller than the grain size, d_g, it is called microstructurally small ($a < d_g$). Other classes that fatigue cracks can be defined as being small include physically small when $a < 1$ mm and chemically small when $a < 10$ mm.

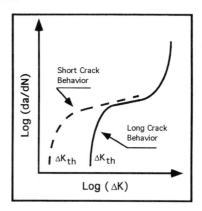

Figure 4.17 Illustration of da/dN Versus ΔK curve for short and long cracks

Tanaka et al. [21] had established a theoretical model to describe the small crack length effect on the threshold stress intensity factor, ΔK_{th}, based on the crack-tip slip band blocked by the grain boundary concept (called the BSB model). The BSB model is based on the assumption that a small crack ceases to advance (the threshold condition) when the crack tip slip band has been arrested by the grain boundary and prevented from growing into the next grain. The effects of crack length and grain size on the threshold stress intensity factor, ΔK_{th} were studied both theoretically and experimentally and a general relationship between ΔK_{th}; crack length, a; and the intrinsic crack length, a_0 was established as:

$$\Delta K_{th} = [a/(a + a_0)]^{1/2} \times (\Delta K_{th})_{R=0} \quad (4.14a)$$

Equation (4.14a) was later modified to account for the stress ratio, R [1]:

$$\Delta K_{th} = [(4/\pi)\tan^{-1}(1 - R)][a/(a + a_0)]^{1/2} \times (\Delta K_{th})_{R=0} \quad (4.14b)$$

where $(\Delta K_{th})_{R=0}$ is the threshold stress intensity factor for $R = 0$. The intrinsic crack length, a_0, defines the boundary between small and long cracks and represents the minimum initial crack size for conducting a meaningful life evaluation using fracture mechanics analysis. Based on the theoretical results and experimental data, an estimated value of 0.004 in. is assigned to a variety of steels and aluminums [29]. As mentioned in Section 4.1, the minimum initial crack size used in the damage tolerance crack growth analysis must be larger than the material's grain size in order not to violate the isotropic continuum concept.

The minimum accepted crack size that can be used for conducting a meaningful crack growth analysis using LEFM can be estimated by employing the Kitagawa and Takahashi diagram [30]. Kitagawa argued that in many structural parts the crack growth process occurs by the initiation and growth of cracks as small as 0.5 to 1.0 mm in length, below which their growth behavior deviates from the LEFM regime. Therefore, it is necessary to develop a method to observe the initiation and growth of such small cracks that may behave differently from the experimentally measured crack growth data associated with large cracks employing the ASTM standards. Kitagawa defined the threshold stress range, Δs_{th}, for a fatigue crack to be associated with a crack growth rate less than 2×10^{-9} mm/cycle ($\approx 8 \times 10^{-11}$ in./cycle). To study the effect of crack length on the threshold stress intensity factor range, ΔK_{th}, the experimental variation of Δs_{th} versus full crack length, $2a$, was plotted (using logarithmic scale) for crack lengths smaller and larger than 0.5 mm. The measured experimental data showed that for crack length larger than 0.5 mm, a good correlation with the LEFM results was obtained (a straight line relationship with slope $-1/2$; see Fig. 4.18). This observation (constant slope of $-1/2$) indicates that the threshold stress intensity factor, ΔK_{th}, is independent of crack length and is a constant for long cracks where $2a > 0.5$ mm. For crack length smaller than 0.5 mm a deviation from the straight line (anomalous behavior associated with small cracks) was observed indicating that the LEFM concept may not be applicable to define the fatigue crack growth threshold behavior. The anomalous behavior can be associated with large scale plasticity at the tip of small cracks ($2c < 0.5$ mm.) where the applied stress amplitude is high and close to the material yield stress.

In the Kitagawa diagram, the boundary between small and large cracks was estimated to be at the onset of deviation from the straight line behavior. The experimental data on HT-80 steel, shown in Reference [30], indicates that the continuation of the trend departing from the straight line becomes gradually asymptotic to the fatigue limit of an uncracked smooth specimen, as shown in Fig. 4.18. Cracks smaller than $a*$ are nondamaging cracks when subjected to an applied cyclic load equal to the endurance limit. Note that the Kitagawa diagram is dependent on the stress ratio, R, and will shift to a lower stress level as R increases (also note that the endurance limit for most material decreases with increasing R value, as discussed in Chapter 2.)

A minimum initial flaw size employed in the aerospace industry for evaluating the life of fracture critical components has been developed based on the capability of crack detection using standard NDI methods. The reader should realize that the crack size assumption by the standard NDI methods is based on the concept of a maximum initial flaw size that may escape detection. If a smaller flaw size is required by the analyst, a special level NDI could be performed on the part. These values are much higher than the limiting value of $a_0 = 0.5$ mm (0.019 in.) observed in the Kitagawa diagram.

Recently, the NASA has accepted the option of replacing the conventional fatigue approach (S–N curve) in estimating the life expectancy of non-fracture critical or low risk parts with fracture mechanics approach that uses a part through crack of depth $a = 0.005$ in. and length $2c = 0.01$ in.. This value is much smaller than the initial flaw size provided by standard NDI methods when performing safe-life analysis for primary space structural components. This method is preferred over the conventional fatigue approach (see chapter 2) because of its feasibility and availability of matereial data as shown in appendix A. The flaw size assumption provided by the NASA (0.005 in.) is larger than the material grain

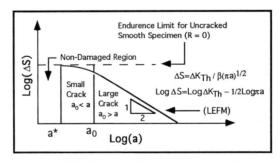

Figure 4.18 Kitagawa diagram for estimating a_0

size and it falls in the boundary between small and large cracks and the concept of isotropic continuum is obeyed.

4.3.2 Newman Crack Closure Approach

The crack closure behavior and its effect on fatigue crack growth rate under constant amplitude loading were discussed by Newman [1, 8, 31]. He developed an analytical model to describe the crack closure phenomenon for both plane stress and plane strain conditions (the thickness influence on the crack closure behavior). The closure model was based on a strip yield model (Dugdale model) that was modified by Newman to incorporate the envelope of all the plastic zones in the wake of an advancing crack. In both plane stress and plane strain conditions, the closure model considers that the material is plastically deformed along the crack surfaces as the crack grows. Newman argued that fatigue cracks remain close during part of the cycle when unloading occurs. When the two crack surfaces are in contact with each other, during the final part of the unloading portion of the cycle, the material may undergo compressive yielding. Upon reversing the cycle as the specimen is reloaded, the crack produces tensile residual stresses at the crack tip that cause the crack to open (zero contact between the two crack surfaces) lower than or equal to the closure stress occurring in the previous portion of the cycle, s_{cl}. The stress at which no surface contact is present upon reloading the specimen is called the crack opening stress, s_{op}. The crack advances during each cycle when the two surfaces are not in contact with each other (where $s_{op} > s_{min}$) and the calculated stress intensity factor range is smaller than the apparent ΔK value when calculating the crack growth rate. An equation was established by Newman that could define the crack opening stress, s_{op}, as a function of stress ratio, R; maximum stress amplitude, s_{max}; and specimen thickness. The specimen thickness effects in closure behavior were defined in terms of a two-dimensional closure model [31] by applying a constraint factor, α, on yielding.

The quantity α is used to describe material yielding in two and three dimensions through the flow stress, σ_o (tensile yielding when σ_o is positive and compressive yielding when σ_o is negative). The flow stress, σ_o, is taken as the average value of the tensile yield and the ultimate of the material. Crack tip yielding occurs depending on the biaxial or triaxial state of stress on that region. Under plane strain conditions, yielding occurs at a stress level much higher than the uniaxial yield stress obtained through laboratory testing. The effective yield stress value, $\alpha\sigma_o$, could be as high as three times the uniaxial stress when a three-dimensional state of stress (plane stress) prevails. Therefore, the material will yield when the stress at the crack tip has reached the quantity $\alpha\sigma_o$. For plane stress and plane strain conditions, the constraint factor, α, takes the values of $\alpha = 1$ and 3, respectively. High α values (2.5 or higher) have been assigned to materials with low K_{Ic}/σ_{Yield} ratio (low fracture toughness value such as high strength steels), whereas high K_{Ic}/σ_{Yield} ratios have $1.5 \leq \alpha \leq 2.0$.

The closure effect on the crack growth rate was described by correlating the effective stress intensity range, ΔK_{eff}, to the apparent stress intensity range, ΔK, through the crack opening stress, s_{op}, and stress ratio, R, as:

$$\Delta K_{eff} = U \Delta K = [(1 - s_{op}/s_{max})/(1 - R)]\Delta K \qquad (4.15)$$

Equation (4.15) simply states that, given a crack growth rate, $(da/dN)_1$, obtained from test data under constant amplitude loading and a stress ratio, R_1 (for simplicity assume $R_1 = 0.2$) with the corresponding stress intensity factor range ΔK; for any other stress ratio, R_2, where $(da/dN)_1 = (da/dN)_2$, the shift in the stress intensity factor range (ΔK_{eff}) can be evaluated through the parameter U (for stress ratio and crack closure adjustment) described by Eq. (4.15); see Fig. 4.19. Experimental data show that for most aluminum alloys, the stress ratio contribution that can shift the crack growth rate curves is negligible for a stress ratio, R, greater than 0.7 [32], see Fig. 4.20. Figure 4.20 shows the experimental data obtained for 2014-T651 aluminum alloy with different stress ratios, $R = 0.1, 0.4$, and 0.7 that were curve fit by employing Eq. (4.5) [1]. Equation (4.5) can be used to describe the crack closure and stress ratio effect due to varying the stress ratio, R, and it provides excellent curve fit.

For a center crack specimen under tension loading, the normalized crack opening stress, s_{op}/s_{max}, describing ΔK_{eff} [which was expressed by Eq. (4.15)] in terms of stress ratio, R, and constraint factor, α, can be written as [8]:

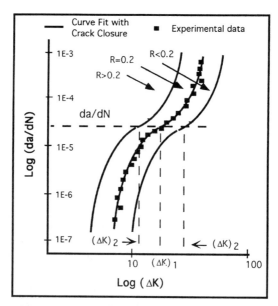

Figure 4.19 Illustration of curve fit with crack closure to crack growth rate data

4.3 Stress Ratio and Crack Closure Effect

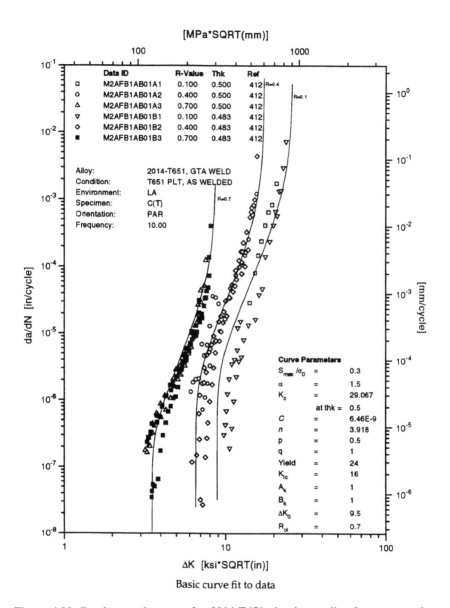

Figure 4.20 Crack growth curves for 2014-T651 aluminum alloy for stress ratios $R = 0.1, 0.4$, and 0.7 [1]

$$S_{op}/S_{max} = [A_o + A_1 R + A_2 R^2 + A_3 R^3] \quad \text{for } R \geq 1 \quad (4.16)$$

and

$$S_{op}/S_{max} = [A_o + A_1 R] \quad \text{for } -1 < R \leq 0 \quad (4.17)$$

The coefficients A_o, A_1, A_2, and A_3 for $S_{op} \geq S_{min}$ are:

$$A_o = (0.825 - 0.34\alpha + 0.05\alpha^2)[\cos(\pi S_{max}/2\sigma_o)]^{1/\alpha} \quad (4.18)$$

$$A_1 = (0.415 - 0.71\alpha) \times (S_{max}/2\sigma_o) \quad (4.19)$$

$$A_2 = 1 - A_0 - A_1 - A_3 \quad (4.20)$$

$$A_3 = 2A_0 + A_1 - 1 \quad (4.21)$$

The flow stress, σ_o, is taken to be the average of the uniaxial yield stress and uniaxial ultimate tensile strength of the material:

$$\sigma_o = (\sigma_{Yield} + \sigma_{Ul})/2 \quad (4.22)$$

The contribution of the crack closure effect on the FNK empirical relationship described by Eq. (4.5) is described by the crack opening function, f. For plasticity induced crack closure, the parameter, f, can be written as [8]:

$$f = K_{op}/K_{max} = \text{Max}(R, A_o + A_1 R + A_2 R^2 + A_3 R^3) \quad \text{for } R \geq 0 \quad (4.23)$$

and,

$$f = K_{op}/K_{max} = (A_o + A_1 R) \quad \text{for } -2 \leq R < 0 \quad (4.24)$$

where K_{op} is the crack opening stress intensity factor below which the crack is closed. For a stress ratio of $R = 0$, the crack closure parameter, f, will have the maximum value given by Eq. (4.23).

Newman and Raju [31] studied the crack closure effect for surface and corner cracks under constant amplitude loading and variable stress ratio, R. They have shown that better crack growth results at the c-tip can be obtained by multiplying ΔK by a crack closure factor (β_R). The value of β_R for the stress ratio $R > 0$ is given by:

$$\beta_R = 0.9 + 0.2R^2 - 0.1R^4 \quad (4.25)$$

and for $R \leq 0$ the factor $\beta_R = 0.9$.

4.3 Stress Ratio and Crack Closure Effect

Table 4.1 Fracture and fatigue crack growth properties for 7075-T651 aluminum alloy

σ_{Yield}	σ_{Ul}	K_{Ic}	K_{Ie}	AK	BK	C	n	p	q	ΔK_0	R_{cl}	α	SR
76,	85,	28,	38,	1.0,	1.0,	0.2E-7,	2.885,	0.5,	1.0,	2.0,	0.7,	1.9,	0.3

The FNK empirical equation describing the crack growth rate [see Eq. (4.5)] is used in the NASA/FLAGRO computer code that was developed to provide an automated procedure for analysis of fracture critical parts of NASA space flight hardware and launch support facilities. In this computer program, the effect of the stress ratio is incorporated through the quantity $(1 - R)^n$ and crack closure behavior through the function f by assuming a constant value of $s_{max}/\sigma_o = 0.3$. This value was selected because it is close to the average value obtained through fatigue crack growth tests using various specimen types. For most aluminum alloys (2000 through 6000 series), α is chosen to have a value of 1.5 and for 7000 series $\alpha = 1.9$. Having the two quantities s_{max}/σ_o and α available, the crack closure effect for any other stress ratio, R, can be evaluated by utilizing Eqs. (4.23) and (4.24). Table 4.1 presents a listing of all the constants that must be provided as input to Eq. (4.5) for 7075-T651 aluminum alloy in order to conduct a comprehensive safe-life analysis. Information contained in Table 4.1 was extracted from the NASA/FLAGRO 2.0 material library. The quantities A_k and B_k are the fit parameters in defining the plane stress fracture toughness, K_c, in terms of plane strain fracture toughness, K_{Ic} [1]. The NASA/FLAGRO computer code has the following capabilities:

1. Safe-life analysis of the structural components subjected to a constant amplitude cyclic load
2. Critical crack size that causes the structure to become unstable
3. Stress intensity factor solutions to different crack geometries and loading conditions
4. Growth rate constants described by Paris law and Eq. (4.5)
5. da/dt life analysis

NASA/FLAGRO 2.0 has a rich material library that has fracture properties data for many aerospace alloys subjected to varying heat treatment conditions. Appendix A provides fracture properties of most aerospace alloys obtained from the NASA/FLAGRO material library (see Table 4.1 for the definition of the symbols). In addition, the program has enhanced stress intensity solutions that cover various crack geometries under different loading conditions. For a more detailed description of the NASA/FLAGRO computer code, the reader may refer to Reference [1]. The improvement for this code is in progress and as the technology advances the program will enhance accordingly. Some areas of improvements that require future attention are:

- Incorporating the da/dt time-dependent environmental crack growth
- Retardation effect
- Cycle by cycle approach for calculating the crack growth, da/dN

Example 4.2

A pressurized cylindrical tank is made of Ti-4Al-4V alloy. Its fracture properties (based on the NASA/FLAGRO material library) are shown in Fig. 4.21. The tank is pressurized and depressurized eight times per year during its service life. It is required that the tank undergo a proof test of 1.5× Maximum Operating Pressure (MOP) prior to its usage. For operating pressure, $P_{op} = 350$ psi; (1) Use a cycle by cycle analysis by using Eq. (4.5) and compare the results of the first three cycles with the NASA/FLAGRO 2.0 computer code output (or any other available code); (2) Determine the final crack length at the end of the year. The original crack length based on penetrant inspection is given as a circular surface crack with $a_o = 0.075$ and $2c = 0.15$ in. The purpose of part (1) of this example problem is to allow the reader to learn the steps necessary for computing the crack growth values and accumulation of growth corresponding to each cycle.

Solution

Using Eq. (4.5), the amount of crack growth associated with each cycle can be computed. The stress intensity range, ΔK, for each cycle is calculated by employing Eq. (3.52) of Chapter 3 for a longitudinal surface crack in pressurized pipe:

$$\beta(a/c, a/t, \theta) = 0.97 * [M_1 + M_2(a/t)^2 + M_3(a/t)^4] * g * f_\phi * f_c * f_i * f_x$$

where:

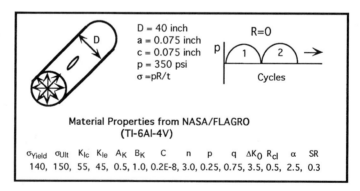

Figure 4.21 Material and crack geometries for the pressurized cylinder in problem 4.2

4.3 Stress Ratio and Crack Closure Effect

$$M_1 = 1.13 - 0.09\,(a/c)$$
$$M_2 = -0.54 + 0.89/(0.2 + a/c)$$
$$M_3 = 0.5 - [1/(0.65 + a/c)] + 14\,(1 - a/c)^{24}$$
$$g = 1 + [0.1 + 0.35\,(a/t)^2](1 - \sin\theta)^2$$
$$f_\phi = [(a/c)^2 \cos^2\theta + \sin^2\theta]^{1/4}$$
$$f_x = [1 + 1.464\,(a/c)^{-1.65}]^{-1/2}$$

The quantity $f_c = [(1 + k^2)/(1 - k^2) + 1 - 0.5\,(a/t)^{1/2}]\,[t/(D/2 - t)]$ where $k = 1 - 2t/D$ and the value of $f_i = 1$ for an internal crack. Note that the quantity $f_x = (\phi)^{-1/2}$, as discussed in Eq. (3.47) of Chapter 3.

The calculated values of M_1, M_2, M_3, f_x, and f_c for $a/t = 0.75$ and $a/c = 1$ using Eq. (3.52) are:

$$M_1 = 1.04,\ M_2 = 0.201,\ M_3 = -0.106,\ f_x = 0.637 \text{ and } f_c = 1.005$$

The correction factor is $\beta = 0.694$, where $\theta = \pi/2$. The stress intensity factor range, ΔK, for the case of $R = 0$ (see Fig. 4.21) can be obtained as:

$$\Delta K = K_{max} = \beta\,\sigma_{max}\,(\pi a)^{1/2}$$
$$= 0.694 \times 1.5 \times 70(3.14 \times 0.075)^{1/2}$$
$$= 35.379 \text{ ksi (in.}^{1/2})$$

Note that the applied stress for the first cycle ($\Delta N = 1$) is equal to the proof stress ($1.5 \times \sigma_{op}$) and the retardation effect due to proof cycle is not included in this analysis. When $\Delta N = 1$ and $R = 0$, Eq. (4.5) can be written as:

$$\Delta a = \frac{C(1-f)^n\,\Delta K^n \left(1 - \dfrac{\Delta K_{th}}{\Delta K}\right)^p}{\left(1 - \dfrac{\Delta K}{K_{1e}}\right)^q}$$

The crack opening function, f, can be calculated from Eq. (4.23) by replacing the quantity s_{max}/σ_o by 0.3 ($f = 0.2745$). Moreover, the threshold stress intensity range $\Delta K_{th} = (\Delta K_{th})_{R=0}$ and is designated by ΔK_0 in the NASA/FLAGRO material library.

Simplifying the crack growth equation by supplying the appropriate values of the constants f, ΔK_0, ΔK_{th}, C, n, p, and q (see the material properties shown in Fig. 4.21), the amount of growth for the first cycle is given by $\Delta a = 0.000106$ in. The original crack depth is advanced by 0.000106 in. and the new crack length at the a-tip is now acting as the original crack depth for the next cycle ($a_o = 0.075 + 0.000106 = 0.075106$ in.)

The amount of growth in the length direction (c-tip) can be calculated in the same way as for the a-tip. Note that the equation of the stress intensity factor for the c-tip was formulated in Chapter 3 [described by Eq. (3.52)] by simply replacing the angle $\theta = 90°$ with $0°$. The correction factor, β, is slightly higher for $\theta = 0°$ due to the dependent quantities g and f_ϕ on the angle θ [see Eq. (3.52)]. The calculated values of g and f_ϕ are 1.296 and 1, respectively. The stress intensity range for the c-tip can be calculated as:

$$\Delta K = K_{max} = \beta \sigma_{max} (\pi c)^{1/2}$$

$$0.9 \times 1.5 \times 70 (3.14 \times 0.075)^{1/2}$$

$$45.88 \text{ ksi (in.}^{1/2})$$

Crack growth for the c-tip can be obtained from Eq. (4.5) in the same manner as was applied for the a-tip direction. However, the fracture toughness value for the c-tip, K_{Ie}, is now replaced by $1.1 K_{Ie}$ [1] (see also Section 3.7.4). The closure effect for the c-tip was described by Eq. (4.25) and for the case of $R = 0$ it takes the value of $\beta_R = 0.9$:

$$\Delta c = \frac{C(1-f)^n (0.9 \Delta K)^n \left(1 - \frac{\Delta K_{th}}{0.9 \Delta K}\right)^p}{\left(1 - \frac{0.9 \Delta K}{1.1 K_{Ie}}\right)^q}$$

The calculated amount of growth at the c-tip is given by $\Delta c = 0.000201$ in. The original crack length at the c-tip is advanced by 0.000201 in. and the new crack length at the c-tip is now acting as the original crack length for the next cycle ($c = 0.075206$ in.).

The amount of crack growth in both the a- and c-tip directions can be calculated for the next cycle in the same manner; however, the aspect ratio a/c for the new analysis must include the actual value, rather than the original value of $a/c = 1$. For example, for the second cycle of pressurization and depressurization, the new aspect ratio $a/c = 0.998 = 0.075106/0.075201$ must be employed. The following table shows the calculated values of the quantities that are used to evaluate the crack growth up to three pressure cycles for both the a- and c-tips. In addition, the values of the amount of crack growth for each cycle are calculated by NASA/FLAGRO computer code and presented for comparison.

Results of the hand analysis for example 4.2

a/c	M_1	M_2	M_3	$g(90)$	f_x	f_ϕ	f_c	β	ΔK	Δa
1	1.04	0.202	−0.12	1.0	0.637	1.0	1.005	0.695	3.5	1.06E−4
0.998	1.04	0.203	−0.11	1.0	0.637	1.0	1.0054	0.697	3.5	1.705E−5
0.998	1.04	0.203	−0.11	1.0	0.638	1.0	1.0053	0.697	3.5	1.707E−5

a/c	M_1	M_2	M_3	$g(0)$	f_x	f_ϕ	f_c	β	ΔK	Δc
1	1.04	0.202	−0.12	1.296	0.637	1.0	1.005	0.901	3.5	2.05E−4
0.998	1.04	0.203	−0.11	1.297	0.637	0.999	1.0054	0.903	3.5	2.88E−5
0.998	1.04	0.203	−0.11	1.297	0.638	0.999	1.0053	0.903	3.5	2.88E−5

Final crack sizes by hand analysis

Cycle 1
Final crack sizes: $a = 0.751060\text{E-}01$, $c = 0.752050\text{E-}01$, $a/c = 0.9986$

Cycle 2
Final crack sizes: $a = 0.75123\text{E-}01$, $c = 0.752338\text{E-}01$, $a/c = 0.9985$

Cycle 3
Final crack sizes: $a = 0.75140\text{E-}01$, $c = 0.752626\text{E-}01$, $a/c = 0.9984$

Results of NASA/FLAGRO (cycle by cycle)

Cycle 1
Final crack sizes: $a = 0.750690\text{E-}01$, $c = 0.751050\text{E-}01$, $a/c = 0.9995$

Cycle 2
Final crack sizes: $a = 0.750833\text{E-}01$, $c = 0.751263\text{E-}01$, $a/c = 0.9994$

Cycle 3
Final crack sizes: $a = 0.750876\text{E-}01$, $c = 0.751476\text{E-}01$, $a/c = 0.9993$

In using NASA/FLAGRO with three cycles at once rather than cycle by cycle, the results are:

Final crack sizes: $a = 0.750975\text{E-}01$, $c = 0.751475\text{E-}01$, $a/c = 0.999335$

Note: The crack growth analysis conducted by the NASA/FLAGRO computer code is not based on a cycle by cycle approach; rather, it is based on incremental growth, $\Delta a = (1/N)a_o$. The advantages associated with this method are cost effectiveness and speed. The NASA/FLAGRO analysis is performed by assuming the amount of incremental growth always to be $1/200$ of the original crack size. The number of cycles, N, associated with $a_o/200$ is calculated by using Eq. (4.5). Therefore, the average amount of growth per cycle is:

$$(\Delta a)_{1\text{cycle}} = (a_o/200)/N$$

Discrepancies between the hand analysis and NASA/FLAGRO are therefore expected for the first few cycles until the calculated Δa value becomes equal to $a_o/200$.

Example 4.3

In the previous example problem (1) determine the number of cycles needed to have leak-before-burst, (2) find the stress value for which leak-before-burst is not achievable (the leak-before-burst criteria were discussed in Chapter 3).

Solution

It would be difficult, if not impossible, to obtain the number of cycles associated with leak-before-burst by employing a hand analysis as demonstrated in the previous example. Using the NASA/FLAGRO computer code, however, the results can easily be provided to the analyst. The following are the results of the NASA/FLAGRO analysis for part 1 of this problem.

Results

Transition to a through crack occurs at cycle no. 948.56.
Where the crack length at the a-tip: $a = 0.9382\text{E-}01$ ($t = 0.10$ in.) and:
crack size at c-tip: $c = 0.105167$, $a/c = 0.892093$
FINAL RESULTS:
Unstable crack growth, max stress intensity exceeds critical value:
$K_{\max} = 88.43$, $K_{cr} = 88.34$
at cycle no. 1756.07, where crack size $c = 0.405322$ in.

The above analysis indicates that the initial crack can grow through the interior wall of the tank ($a \approx t = 0.1$ in.) without failure when it is subjected to fluctuating load due to pressurization and depressurization cycles. It requires approximately 948 cycles for the crack to become a through crack. Moreover, it takes an additional 808 cycles (1756-948 cycles) for the instability to occur.

For the tank to burst before leaking, the level of pressure must be high enough to have failure before the surface crack becomes a through crack. Two cases are illustrated in this example problem to demonstrate failure before tank leakage;

(1) to include the proof test as the first cycle in the fatigue crack growth analysis and (2) to exclude the proof test cycle. To obtain the maximum stress value for which leak-before-burst is prevented, several crack growth runs based on different applied fluctuating stresses were performed. The maximum operating stress that can be used in order to have a valid proof test (without yielding the material) is:

$$(\sigma_{max})_{operating} = 0.9 \times \sigma_{Yield}/1.5 = 84 \text{ ksi}$$

Note that, in conducting the proof test, no permanent deformation is permitted on the tank. For this reason, it is reasonable to allow the proof test stress to go as high as 90% of the tensile yield of the material. Based on the above criteria, the maximum operating pressure will be 84 ksi.

Results

Transition to a through crack at cycle no. 423.38:
Where the crack length at the a-tip: $a = 0.913\text{E-}01$ ($t = 0.1000$) and:
Crack size: $c = 0.101266$, $a/c = 0.902003$
FINAL RESULTS:
Unstable crack growth, max stress intensity exceeds critical value:
$K_{max} = 88.58$, $K_{cr} = 88.34$
at cycle no. 776.28, where crack size $c = 0.302632$

From the above results, one can conclude that for the tank with thickness $t = 0.1$ in. subjected to stress level of 84 ksi, the leak-before-burst condition cannot be avoided. However, by eliminating the proof cycle (not considering its effect in the crack growth rate analysis) and allowing the maximum operating stress in the above analysis to go as high as 135 ksi $< \sigma_{yield} = 140$ ksi, failure of the tank is expected to occur before the surface crack becomes a through crack, $K_{Ie} \geq K_{max}$ (see Eq. 3.55 of Chapter 3 for instability condition). As indicated in the results of the analysis below, it takes approximately 17 cycles for the tank to burst at a stress level corresponding to 135 ksi.

Results

Unstable crack growth, max stress intensity exceeds critical value:
$K_{max} = 60.64$, $K_{cr} = 60.50$
at cycle no. 16.80 (the end of the second year)
Crack sizes: $a = 0.801573\text{E-}01 < t = 0.1$ in., $c = 0.839072\text{E-}01$, $a/c = 0.955309$

Example 4.4

An embedded crack was detected by x-ray inspection in the weld region of a pressurized liquid oxygen tank to be used for a space vehicle, as shown in Fig. 4.22.

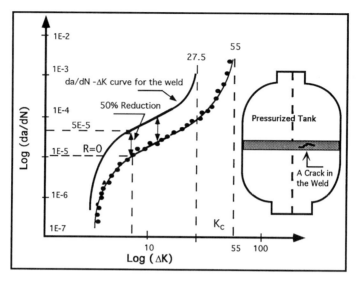

Figure 4.22 Crack growth curve for the welded region described in Example 4.4

The detected crack was shown to be very close to the surface of the weld and conservatively considered as a surface crack having length $2c = 0.32$ in. The crack is oriented in the circumferential direction in the shell region of the tank with $\sigma_{Applied} = PR/2t$ (where P is the maximum operating pressure, R is the tank radius, and t is the thickness associated with the weld region in which the crack was detected). The $da/dN - \Delta K$ data curve for this alloy is shown in Fig. 4.22. Use the Forman equation and compare the results with the NASA/FLAGRO computer code (or any other available fatigue crack growth computer code) to determine the number of pressurization cycles that the tank can withstand before failure. Assume the weld properties are 50% lower than those of the parent material (stress ratio, $R = 0$, $P = 35$ psi, $(\sigma_{Yield})_{parent} = 40$ ksi, $(\sigma_{Ult})_{parent} = 60$ ksi, tank radius, $R = 72$ in., and $t = 0.1$ in.).

Solution

To obtain the total number of pressurization cycles in the weld region of the liquid oxygen tank, a 50% reduction in properties must be considered (as shown in Fig. 4.22). Furthermore, region III of the curve is approximated by using the asymptotic line associated with 27.5 ksi (in.$^{1/2}$) (this is equal to 50% of the parent material fracture toughness). The Forman constants c and n for the weld material [from Eq. 4.4)] can be obtained by selecting two points on the da/dN curve, as shown in Fig. 4.22:

4.3 Stress Ratio and Crack Closure Effect

$$0.0001 = [c(10)^n]/(27.5 - 10)$$
$$5 \times 0.00001 = [c(8.5)^n]/(27.5 - 8.5)$$

Solving for c and n:

$$c = 3.11 \times 10^{-7} \text{ and } n = 3.75$$

The Forman equation for the weld material can be written as:

$$da/dN = \frac{3.11 \times 10^{-7}(\Delta K)^{3.75}}{27.5 - \Delta K}$$

The number of cycles to failure ΔN can be evaluated by:

$$\Delta N = \int_{c_i}^{c_f} [8.84 \times 10^7 (\Delta K)^{-3.75}] dc - \int_{c_i}^{c_f} [0.321 \times 10^7 (\Delta K)^{-2.75}] dc$$

where $c_i = 0.16$ in. and final crack length, c_f, is computed by:

$$K_c = \beta\sigma(\pi c_f)^{1/2} \rightarrow c_f = (27.5/25.2)^2/\pi = 0.379 \text{ in.}$$

The number of cycles to failure in terms of final crack length is:

$$\Delta N = \int_{0.32}^{0.379} [8.84 \times 10^7 (\Delta K)^{-3.75}] dc$$
$$- \int_{0.32}^{0.379} [0.321 \times 10^7 (\Delta K)^{-2.75}] dc$$

$$\Delta N = \int_{0.32}^{0.379} [8.84 \times 10^7 [25.2\sqrt{\pi c}]^{-3.75}] dc$$
$$- \int_{0.32}^{0.379} [0.321 \times 10^7 [25.2\sqrt{\pi c}]^{-2.75}] dc$$

$$\Delta N = \int_{0.32}^{0.379} [57.4 c^{-3.75}] dc - \int_{0.32}^{0.379} [93.1 c^{-2.75}] dc = 83 \text{ Cycles}$$

To use the NASA/FLAGRO computer code for determination of the number of cycles to failure, the fatigue growth rate constants used to establish the da/dN, ΔK curve must be computed from the data points shown in Fig. 4.22.

Material properties:
UTS: YS: K1e: K1c: Ak: Bk: Thk: Kc
33.0: 27.5: 30.2: 27.5: 1.0: 0.0: 0.10: 27.5

where $K_{1e} = 1.1\ K_{1c}$ (see Chapter 3). The crack geometry was assumed to be a circular surface crack with total crack length $2c = 0.32$ and crack depth $a = 0.16$ in. From da/dN, ΔK test data with 50% reduction in the properties:

da/dn	:	ΔK
0.1000E-02	:	27.500
0.4000E-03	:	25.000
0.2000E-03	:	18.000
0.5500E-04	:	9.000
0.2200E-04	:	4.500

Final results:
Unstable crack growth, max. stress intensity exceeds critical $K_{cr} = 27.5$ value:
$K_{max} = 27.53 \qquad K_{ref.} = 0.0000 \qquad K_{cr} = 27.50$
at cycle no. 691.9 crack size $c = 0.369646$

By comparing the number of cycles to failure for the two approaches, it can be seen that it takes only 83 cycles for the Forman and 692 cycles when the NASA/FLAGRO crack growth equation is used. It can be concluded that the Forman approach yields a more conservative result when the life of a structural part is calculated.

4.4 Variable Amplitude Stress and the Retardation Phenomenon

Fatigue crack growth rate test data needed for life analysis (da/dN versus ΔK) are produced in the laboratory under the condition of constant amplitude loading. All of the fracture properties that are compiled in the NASA/FLAGRO material library [1] (shown in Appendix A) are generated under a constant amplitude loading condition. However, in real situations, most structural parts are subjected to variable amplitude loading throughout their service life. Crack growth test data have shown that, under variable amplitude loading, there is delay (retardation) or acceleration in the amount of crack growth due to high or low loads. Therefore, the effect of high and low loads on the amount of crack growth must be addressed. For example, the loads that the wing of a transport aircraft encounters during its service usage are complex and the effect of stress interaction due to variable amplitude loading on crack growth rate is an important problem in aircraft design.

Experimental data have indicated that a high tensile load, followed by a constant amplitude load, will reduce or retard the rate of growth, as illustrated in Fig.

4.4 Variable Amplitude Stress and the Retardation Phenomenon

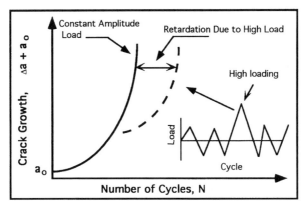

Figure 4.23 Illustration of the decrease in the rate of loading due to the high load followed by low constant amplitude load

4.23. This phenomenon is called retardation [33]. Retardation or delay in the rate of crack growth results from the plastic deformation that occurs at the crack tip. The tensile overload produces large tensile plastic deformation (called the overload affected zone) and, upon removal of this load, the material in the vicinity is elastically unloaded. The plastic zone surrounding the crack tip, however, experiences compressive stresses. Crack growth by subsequent smaller cycles have a retarded rate across the affected zone. The higher the magnitude of the tensile overload, the larger the retardation effect when a subsequent low cycle amplitude is applied.

Even though the plastic deformation under constant amplitude loading has a smaller dimension, it does not have any retardation effect at the crack tip. The constant amplitude plastic deformation contributes to the closure phenomenon described in Section 4.3 where crack surfaces close at a nonzero load level ($S_{min} < S_{cl}$).

In general, structural parts undergo random loading during their service life. The load environment contains not only the tensile overloading, as shown in Fig. 4.23, but also other types of overloading, such as compressive and tensile-compressive overload (see Fig. 4.24), which affect the rate of crack growth. Negative or compressive overloads may have the opposite effect and tend to accelerate the crack growth rate [34, 35] (it is worth mentioning that for precracking the brittle material, when it is difficult to precrack under normal tensile mode I loading procedures, the high compressive load concept can be used to initiate the crack or to extend the preexisting crack in the material). When a tensile overload is followed by a compressive overload, the effect of the retardation on the rate of growth is reduced or may be totally eliminated.

To adjust the crack growth rate due to tensile overload, the empirical crack growth equations, described in Section 4.2, must be corrected. Various models

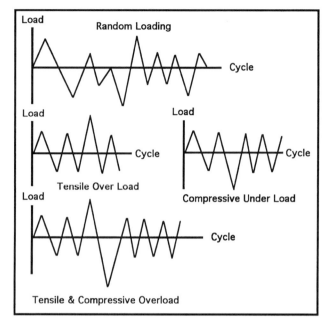

Figure 4.24 Illustration of the types of loading that occur in service life

are available that describe the retardation phenomenon. Two well-known mathematical models, called the Wheeler and Willenborg models, based on yield zone [36, 37] are presently available and will be discussed in Sections 4.4.1 and 4.4.2, respectively.

4.4.1 Wheeler Retardation Model

In this model, the crack growth equation is corrected to account for the retardation effect. The effect of the tensile overload on the rate of growth is through the crack growth reduction factor parameter, C_p [36]. The retarded crack growth rate, $(da/dN)_r$, in terms of the apparent crack growth rate, da/dN, can be written as:

$$(da/dN)_r = C_p (da/dN) = C_p f(\Delta K) \tag{4.26}$$

The retardation factor, C_p, is related to the size of the plastic zones created by the tensile overload [overload-affected zone, $(r_p)_{OL}$], crack size at the onset of overload, a_{OL}, and the plastic zone size for the current or subsequent cycles, $(r_p)_n$, as illustrated in Fig. 4.25:

$$C_p = \left(\frac{(r_p)_n}{a_{OL} + (r_p)_{OL} - a_n} \right)^m \tag{4.27}$$

4.4 Variable Amplitude Stress and the Retardation Phenomenon

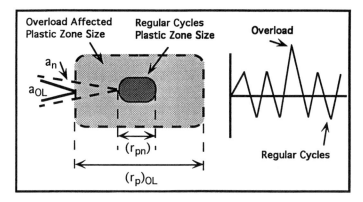

Figure 4.25 Illustration of plastic zones and the retardation effect as the result of overload

where m is an empirical constant that can be tailored from variable amplitude test data to allow for reasonably accurate life predictions. Test data have indicated that the constant m depends on the material, crack size, and the level of the applied overload. The quantity $(a_n - a_{OL})$ is the amount of growth associated with n cycles consumed in the overload affected zone. The plastic zone sizes, $(r_p)_{OL}$ and $(r_p)_n$, shown in Fig. 4.25, can be written in terms of the stress intensity factor and the tensile yield of the material by:

$$(r_p)_{OL} = (1/2\pi)[(K)_{OL}/\sigma_{Yield}]^2 \qquad (4.28a)$$

$$(r_p)_n = (1/2\pi)[(K)_n/\sigma_{Yield}]^2 \qquad (4.28b)$$

From Eq. 4.27 (see also Fig. 4.25), it can be seen that the Wheeler retardation model described by the parameter C_p is effective as long as the plastic zone size associated with the nth cycle, $(r_p)_n$, is within the larger zone $(r_p)_{OL}$. That is, when the current crack size reaches the end of the yield zone, $a_n = (r_p)_{OL}$, and the crack growth reduction factor parameter $C_p = 1$.

4.4.2 Willenborg Retardation Model

The Willenborg model considers the effect of retardation on the rate of crack growth through the extent of the plastic zone formed by the overload [37]. The correction to the crack growth is handled through the use of an effective stress intensity factor that can be expressed as:

$$(K_n)_{eff} = K_n - K_{red} \qquad (4.29)$$

where the reduction stress intensity factor, K_{red} (also called the residual stress intensity factor, owing to the compressive residual stress state caused by differences in load levels) as the result of overload is given as [37]:

$$K_{red} = (K_{OL})_{max} [1 - (a_n - a_{OL})/r_{pOL}]^{1/2} - K_{max,n} \quad (4.30)$$

in which $(K_{OL})_{max}$ is the maximum stress intensity factor due to the overload and $K_{max,n}$ is the maximum stress intensity factor for the subsequent smaller cycle (see Fig. 4.25). When the current crack size has reached the boundary of the overload yield zone $(a_n - a_{OL} = r_{pOL})$, the retardation effect becomes zero, $K_{red} = 0$. Any overload greater than the previous overload creates a new retardation effect that is independent from the preceding one.

Equation (4.29) in terms of maximum $(K_{max})_{eff}$ and minimum $(K_{min})_{eff}$ effective stress intensity factors can be written as:

$$(K_{max,n})_{eff} = \beta(\sigma_{max,n} - \sigma_{red})(\pi a_n)^{1/2} \quad (4.31)$$

$$(K_{min,n})_{eff} = \beta(\sigma_{min,n} - \sigma_{red})(\pi a_n)^{1/2} \quad (4.32)$$

and the effective stress intensity range is:

$$(\Delta K_n)_{eff} = (K_{max,n})_{eff} - (K_{min,n})_{eff} \quad (4.33)$$
$$= \Delta K_n$$

From Eqs. (4.30), (4.31), and (4.32), one can conclude that the effective stress intensity factor range, $(\Delta K_n)_{eff}$, for a complete cycle is equal to the apparent stress intensity factor, ΔK_n. That is, in the Willenborg retardation model, both quantities, the effective stress range $(\Delta \sigma)_{eff}$ and the apparent stress range, $\Delta \sigma$ are the same:

$$(\Delta \sigma)_{eff} = (\sigma_{max,n})_{eff} - (\sigma_{min,n})_{eff} = \sigma_{max,n} - \sigma_{min,n} \quad (4.34)$$

where the effective stress ratio, R_{eff}:

$$R_{eff} = (K_{min})_{eff}/(K_{max})_{eff} \quad (4.35)$$

Because the effective and apparent stress intensity ranges are equal in the Willenborg retardation model, the retardation effect is affected by the effective stress ratio described by Eq. (4.35). By calculating $(\Delta K)_{eff}$ or ΔK and R_{eff}, the amount of growth, da/dN, in the retardation zone can be computed by any crack growth equation in which the stress ratio, R, appears. For example, consider the Forman crack growth relation described by Eq. (4.4):

4.4 Variable Amplitude Stress and the Retardation Phenomenon

$$da/dN = \frac{c(\Delta K_{\text{eff}})^n}{(1 - R_{\text{eff}})K_c - \Delta K_{\text{eff}}} \quad (4.36)$$

The Willenborg retardation effect can be handled cycle by cycle through Eq. (4.36). This equation is effective as long as R_{eff} is positive. One limitation of the Willenborg retardation model [described by Eq. (4.30)] is that, for the case of $(K_{\text{OL}})_{\text{max}} = 2 \times K_{\text{max}, n}$, at the application of the overload where $a_i = a_{\text{OL}}$, complete crack arrest is obtained and crack growth ceases completely. That is:

when $a_n = a_{\text{OL}} \to K_{\text{red}} = (K_{\text{OL}})_{\text{max}} - K_{\text{max}, n} = K_{\text{max}, n} \to (K_n)_{\text{eff}} = K_n - K_{\text{red}} = 0$

To account for an overload greater than twice the previous load, the Willenborg model was revised by Gallagher and Hughes [38] and is known as the "Generalized Willenborg Model." Gallagher and Hughes introduced a parameter ϕ into Eq. (4.30), such that:

$$K_{\text{red}} = \phi\{(K_{\text{OL}})_{\text{max}} [1 - (a_n - a_{\text{OL}})/r_{p\text{OL}}]^{1/2} - K_{\text{max}, n}\} \quad (4.37)$$

where:

$$\phi = \frac{1 - \dfrac{K_{\text{max, th}}}{K_{\text{max}, n}}}{(S_{\text{OL}})_{\text{so}} - 1} \quad (4.38)$$

The quantity $K_{\text{max, th}}$ is the threshold stress intensity factor associated with zero crack growth and $(S_{\text{OL}})_{\text{so}}$ is the shutoff overload ratio that can cause complete retardation when the crack ceases to grow ($da/dN = 0$):

$$(S_{\text{OL}})_{\text{so}} = \text{Overload/Underload} = (K_{\text{OL}})_{\text{max}}/K_{\text{max}, n} \quad (4.39)$$

Example 4.5

An aircraft component is made of 7075-T651, L-T, 75F aluminum alloy (constant amplitude crack growth data as shown in Fig. 4.26 [1]) and is subjected to a variable load environment. The fluctuating load is repeated every three cycles and has maximum peak stresses of 18, 14, and 10 ksi with $R = 0$, as illustrated in Fig. 4.27. Visual inspection shows that the maximum detected flaw size is a center crack with length $2a = 1.0$ in. The thickness and width of the part are 0.25 and 10.0 in., respectively. Use the Forman equation to calculate the amount of crack growth for the following cases: (1) no retardation effect; (2) apply the retardation effect using the Wheeler model (assume $m = 1.4$); (3) use the Willenborg model, and generalized Willenborg models. Conduct the above mentioned analysis for

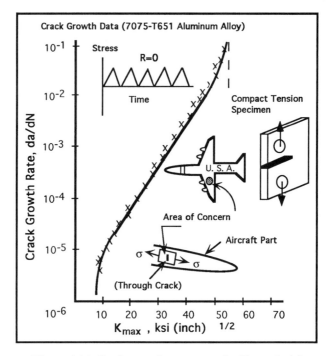

Figure 4.26 Crack growth rate curve for Example 4.5

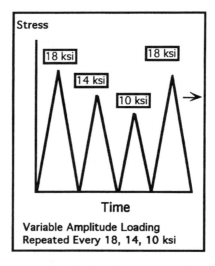

Figure 4.27 Load spectrum for the aircraft component described in Example 4.5

4.4 Variable Amplitude Stress and the Retardation Phenomenon

the first six cycles (cycles associated with 18, 14, 10, and 18, 14, 10 ksi). The tensile yield of the material is given as 65 ksi [1].

Solution

The equation for the stress intensity factor of a through center crack was formulated in Section 3.2.5 of Chapter 3 [Eq. (3.19)]:

$$K_{max} = \beta\sigma(\pi a)^{1/2}$$

For the first cycle, with $\sigma = 18$ ksi:

$$K_{max} = 18 \times (\pi \times 0.5)^{1/2} = 22.55 \text{ ksi (in.}^{1/2}) \quad (\beta = 1)$$

The Forman crack growth equation for the case of $R = 0$ can be written as:

$$da/dN = \frac{c(K_{max})^n}{K_c - K_{max}}$$

The constants c and n can be obtained by selecting two points in the linear region of the crack growth data (Fig. 4.26), as demonstrated in the example problem of Section 4.2.

$$c = 1.13\text{E-}8, \text{ and } n = 4.0$$

The Forman equation, in terms of constants c and n, becomes:

$$da/dN = \frac{1.13\text{E-}8(K_{max})^{4.0}}{K_c - K_{max}}$$

1. No retardation effect

The amount of growth for the first cycle ($\sigma = 18$ ksi) with $a_o = 0.5$ (growth at one tip) is:

$$(\Delta a)_1 = da = 1.13\text{E-}8 \, (22.55)^{4.0}/(55 - 22.55) = 0.0000900 \text{ in.}$$

For the second cycle:

$$K_{max} = 14 \times (\pi \times 0.50009)^{1/2} = 17.54 \text{ ksi (in.}^{1/2}) \quad \text{for } \sigma = 14 \text{ ksi}$$
$$(\Delta a)_2 = da = 1.13\text{E-}8 \, (17.54)^{4.0}/(55 - 17.54) = 0.0000285 \text{ in.}$$

For the third cycle:

$K_{max} = 10 \times (\pi \times 0.500118)^{1/2} = 12.53$ ksi (in.$^{1/2}$)   for $\sigma = 10$ ksi

$(\Delta a)_3 = da = 1.13\text{E}-8 \, (12.53)^{4.0}/(55 - 12.53) = 0.0000065$ in.

For the fourth cycle:

$K_{max} = 18 \times (\pi \times 0.5001245)^{1/2} = 22.5567$ ksi (in.$^{1/2}$)   for $\sigma = 18$ ksi

$(\Delta a)_4 = da = 1.13\text{E}-8 \, (22.5567)^{4.0}/(55 - 22.5567) = 0.0000902$ in.

For the fifth cycle:

$K_{max} = 14 \times (\pi \times 0.500215)^{1/2} = 17.5472$ ksi (in.$^{1/2}$)   for $\sigma = 14$ ksi

$(\Delta a)_5 = da = 1.13\text{E}-8 \, (17.5472)^{4.0}/(55 - 17.5472) = 0.0000286$ in.

For the sixth cycle:

$K_{max} = 10 \times (\pi \times 0.500244)^{1/2} = 12.533$ ksi (in.$^{1/2}$)   for $\sigma = 10$ ksi

$(\Delta a)_6 = da = 1.13\text{E}-8 \, (12.533)^{4.0}/(55 - 12.533) = 0.00000656$ in.

No retardation effect

Original crack length (in.)	Final crack length (in.)
0.5000000	0.5000901
0.5000901	0.5001180
0.5001180	0.5001245
0.5001245	0.5002150
0.5002150	0.5002440
0.5002440	0.5002506

2. Wheeler retardation model

To apply the Wheeler model, the quantities associated with the retardation factor, C_p, should be calculated. In this example problem, the first cycle with 18 ksi is considered as the overload and will retard the growth of the subsequent cycles:

$$C_p = \left(\frac{r_{pn}}{a_{OL} + (r_p)_{OL} - a_n} \right)^m$$

where

$$(r_p)_{OL} = (1/2\pi)[(K)_{OL}/\sigma_{Yield}]^2$$
$$\text{and } r_{pn} = (1/2\pi)[(K_{max})_n/\sigma_{Yield}]^2$$

4.4 Variable Amplitude Stress and the Retardation Phenomenon

$$(r_p)_{OL} = (1/2\pi)(22.55/65)^2 = 0.0191 \text{ in.,}$$
$$\text{and } r_{pn} = (1/2\pi)[17.54/65]^2 = 0.01159 \text{ in.}$$

The amount of growth associated with the first cycle is the same as in the previous case (no interaction effect) where $C_p = 1$:

$$K_{max} = 18 \times (\pi \times 0.5)^{1/2} = 22.55 \text{ ksi (in.}^{1/2}) \quad \text{for } \sigma = 18 \text{ ksi}$$
$$(\Delta a)_1 = da = 1.13\text{E}-8 \ (22.55)^{4.0}/(55 - 22.55) = 0.0000900 \text{ in.}$$

To obtain $(\Delta a)_2$ and $(\Delta a)_3$, the quantity C_p for each case must be calculated. The values of a_{OL} and a_1 are 0.5 and 0.50009 in., respectively. The value of C_p for $(\Delta a)_2$ is:

$$C_p = [0.01159/(0.5 + 0.0191 - 0.50009)]^{1.4} = 0.5$$
$$(\Delta a)_2 = C_p f(\Delta K) = 0.5 \times 1.13\text{E}-8 \ (17.54)^{4.0}/(55 - 17.54)$$
$$= 0.00001425 \text{ in.}$$

and for the case of $(\Delta a)_3$:

$$K_{max} = 10 \times (\pi \times 0.5001043)^{1/2} = 12.5312 \text{ ksi (in.}^{1/2})$$
$$(r_p)_{OL} = (1/2\pi)(22.55/65)^2 = 0.0191 \text{ in., and } r_{pn} = (1/2\pi)[12.53/65]^2$$
$$= 0.005917 \text{ in.}$$
$$C_p = [0.005917/(0.5 + 0.0191 - 0.5001043)]^{1.4} = 0.195$$
$$(\Delta a)_3 = 0.195 \times 1.13\text{E}-8 \ (12.5312)^{4.0}/(55 - 12.5312)$$
$$= 0.000001268 \text{ in.}$$

and the amount of growth associated with the fourth cycle (no retardation) is:

$$(\Delta a)_4 = da = 1.13\text{E}-8 \ (22.5567)^{4.0}/(55 - 22.5567) = 0.0000902 \text{ in.}$$

where

$$K_{max} = 18 \times (\pi \times 0.50010557)^{1/2} = 22.5563 \text{ ksi (in.}^{1/2})$$

Accordingly, the values of $(\Delta a)_5$ and $(\Delta a)_6$ can be calculated as shown in the previous cases of $(\Delta a)_2$ and $(\Delta a)_3$ by considering the effect of retardation as a result of 18 ksi overload:

$$(r_p)_{OL} = (1/2\pi)(22.5563/65)^2 = 0.01917,$$

$$\text{and } r_{pn} = (1/2\pi)[17.545/65]^2 = 0.01160 \text{ in.}$$

where

$$K_{max} = 14 \times (\pi \times 0.50019577)^{1/2} = 17.545 \text{ ksi (in.}^{1/2})$$
$$C_p = [0.01160/(0.5 + 0.01917 - 0.50019577)]^{1.4} = 0.611$$
$$(\Delta a)_5 = 0.611 \times 1.13\text{E}-8 \, (17.545)^{4.0}/(55 - 17.54) = 0.00001746$$

and in the same way, the quantity $(\Delta a)_6$ can be calculated:

$$K_{max} = 10 \times (\pi \times 0.500213)^{1/2} = 12.5326 \text{ ksi (in.}^{1/2})$$
$$(r_p)_{OL} = (1/2\pi)(22.5563/65)^2 = 0.01917 \text{ in,}$$
$$\text{and } r_{pn} = (1/2\pi)[12.5326/65]^2 = 0.0059197 \text{ in.}$$
$$C_p = [0.005919/(0.50010557 + 0.01917 - 0.500213)]^{1.4} = 0.1945$$
$$(\Delta a)_6 = 0.1945 \times 1.13\text{E}-8 \, (12.5326)^{4.0}/(55 - 12.5326) = 0.000001276 \text{ in.}$$

Wheeler retardation effect

Original crack length (in.)	Final crack length (in.)
0.50000000	0.50009010
0.50009010	0.50010430
0.50010430	0.50010557
0.50010557	0.50019577
0.50019577	0.50021300
0.50021300	0.50021430

3. Willenborg retardation model

The Willenborg correction to the crack growth is handled through the use of an effective stress intensity factor and the effective stress ratio. From Eq. (4.29):

$$(K_{max})_{eff} = K_{max} - K_{red}, \quad (K_{min})_{eff} = K_{min} - K_{red} \quad \text{(where } K_{min} = 0\text{)}$$

and

$$K_{red} = (K_{OL})_{max}[1 - (a_i - a_{OL})/r_{pOL}]^{1/2} - K_{max,i}$$

From the above equations, it can be deduced that:

$$(K_{max})_{eff} - (K_{min})_{eff} = K_{max}$$

which states that for the case of $R = 0$, both the Willenborg and the "generalized" Willenborg model cannot be effective to correct the retardation effect. For any other stress ratio larger than zero, the Willenborg retardation models can be used to address the retardation effect. Therefore, case 3 is identical to case 1 of this example problem. The following summary table summarizes the calculated crack growth for the first six cycles associated with 18, 14, 10, and 18, 14, 10 ksi).

Summary table for example 4.5

No retardation effect		Wheeler retardation effect	
Original crack length (in.)	Final crack length (in.)	Original crack length (in.)	Final crack length (in.)
0.5000000	0.5000901	0.50000000	0.50009010
0.5000901	0.5001180	0.50009010	0.50010430
0.5001180	0.5001245	0.50010430	0.50010557
0.5001245	0.5002150	0.50010557	0.50019577
0.5002150	0.5002440	0.50019577	0.50021300
0.5002440	0.5002506	0.50021300	0.50021430

4.5 Cycle by Cycle Fatigue Crack Growth Analysis

To prevent catastrophic failure due to the presence of undetected flaws that would result in the loss of life and the structure, it is important for the engineer to design components that provide structural integrity, even in the presence of undetected flaws. Therefore, it is essential to know whether the preexisting flaw will grow to a critical length during its service life.

Owing to the complexities in the crack growth analysis that emerge as a result of a large number of cycles (containing constant and variable amplitude load) together with complicated load and crack geometries, automated flaw growth programs are usually employed. Ideally, a flaw growth computer program should have libraries for (1) standard NDE flaw sizes, (2) stress intensity factor solutions for different crack geometries and loading conditions, (3) empirical fatigue crack growth equations, and (4) material properties. In addition, it should be efficient and easy to run ("user friendly").

Most crack growth computer programs that are tailored for the aircraft industry have the capability to perform variable amplitude fatigue life evaluation by employing one or both of the retardation models discussed in Section 4.4. It should be noted that, based on NASA requirements stated in NHB 8071-1, "Fracture Control Requirements for Payloads Using the National Space Transportation System (NSTS)," the retardation effect is not allowed to be considered in evaluating the life of fracture critical or high-risk components that will be used as payloads in the NSTS transporter. The cycle by cycle crack growth procedure generally consists of the following steps:

Step 1. A standard NDE initial crack size, a_1, is assumed. As an example, for standard penetrant inspection, the initial crack size is a part through crack with depth $a = 0.075$ and $2c = 0.15$ in., see Table 5.5 of Chapter 5.

Step 2. The stress intensity range for the first cycle can be determined, $\Delta K_1 = \beta \Delta \sigma_1 (\pi a)^{1/2}$, where $\Delta \sigma_1$ is the stress range in the first cycle and β is the appropriate correction factor for crack geometry and loading conditions. Note that the crack growth analysis performed for a part through crack should consider the growth for both the depth (a-tip) and the length direction (c-tip).

Step 3. Determine the amount of growth for the first cycle, Δa_1, by using the appropriate crack growth equation that considers the effect of the stress ratio, R [(for example, the Forman relationship described by Eq. (4.4)] or directly from the da/dN versus ΔK curve. The amount of growth for the first cycle, Δa_1, is:

$$\Delta a_1 = (da/dN)_1 \times 1$$

The new crack length will be $a_2 = a_1 + \Delta a_1$. If the crack geometry in consideration is a part through crack, it is preferable to have the da/dN versus ΔK test data for the crack geometry under study.

Step 4. The size of the plastic zone, r_p, in the first cycle is calculated by Eq. (3.33) described in Chapter 3. If the first cycle is considered as the overload cycle then the extent of the plastic zone is:

$$Z_1 = ((r_p)_{OL})_1 + a_1 \qquad \text{where } a_1 = a_{OL}$$

where for plane stress condition:

$$((r_p)_{OL})_1 = 1/2\pi (K_{max1}/\sigma_{Yield})^2$$

and for plane strain condition:

$$((r_p)_{OL})_1 = 1/6\pi (K_{max1}/\sigma_{Yield})^2$$

Step 5. The new crack length is now a_2 and the stress intensity range for the second cycle $\Delta K_2 = \beta \Delta \sigma_2 (\pi a)^{1/2}$. If the new cycle has a load magnitude larger than the previous cycle, the new overload plastic zone must be calculated. On the other hand, if the subsequent cycle has a smaller load magnitude, then the effect of retardation has to be considered. The extent of the yield zone for the new cycle is:

$$Z = a_2 + (r_p)_2$$

Step 6. If the extent of the plastic zone $Z < Z_1$, the correction to retardation effect according to the Willenborg [Eqs. (4.33) and (4.35)] or Wheeler model [cal-

4.5 Cycle by Cycle Fatigue Crack Growth Analysis

culate the quantity C_p from Eq. (4.27)] is needed. For the Willenborg model the new crack length is:

$$a_3 = a_2 + \Delta a_2$$

where

$$\Delta a_2 = (da/dN)_2 \times 1$$

The quantity $(da/dN)_2$ is obtained by using the ΔK_{eff} and R_{eff} in the Forman crack growth equation.

When using the Wheeler model, the stress intensity factor, $(\Delta K)_2$, is calculated and the retardation factor, C_p, is incorporated to calculate the new crack length:

$$a_3 = a_2 + \Delta a_2 + C_p (da/dN)_2 \times 1$$

Step 7. If the extent of the plastic zone $Z > Z_1$ (no retardation effect), determine the quantity $\Delta a_2 = (da/dN)_2 \times 1$ and the new crack size, a_3, as:

$$a_3 = a_2 + (da/dN)_2 \times 1$$

For the subsequent cycles steps 6 and 7 can be repeated in such away that the nth cycle with new crack length a_n replaces a_2 and the subsequent cycle a_{n+1} replaces a_3.

Table 4.2 shows the current state of the art computer codes used in various aerospace programs.

Table 4.2 Fatigue crack growth computer codes used in several aerospace industry

Programs	Computer code
C17 Program (Douglas Aircraft)	Air Force Cracks Program
High-Speed Transporter (Douglas Aircraft)	NASGROW
International Space Station	NASA/FLAGRO
(McDonnell Douglas Aerospace)	NASA/FLAGRO
MSC/PDA Supported Programs	PFATIGUE/nCode
F-18 Aircraft (Northrop)	NORCRAK
Douglas Aircraft (MD-95 Project)	CRACK95(DAC)
NASA'S Crack Growth Analysis (Using Crack Closure Concept)	FASTRAN

4.6 Structural Integrity Analysis of Bolted Joints Under Cyclic Loading

There are several types of joints that are used to join structural parts. The most common types have bolted, welded, or pinned connections. A joint can be viewed as a source of stress concentration and preventative measures must be taken to minimize structural failure. In this section, bolted joints (also called mechanically fastened joints) are discussed briefly and the emphasis is given to the integrity of the bolts and the pads (pads are referred to as abutments, the two pieces being fastened together) when they are subjected to a fluctuating load environment. Many structures, such as buildings, bridges, space vehicle, and airplanes, rely heavily upon bolted joints for their structural integrity. The factors that are important to consider in designing and analyzing a bolted joint are the bolt material, pad material, bolt pattern, and the degree of preload.

4.6.1 Preloaded Bolt Subjected to Cyclic Loading

Fatigue failure of bolts usually occurs at the threaded location where bolt and nut are engaged [39, 40]; however, in some cases (less probable) bolt failure is observed in the shank to bolt head area, as shown in Fig. 4.28(a). Experimental data on threaded bolts subjected to rotating–bending tests suggest that a larger root radius reduces the stress concentration at the root and increases the fatigue life of the bolt [41]. A large root radius allows smoother flow of stress in that

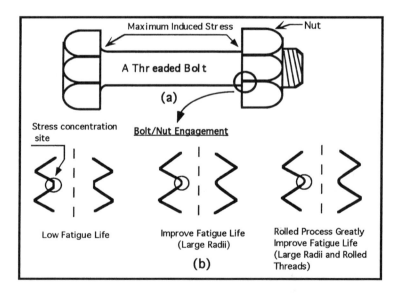

Figure 4.28 Probabale locations where a threaded bolt fails

4.6 Structural Integrity Analysis of Bolted Joints Under Cyclic Loading

region, thus reducing the degree of stress concentration, K_t. Other experimental studies have show that thread processing (by rolling the threads) improves the fatigue life and this method is preferred over machining, Fig. 4.28(b) [42, 43]. Rolling threads will introduce compressive residual stresses at thread roots which help to prevent fatigue failure of bolts in a joint (thread rolling introduces a larger root radius and smoother stress flow). Threaded bolts used in fastening two fracture critical pieces of hardware which is considered as a high risk joint are required to undergo a rolling process in order to avoid fatigue failure at the roots during their service life, see figure 4.28(b). In addition to the rolling process for reducing the stress concentration in the threaded region, it is also a good practice to distribute the load among threads for avoiding crack initiation and improving the fatigue life of the bolted joint. The designer should be aware that the first few threads which are engaged carry the major part of the load.

A tensile preloaded bolt is also considered desirable in reducing the fatigue failure and increasing the life of a bolt (by decreasing the load amplitude at the expense of increasing the mean stress) when it is exposed to cyclic loading as illustrated in Fig. 4.29. Preloading a bolt in a joint is the result of an applied torque for the purpose of tightening the joint and to prevent any gapping between the two fastened plates. Consider a bolt in a bolted joint with maximum fluctuating stress of $0.3\sigma_{UL}$ with a stress ratio $R = 0$. The calculated life of this bolt is higher in the presence of preload. Higgins [42] in his experimental work with high strength steel bolts concluded that the preloading can increase the fatigue life considerably than otherwise possible with similar joints. The following bolt analysis introduces some of the factors that are involved when assessing the fatigue life of a bolt.

When a bolt is subjected to an alternating tensile applied load, $(P_t)_{\text{applied}}$, and a preload, P_{preload} (see Fig. 4.30), the total bolt load, P_b, can be obtained [44] through the following relationship:

$$P_b = P_{\text{preload}} + [K(P_t)_{\text{applied}}] \times \text{Factor of Safety} \qquad (4.40)$$

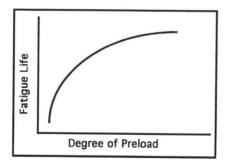

Figure 4.29 The effect of preload on fatigue life of the part

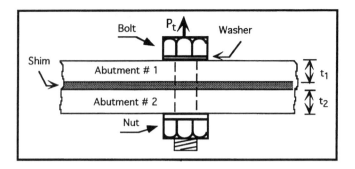

Figure 4.30 A bolted joint with applied tensile load

where

$$P_{preload} = \frac{T}{D\mu} \quad (4.40a)$$

T is the mean wrenching applied torque and D is the nominal bolt diameter. The quantity μ is the friction coefficient (at the nut to bolt assembly) and it's value is dependent on the degree of lubricant at the nut to bolt assembly. The joint stiffness factor, K:

$$K = C/[L(1 + A_m E_m / A_b E_b)] \quad (4.41)$$

where:

$$L = t_1 + t_2 + t_{shim}, \quad C = (t_1 + t_2)/2 + t_{shim}$$

and A_m and A_b are the compressed joint area and the bolt shank area, respectively. E_b is the modulus of elasticity in tension of the bolt and E_m is the modulus of elasticity in compression of the abutment material. The equations describing $A_m, A_b,$ and E_m are:

$$A_m = \frac{\pi}{4}(B^2 - D_{hole}^2)$$

where

$$B = D_{head} + L(\tan 30)/2$$

4.6 Structural Integrity Analysis of Bolted Joints Under Cyclic Loading

and D_{head} is approximately $1.6 D_{\text{bolt}}$. Furthermore,

$$A_b = \frac{\pi}{4} D_{\text{bolt}}^2$$

and

$$E_m = (t_1 + t_2 + t_{\text{shim}})/[t_1/E_1 + t_2/E_2 + t_{\text{shim}}/E_{\text{shim}}]$$

Let us for simplicity assume that the bolt shown in Fig. 4.30 is subjected to a fluctuating stress $(\sigma_t)_{\text{applied}} = 0.3\sigma_{\text{UL}}$ with stress ratio of $R = 0$ and $\sigma_{\text{preload}} = 0.5\sigma_{\text{UL}}$. Other pertinent information related to the bolt, plate geometry, and material properties needed to calculate the joint stiffness factor, K, are: $D_{\text{bolt}} = 0.25$ in., $t_{\text{shim}} = 0$, $t_1 = t_2 = 0.319$ in., $E_1 = E_2 = 10.8E + 6$ psi, $E_b = 29.1E + 6$ psi, $\sigma_{\text{UL}} = 1.1E + 5$ psi, and $D_{\text{head}} = 1.6D_{\text{hole}}$, with factor of safety = 1.4.

$L = t_1 + t_2 + t_{\text{shim}}$	0.638 in.
$C = t_1/2 + t_2/2 + t_{\text{shim}}$	0.319 in.
$t_{\text{stack}} = L$	0.638 in.
$B = D_{\text{head}} + 0.5 t_{\text{stack}} (\tan 30)$	0.542 in.
$A_m = \pi/4 (B^2 - D_{\text{hole}}^2) =$	0.217 in^2.
$A_b = \pi/4 (D_{\text{bolt}})^2$	0.049 in^2.
$K = C/[L(1 + A_m E_m/A_b E_b)]$	0.189

Equation (4.40) for maximum alternating bolt stress becomes:

$$\sigma_b = 0.5\sigma_{\text{UL}} + 0.189 \times 1.4 \times 0.3\sigma_{\text{UL}} \tag{4.42a}$$

In terms of ultimate stress, σ_{UL}:

$$\sigma_b = 0.5\sigma_{\text{UL}} + 0.0793\sigma_{\text{UL}} \tag{4.42b}$$

The second quantity in Eq. (4.42) is fluctuating between zero and $0.0793\sigma_{\text{UL}}$. Figure 4.31 illustrates the cyclic loading that the bolt will experience when it is subjected to mean stress slightly higher than the preload induced stress, $\sigma_M = 59.34$ ksi > preload = 55 ksi. It can be concluded that, in the presence of preload, the mean stress on the bolt increases, but the amplitude of the fluctuating stress, $\sigma_a = 4.34$ ksi < $(0.3/2)\sigma_{\text{UL}} = 16.5$ ksi, decreases, which results in increasing the fatigue lifetime of the bolt.

Example 4.6 illustrates the effect of preload on service life of a bolt. Fatigue crack growth analysis of pads in a joint is discussed in Section 4.6.2. Example 4.7 shows the analysis approach to evaluating the integrity of pads in a bolted joint by

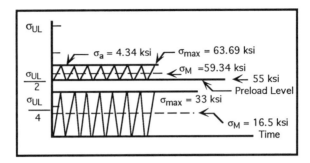

Figure 4.31 Illustrating the effect of preload on a bolt subjected to fluctuating load

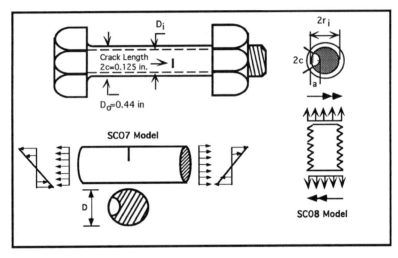

Figure 4.32 A surface crack shown on the shank area of a bolt described in Example 4.6

assuming the existence of a corner crack emanating from the holes. The final results, in terms of the NASA/FLAGRO computer code output, are included.

Example 4.6

A fracture critical bolt (shown in Fig. 4.32) is subjected to a fluctuating load of $(P_t)_{applied} = 5000$ lb with stress ratio $R = 0$. The bolt is made of A286 high strength steel with fracture properties shown in Table 4.3. The threads are rolled

Table 4.3 Material properties from NASA/FLAGRO, A286, 200 ksi, (forged rod, L–R)

σ_{Yield}	σ_{Ul}	K_{Ic}	K_{Ie}	AK	BK	C	n	p	q	ΔK_0	R_{cl}	α	SR
190	200	100	140	1.0	0.5	0.3E-8	2.1	0.25	.25	3.5	0.2	3.0	0.3

4.6 Structural Integrity Analysis of Bolted Joints Under Cyclic Loading

to minimize the stress concentration[1]. The standard eddy current inspection performed on the bolt indicated that the initial flaw size in the shank area is a part through crack having half crack length $c = 0.075$ in. and crack depth $a = 0.064$ in.; see Fig. 4.32. Evaluate the life of the bolt for the preload of $0.5\sigma_{UL}$ where the joint stiffness factor, $K = 0.25$. Compare the results of the analysis with no preload case. The factor of safety for the load environment is 1.4.

Solution

From Eq. (4.40), the bolt load can be written as:

$$P_b = P_{preload} + [K(P_t)_{applied}] \times 1.4$$

where $K(P_t)_{applied} = 1250$ lb and $\sigma_{preload} = 0.5\sigma_{UL}$. The load variations that the bolt is subjected to for the two above cases (with and without preload) are shown in Fig. 4.33. The crack geometry used for crack growth analysis is a thumbnail crack as shown in Fig. 4.32 (designated by SC07 in the NASA/FLAGRO computer code, see also the note provided at the end of this example problem) [1].

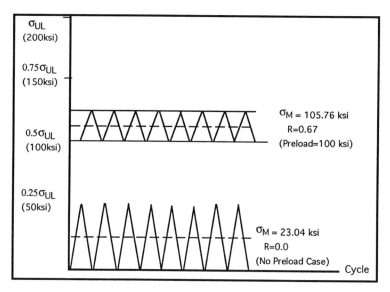

Figure 4.33 The Cyclic load environment for two cases: (1) preload and (2) no preload

---

[1] The rolling process will introduce compressive residual stresses at the threaded roots. For this reason, the crack growth analysis in this example problem is performed in the shank area where the probability of failure is higher (see the note provided at the end of this example problem).

The crack depth length, a, in terms of crack length $2c$, was expressed in terms of the diameter $D = 2r = 0.44$ in. [1] as:

$$a = \frac{D}{2}[\tan(2c/D) + 1 - (\tan^2(2c/D) + 1)^{1/2}]$$

The fracture mechanics analysis for both the preloaded and no preload (where the joint stiffness factor $K = 1$) cases are conducted and the final results in terms of NASA/FLAGRO output are shown here. The final results clearly indicate that longer life is expected in the presence of the preload.

FATIGUE CRACK GROWTH ANALYSIS
(computed: NASA/FLAGRO Version 2.03, January 1995)

PROBLEM TITLE
BOLT ANALYSIS WITHOUT PRELOAD

GEOMETRY
MODEL: SC07-Part-circular Surf. crk on cylinder circ. plane
Cylinder Diameter, D = 0.4400

FLAW SIZE
a (init.) = 0.6461E-01
c (init.) = 0.75E-01

MATERIAL
MATL 1; A286 (200ksi Bolt material)
Forg. rod, L-R
Material Properties;

	UTS	YS	K1e	K1c
	200.0	190.0	140.0	100.0

Material Crack Growth Eqn Constants

C	n	p	q	DKo	Rcl	Alpha	Smax/SIGo
0.3D-08:	2.1	0.25	0.25	3.50	0.20	3.00	0.30

FINAL RESULTS
Net-section stress exceeds the flow stress.
(Flow stress = average of yield and ultimate)
At cycle no. 91536.03
Crack size a = 0.203986
Corresponding semicrack length, c = 0.750000E-01

Note: The final results (net-section stress exceeds the flow stress) do not represent the final failure where the total separation occur (failure of the bolt in two parts).

4.6 Structural Integrity Analysis of Bolted Joints Under Cyclic Loading

When the total length of the advancing crack is in such a way that the net section $(W - a)$ is yielding, the linear elastic fracture mechanics cannot further be used to describe the crack tip stress field and the corresponding crack growth rate.

Based on the above result, the number of cycles to failure for the bolt when it is not preloaded is shown to be 91,536. However, the life of the bolt is expected to increase by applying $\sigma_{\text{preload}} = 0.5\sigma_{\text{UL}}$ to the bolt. This is true because upon the application of preload, some of the applied load is absorbed by the jointed plates in compression and only a portion of it will be affected by the bolt, $K(P_t)_{\text{applied}}$, where $K = 0.25$. The following is the final result of the analysis conducted by NASA/FLAGRO where the magnitude of applied fluctuating stress is reduced by the joint stiffness factor $K = 0.25$ ($\sigma_{\text{max}} = 111.5$, and $\sigma_{\text{min}} = 100$ ksi).

FINAL RESULTS (WITH PRELOAD):
Net-section stress exceeds the flow stress.
(Flow stress = average of yield and ultimate)
At cycle no. 832977.38:
Crack size $a = 0.132176$
Corresponding semicrack length, $c = 0.750000\text{E-}01$

The fracture analysis of the bolt with preload clearly indicates that the number of cycles to failure has increased by nine times as compared to the previous analysis where preload was not incorporated.

Final Note: If a thorough inspection is possible to ensure that the rolling threads are properly performed and fully effective, then the area of concern would be in the shank and bolt head-to-shank region. However, because 100% inspection at the roots are not possible, it is recommended that bolt fracture analysis be performed in the threaded region only. Note that in analyzing the threaded region (using SC08 crack model in the NASA/FLAGRO computer code or any other available code, see Fig. 4.32), the far field stress will be based on the minor diameter which will give higher preload stress compared with the shank region.

4.6.2 Fatigue Crack Growth Analysis of Pads in a Bolted Joint

Generally, in joints that contain more than one bolt, the failure of one bolt due to cyclic loading will not necessarily cause the total destruction of the joint. If by analysis it can be shown that the remaining bolts can withstand the redistributed load during its service life, the system has a fail-safe design. Even though the bolts in a given joint might be classified as having a fail-safe design and the failure might be prevented by redistribution of load to the remaining bolts, the fatigue cracking may initiate at other locations in the joint and cause a detrimental effect on the structural parts fastened together. For example, a crack may initiate from the hole in the fastened plates (as illustrated in Fig. 4.34) where the two plates are

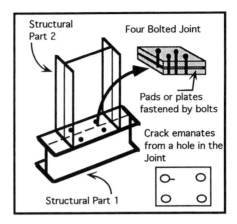

Figure 4.34 A four bolt joint where the two plates are fastened together

bolted together by a four-bolted joint pattern. The initiated crack from one of the holes in the joint may grow to its critical size due to the fluctuating load environment, and cause the complete separation of the two structural parts.

When a bolted joint such as the one shown in Fig. 4.34 is subjected to a fluctuating load environment, the tension and shear forces applied to the bolts will induce localized bending and bearing stresses on the plate. The calculated induced stresses are useful information for determining whether the joint can maintain its integrity during its usage period. Figure 4.35 illustrates the possibility of a crack emanating from a hole in a joint as a result of localized cyclic bearing and bending stresses. The following analysis briefly describes the methodology used in evaluating the structural integrity of a bolted joint when it is subjected to a resul-

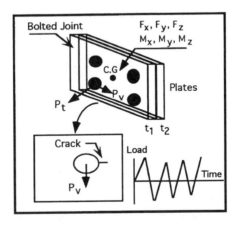

Figure 4.35 The axial and shear induced loads at a bolt

4.6 Structural Integrity Analysis of Bolted Joints Under Cyclic Loading

tant force and moments (F_x, F_y, F_z, M_x, M_y, M_z), applied at the centroid of the bolt pattern, shown in Fig. 4.35. Using the bolt load analysis approach, described by Bruhn in Reference [45], the maximum resultant bolt tension load, P_t, and shear, V, can be calculated. The bearing stress that is effective for growing the crack (see Fig. 4.35) is given by:

$$\sigma_{br} = V/Dt \qquad (4.43)$$

where D and t are the diameter of the hole and plate thickness, respectively. Moreover, the bending stress induced in the pad, due to the axial load on the bolt, P_t, also contributes to growing the crack from a hole and can be calculated as a bath tub channel fitting (case a), angle fitting (case b), or from the plate bending approach (case c), depending on the geometry of the joint [46] (see Fig. 4.36).

For the simple case of plate bending (case c), the induced plate bending stress, $\sigma_{bending}$, to be employed in the NASA/FLAGRO computer code can be estimated for the region designated as $ABCD$ by the following relationship (see Fig. 4.37):

$$\sigma_{bending} = 6M/bt^2 \qquad (4.44)$$

Example 4.7

A bolted joint with a four-bolt pattern (situated in the flange of an I-beam), shown in Fig. 4.38, is subjected to forces and moments at its centroid location. The bolt

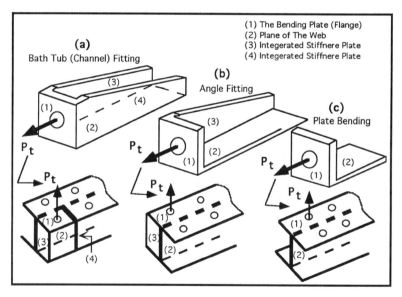

Figure 4.36 The three types of fitting design used in designing a joint: (a) channel, (b) angle and (c) plate bending design [46]

240 Chap. 4 Fatigue Crack Growth

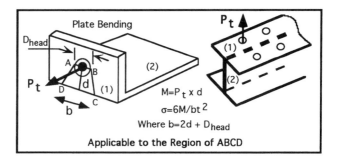

Figure 4.37 Evaluating bending stress in a plate subjected to tensile bolt load

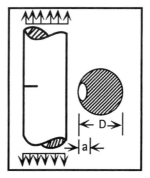

Figure 4.38 Surface crack in a preloaded bolt subjected to cyclic loading (part 1 of Example 4.7)

Table 4.4 Material properties from NASA/FLAGRO, 2219–T851 L–T

σ_{Yield}	σ_{Ul}	K_{Ic}	K_{Ie}	AK	BK	C	n	p	q	ΔK_0	R_{cl}	α	SR
53	65	100	46	1.0	1.0	0.12E-7	3.2	0.50	1.0	2.5	0.7	1.5	0.3

is made of A286 high strength steel (Fig. 4.38) with fracture properties for bolts and plated shown in Tables 4.3 and 4.4 respectively. The analysis indicated that the maximum fluctuating bolt tension load $P_t = 2500$lb with resultant shear load $V = 2000$ lb ($R = 0$). To improve fatigue life and avoid gapping, the bolts are required to be preloaded as high as 0.35 σ_{Ul}. The load spectrum for one service life is given as:

4.6 Structural Integrity Analysis of Bolted Joints Under Cyclic Loading

Load Spectrum Table for Example 4.7

Steps	Cycles	Cyclic stress (%Limit)	
		Minimum	Maximum
1	100	0.0	100
2	300	0.0	90
3	400	0.0	80
4	700	0.0	70
5	1000	0.0	60
6	2000	0.0	50
7	2800	0.0	40
8	5500	0.0	30
9	9000	0.0	20
10	15019	0.0	10
11	28853	0.0	7

(1) Determine if the bolts can survive four service lives by assuming the existence of a part through crack in the shank location with $a = 0.04$ and $2c = 0.1$ in. and (2) Assume a corner crack of length $c = 0.05$ and depth $a = 0.05$ in. is emanating from the hole in the joint (see Fig. 4.39). Determine if the joint can survive the above environment (minimum of four service lives) with far field tension

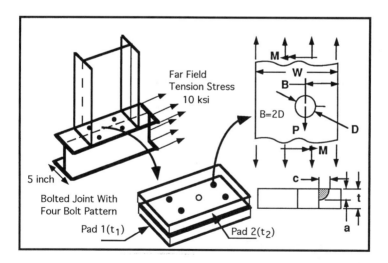

Figure 4.39 A bolted joint and crack geometry for a crack emanating from a hole (Part 2 of Example 4.7)

stress $\sigma_{\text{Tension}} = 10$ ksi ($R = -1$), as shown in Fig. 4.39. The joint stacking thickness and other related information are: $t_1 = t_2 = 0.4$ in., $t_{\text{shim}} = 0$, $D_{\text{bolt}} = 0.5$ in., $D_{\text{head}} = 0.8$ in. and $E_1 = E_2 = 10.8\text{E} + 6$ psi, $E_b = 29.1\text{E} + 6$ psi, $\sigma_{\text{UL}} = 1.99\text{E} + 5$ psi, $d = 2D$.

Solution

The stiffness factor, K, for this joint can be calculated by employing Eq. (4.38):

$L = t_1 + t_2 + t_{\text{shim}}$	0.800 in
$C = t_1/2 + t_2/2 + t_{\text{shim}}$	0.400 in
$t_{\text{stack}} = L$	0.800 in
$B = D_{\text{head}} + 0.5\, t_{\text{stack}}\, (\tan 30)$	1.014 in
$A_m = \pi/4\, (B^2 - D_{\text{hole}}^2)$	0.611 in^2
$A_b = \pi/4\, (D_{\text{bolt}})^2$	0.196 in^2
$K = C/[L(1 + A_m E_m/A_b E_b)]$	0.232

The alternating tensile stress in the bolt can be calculated with the presence of induced preload stress of $\sigma_{\text{preload}} = 0.3 \times 200 = 60$ ksi. From Eq. (4.40), the total bolt stress can be written as:

$$\sigma_{\text{bolt}} = \sigma_{\text{preload}} + K\sigma_{\text{applied}}$$

where the maximum and minimum alternating bolt stresses (based on $P_t = 2500$ lb) are 62.96 and 60 ksi, respectively. The final results pertaining to fracture analysis of bolts subjected to the fluctuating load environment (described in the load spectrum table for Example 4.7) are shown for four service lives as:

FINAL RESULTS:
All stress intensities are below the fatigue threshold.
NO growth in four service lives
Crack size $a = 0.455928\text{E}-01$
Corresponding semicrack length, $c = 0.500000\text{E}-01$

The final result on the fracture analysis of bolts indicates that all the calculated stress intensity factors are below the threshold stress intensity factor, ΔK_{th} (described in Section 4.3.1.1) and thus no growth is expected when the bolt is subjected to the fluctuating load environment shown below. The induced tensile stress in the bolt due to the 2500 lb fluctuating load is not effective in driving the crack above the threshold value.

4.6 Structural Integrity Analysis of Bolted Joints Under Cyclic Loading

Load Spectrum

Steps	Cycles	% Limit Load Min	% Limit Load Max
1	100.00	60.00	62.96
2	300.00	60.00	62.66
3	400.00	60.00	62.36
4	700.00	60.00	62.07
5	1000.00	60.00	61.78
6	2000.00	60.00	61.48
7	2800.00	60.00	61.18
8	5500.00	60.00	60.88
9	9000.00	60.00	60.59
10	15019.00	60.00	60.29
11	28853.00	60.00	60.20

To examine the integrity of the pad in the joint, the bending stress, $\sigma_{bending}$, and bearing stresses, σ_{br}, should be calculated to check if the corner crack is stable during its four service lives. From Eq. (4.44):

$$\sigma_{bending} = 6M/bt^2$$

where $M = P_t d = 2500 \times 2D = 2500$ in-lb. From Fig. 4.37: $b = (4D + D_{head}) = 2.8$ in.

$$\sigma_{bending} = 6 \times 2500/(2.8 \times 0.4^2) = 33.48 \text{ ksi}$$

From Eq. (4.43), the bearing stress, σ_{br}:

$$\sigma_{br} = V/Dt$$

$$\sigma_{br} = 2000/(0.5 \times 0.4) = 10.0 \text{ ksi}$$

FINAL RESULTS:
Unstable crack growth, max stress intensity exceeds critical value:
$K_{max} = 56.44$ $K_{ref} = 0.0000$ $K_{cr} = 50.60$:
At the end of two service lives:
Crack sizes: $a = 0.381844$, $c = 0.294645$, $a/c = 1.29594$

The above results indicate that unstable crack growth occurs when the pads in the bolted joint are exposed to two service lives (associated with the number of cycles used in the two load spectrum). In summary, the pads in the bolted joint could not survive the minimum of four service lives. The designer may be advised to make one of the following changes in order to have a safe joint:

1. Reduce applied loads by redesigning the joint.
2. Use special NDE inspection to obtain a smaller flaw size, as discussed in Section 5.4 of Chapter 5.
3. Consider changing the pad material.

Example 4.8

A low risk structural component carries a fluid line (a clamp is holding the fluid line as shown in Fig. 4.40) and is attached to a primary piece of hardware (a high risk part) by a four-bolt pattern (bolt diameter $D = 0.25$ in. and plate thickness $t = 0.1$ in.). The fluid line and the clamp assembly have a total mass of 5 lb-m and are subjected to 50 ft./sec^2 acceleration (in the x and y directions) in a fluctuating load environment. Use NASA/FLAGRO or any other computer code to determine the integrity of the interface during its service life (one service life is equal to four times the number of cycles shown in the load spectrum table). Both the primary and secondary structures are made of 2219-T851 aluminum alloys with fracture properties as shown in Table 4.4. Assume a corner crack of length 0.05 in. with $a/c = 1$ preexisting at the hole.

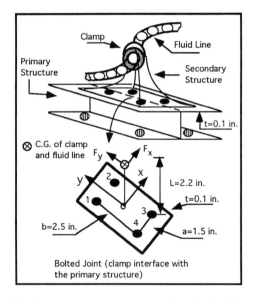

Figure 4.40 Loads and crack geometry descibed in Example 4.8

4.6 Structural Integrity Analysis of Bolted Joints Under Cyclic Loading

Load Spectrum for Example 4.8

Steps	Cycles	%Limit Load	
		Max %	Min%
1	100.00	100.00	−100.00
2	500.00	90.00	− 90.00
3	1000.0	80.00	− 80.00
4	2000.0	70.00	− 70.00
5	3000.0	60.00	− 60.00
6	4000.0	50.00	− 50.00
7	5000.0	40.00	− 40.00
8	10000.0	30.00	− 30.00
9	12000.0	20.00	− 20.00
10	15000.0	10.00	− 10.00

Solution

The forces induced in the x and y directions due to 50 ft./sec^2 acceleration are $F_y = F_x = 5 \times 50 = 250$ lb. These forces carry moments M_x and M_y at the centroid of the four-bolt pattern as shown in Fig. 4.41. The reactions at the bolted joints due to $F_x, F_y, M_x,$ and M_y can be calculated:

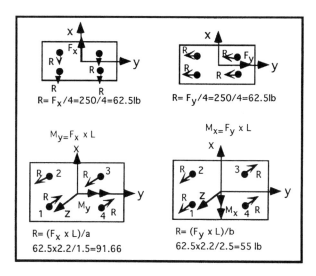

Figure 4.41 Bolt load due to forces and moments applied to the centroid of the bolt pattern

The maximum tension load is taken by bolt number 1 where the reactions from M_x, and M_y contributions are added together (91.66 + 55 = 146.66 lb). The bending stress induced on the plate ($t = 0.10$ in.) due to tension on the bolt can be determined by using an equation describing the bending stress for case (c) of Fig. 4.36; see also Fig. 4.37. From Eq. (4.44) (see also Fig. 4.42):

$$M = 146.66 \times (1.5/2) = 110 \text{ in.-lb}$$

$$b = 2 \times 0.75 + 1.6 \times 0.25 = 1.9 \text{ in., and } t^2 = 0.010 \text{ in.}^2$$

$$\sigma_{\text{bending}} = 6M/bt^2 = (6 \times 110)/(1.9 \times 0.010) = 34736 \text{ psi}$$

The bearing stress due to shear forces will also contribute to crack growth. The resultant of the shear forces for the bolt at location 4 is (see Eq. (4.43) and also Fig. 4.43):

$$R = V = (91.66^2 + 55^2)^{1/2} = 107 \text{ lb}$$

$$\sigma_{\text{br}} = V/Dt = 107/(0.25 \times 0.1) = 4276 \text{ psi}$$

Fatigue crack growth analysis
(computed: NASA/FLAGRO version 2.03, January 1995.)
U.S. customary units [inches, ksi, ksi sqrt(in)]

GEOMETRY
MODEL: CC02-Corner crack from hole in plate (2D)

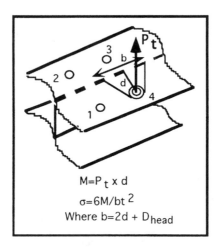

Figure 4.42 Bolt tension load and plate bending as described in Example 4.8

4.6 Structural Integrity Analysis of Bolted Joints Under Cyclic Loading

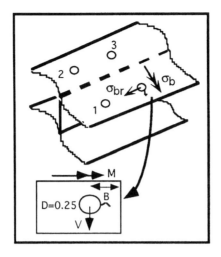

Figure 4.43 Shear resultant and the induced bearing stress as described in Example 4.8

Plate thickness, $t = 0.1000$
Plate width, $W = 2.0000$
Hole diameter, $D = 0.2500$
Hole-center-to-edge dist., $B = 0.5000$
Poisson ratio $= 0.30$

FLAW SIZE
a (init.) $= 0.5000\text{E-}01$
c (init.) $= 0.5000\text{E-}01$
a/c (init.) $= 1.000$

MATERIAL PROPERTIES

UTS	YS	K1e	K1c	Ak	Bk	Thk	Kc
65.0	53.0	46.0	33.0	1.00	1.00	0.100	65.7

Crack Growth Eqn Constants

C	n	p	q	DK_o	R_{cl}	α	$S_{max}/S_1 G_o$
0.119E-7	3.156	0.50	1.00	3.00	0.70	1.50	0.30

ANALYSIS RESULTS
FINAL RESULTS
Critical crack size has NOT been reached.
At cycle no. 15000.00 of load step no. 10
of block no. 4 (at the end of 4 servise lives)
Crack sizes: $a = 0.665341\text{E-}01$, $c = 0.183923$, $a/c = 0.361750$

The result of the analysis indicated that the part can survive the load environment with final crack length $a = 0.0665$ in. (initial crack length was 0.05 in.) and $a/c = 0.36$.

4.7 Material Anisotropy and Its Applications in Beam Analysis

Almost all of the beams used in structural hardware are oriented in such a way that the elongated grain direction (*L*-direction) is along the length of the beam. The long transverse, *T*, and short transverse, *S*, directions are associated with the width of the flange and the web, respectively (as illustrated in Fig. 4.44 for a beam with *I* cross sectional area). Fracture toughness test data on most alloys have shown lower values in the *T–L* and *S–L* compared to the *L–T* and *L–S* directions, as indicated in Table 4.5 for a few selected materials extracted from the NASA/FLAGRO material library [1] (see Section 3.3.1 of Chapter 3 for standard nomenclature relative to directions of mechanical working (grain direction) for rectangular sections). For 2219-T851 aluminum alloy, the fracture toughness values for the *L–T* and *T–L* directions are almost equal; however, for other materials, such as 2124-T851 aluminum alloy, the difference is no longer small. The analyst must use the appropriate value of fracture toughness for the life evaluation of the part under study.

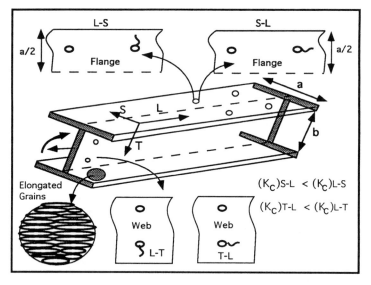

Figure 4.44 The material anisotropy, crack orientations, and fracture toughness for an I-beam

Table 4.5 Fracture toughness properties for several aluminum alloys with respect to their material grain orientaions

2219-T851 (Pit & Sht.)	K_{1c}	K_{1e}	YS	UTS
(L–T)	33	46	53	65
(T–L)	31	43	50	66
2219-T87 (Pit & Sht.)	K_{1c}	K_{1e}	YS	UTS
(L–T)	30	42	57	68
(T–L)	27	38	58	69
2219-T851 (Pit & Sht.)	K_{1c}	K_{1e}	YS	UTS
(L–T)	30	42	63	71
(T–L)	23	31	64	72
(S–T), (S–L)	21	28	60	69

When a beam is mechanically bolted to another structure (bolted joints at the flange or web locations), it is important to select the correct material orientation when evaluating the usage life of the part under study. In most cases, when the beam is subjected to far-field induced bending and tension stresses (as shown in Fig. 4.44), the L–S and L–T crack orientations (for the flange and the web, respectively) are considered to be the appropriate orientation to use for fracture analysis of the joints, even though the $(K_c)T$–L and $(K_c)S$–L values are lower than $(K_c)L$–T and $(K_c)L$–S, respectively. For this reason, in the case of a flange, as well as the web, it is recommended that the analyst orient the assumed initial flaw (crack emanating from a hole) perpendicular to the far field stress when conducting safe-life analysis (Fig. 4.44). It should be noted that the S–L and T–L crack orientations are not affected by the far field induced stresses. However, when dealing with localized induced stresses such as bearing stress at the hole of a joint (no contribution due to far field stress), the material orientation must be considered. In using the L–T or L–S fracture toughness values, less conservative results (a larger number of cycles to failure) can be expected.

References

1. Fatigue Crack Growth Computer Program "NASA/FLAGRO," developed by R. G. Forman, V. Shivakumar, and J. C. Newman. JSC-22267A, January 1993.
2. W. G. Clark Jr. and S. J. Hudak Jr., "Variability in Fatigue Crack Growth Rate Testing," J. Test. Evaluat., Vol. 3, No. 6, 1975, pp. 454–476.

3. D. W. Hoeppner and W. E. Krupp, "Prediction of Component Life by Application of Fatigue Crack Growth Knowledge," Lockheed California Company, Burbank, CA, pp. 8–9.
4. P. C. Paris, M. P. Gomez, and W. E. Anderson, "A Rational Analytic Theory of Fatigue", Trend Engin. Univ. Wash., Vol. 13, No. 1, 1961, pp. 9–14.
5. E. K. Walker, "The Effect of the Stress Ratio During Crack Propagation and Fatigue for 2024-T3 and 7075-T6, The Effect of Environment and Complex Load History on Fatigue Life Life," ASTM-STP 462, ASTM, pp. 1–15, Philadelphia, 1967.
6. R. G. Forman, V. E. Kearney, and R. M. Engle, "Numerical Analysis of Crack Propagation in Cyclic Load Structures," J. Bas. Engin. Trans. ASME, Ser. D, 89, 1967, p. 459.
7. R. G. Forman and S. R. Mettu, "Behavior of Surface and Corner Cracks Subjected to Bending and Tensile Loads in Ti-6Al-4V Alloy," Fracture Mechanics: Twenty Second Symposium, Vol. 1, ASTM STP 1131, edited by H. A. Earnest, A. Saxena, and D. L. McDowell, American Society for Testing and Materials, Philadelphia, 1992, pp. 519–546.
8. J. C. Newman Jr., "A Crack Opening Stress Equation for Fatigue Crack Growth," Int. J. Fract., Vol. 24, No. 3, March 1984, pp. R131–R135.
9. J. K. Donald, and D. W. Schmidth "Computer-Controlled Stress Intensity Gradient Techniques for High Rate Fatigue Crack Growth Testing," J. Test. Evaluat., Vol. 8, No. 1, January 1980, pp. 19–24.
10. A. Saxena and S. J. Hudak Jr., "Review and Extension of Compliance Information for Common Crack Growth Specimens," Int. J. Fract., Vol. 14, No. 5, October 1978.
11. J. C. Newman Jr., "Stress Analysis of Compact Specimens Including the Effects of Pin Loading," Fracture Analysis (8th Conference), ASTM STP-560, ASTM, 1974, pp. 105–121.
12. C. E. Fedderson, "Wide Range Stress Intensity Factor Expression for ASTM Method E 399 Standard Fracture Toughness Specimens," Int. J. Fract., Vol. 12, June 1976, pp. 475–476.
13. W. Elber, "Fatigue Crack Propagation," Ph.D. thesis, University of New South Wales, Australia, 1968.
14. W. Elber, "Fatigue Crack Closure Under Cyclic Tension," Engin. Fract. Mech., Vol. II, No. 1, July 1970.
15. W. Elber, "The Significance of Fatigue Crack Closure," Damage Tolerance in Aircraft Structure, ASTM STP 486, 1971, p. 230.
16. J. R. Rice, "Mechanics of Crack Tip Deformation and Extension by Fatigue," Fatigue Crack Propagation, ASTM STP 415, 1967, p. 247.
17. J. Schijve, "The Stress Ratio Effect on Fatigue Crack Growth in 2024-T3 Aluminum Clad and the Relation to Crack Closure," Delft University of Technology Department of Aerospace Engineeing, Memorandum M-336, August 1979.
18. J. Backlund, A. F. Blom, and C. J. Beevers, "Fatigue Threshold, Fundamentals and Engineering Applications," in Proceedings of an International Conference held in Stockholm, June 1–3, 1981.
19. M. Klesnil and P. Lukas, "Effect of Stress Cycle Asymmetry on Fatigue Crack Growth," Mater. Sci. Engin., Vol. 9, 1972, pp. 231–240.

20. J. M. Basom, "Fatigue Behavior of Pressure Vessels Steel," WRC Bulletin 194, Welding Research Council, New York, May 1964.
21. K. Tanaka, Y. Nakai, and M. Yamashita, "Fatigue Growth Threshold of Small Cracks," Int. J. Fract., Vol. 17, No. 5, October 1981, pp. 519–533.
22. N. E. Frost, Proceedings of Institution of Mechanical Engineers, Vol. 173, 1959, pp. 811–835.
23. N. E. Frost, Alternating Stress Required to Propagate Cracks in Copper and Nickel-Chromium Alloy Steel Plates, J. Mech. Engin. Sci., Vol. 5, 1963, pp. 15–22.
24. H. Ohuchida, S. Usami, and A. Nishioka, Trans. Japan Soc. Mech. Engin., Vol. 41, 1975, pp. 703–712.
25. S. Usami and S. Shida, Fatigue Engin. Mater. Struct., Vol. 1, 1979, pp. 471–482.
26. Small Fatigue Cracks, edited by R. O. Ritchie and J. Lankford, Proceedings of the Second Engineering Foundation International Conference/Workshop, Santa Barbara, CA, January 5–10, 1986." A publication of the Metallurgical Society, Inc.
27. Short Fatigue Crack, edited by K. J. Miller and E. R. de los Rios, ESIS, Publication 13.
28. The Behavior of Short Fatigue Cracks, edited by K. J. Miller and E.R. de los Rios, EGF, Publication 1 (Collection of Papers and References in Crack Initiation).
29. T. C. Landley, "Near Threshold Fatigue Crack Growth: Experimental Methods, Mechanism, and Applications," in Subcritical Crack Growth Due to Fatigue, Stress Corrosion, and Creep, edited by L. H. Larsson, Elsevier Applied Science, 1985, pp. 167–213.
30. H. Kitagawa and S. Takahashi, "Applicability of Fracture Mechanics to Very Small Cracks or the Cracks in the Early Stage," in Proceedings of the Second International Conference on Mechanical Behavior of Materials, Boston, MA, 1976, pp. 627–631.
31. J. C. Newman Jr. and I. S. Raju, "Prediction of Fatigue Crack Growth Patterns and Lives in Three-Dimensional Cracked Bodies," presented at the Sixth International Conference on Fracture, New Dehli, India, December 1984.
32. R. A. Schmidt and P. C. Paris, "Threshold for Fatigue Crack Growth Propagation and Effects of Load Ratio and Frequency," in Progress in Flaw Growth and Fracture Toughness Testing, ASTM STP-536, American Society for Testing and Materials, Philadelphia, pp. 79–94.
33. J. Schijve and D. Broek, "Crack Propagation Based on a Gust Spectrum with Variable Amplitude Loading," Aircraft Engin., Vol. 34, 1962, pp. 314–316.
34. J. Schijve, "The Accumulation of Fatigue Damage in Aircraft Materials and Structures," AGARDograph No. 157, 1972.
35. J. Schijve, "Cumulative Damage Problems in Aircraft Structures and Materials," The Aeronaut. J., Vol. 74, 1970, pp. 517–532.
36. O. E. Wheeler, "Spectrum Loading and Crack Growth," J. Basic Engin., Vol. 94D, 1972, pp. 181–184.
37. J. D. Willenborg, R. M. Eagle, and H. A. Wood, "A Crack Growth Retardation Model Using an Effective Stress Concept," AFFDL-TM-71-1 FBR, 1971.
38. J. P. Gallagher and T. F. Hughes, "Influence of the Yield Strength on Overload Affected Fatigue-Crack-Growth Behavior of 4340 Steel," AFFDL-TR-74-27, 1974.

39. S. M. Arnold, "Effect of Screw Threads on Fatigue," Mech. Engin., Vol. 65, July, 1943, pp. 497–505.
40. A. Schwartz Jr., "New Thread Form Reduces Bolt Breakage," Steel, Vol. 127, September, 4, 1950, pp. 86–94.
41. A. M. Smith, "Screw Threads, the Effect of Method of Manufacture on the Fatigue Strength," Iron Age, Vol. 146, August 22, 1940, pp. 23–28.
42. J. O. Almen, "On Strength of Highly Stressed, Dynamically Loaded Bolts and Studs," SAE J., Vol. 52, No. 4, April, 1944, pp. 151–158.
43. T. R. Higgins, "Bolted Joints Found Better Under Fatigue," Engin. News-Record, Vol. 147, August 2, 1951, pp. 35–36.
44. "Preloaded Bolts and Screws," Lockhead Aircraft Corporation, Report No. 2072, March 1, 1950, pp. 1–7.
45. E. F. Bruhn, "Analysis and Design of Flight Vehicle Structure," Jacobs Publishing, pp. D1–D14.
46. "Tension Type Fittings," Lockhead Report No. 2072, Lockhead California Company, November 1, 1968, pp. 1–12.

Chapter 5

Fracture Control Program and Nondestructive Inspection

5.1 Introduction

Full assurance of maintaining a trouble-free damage tolerant space vehicle during its operational life requires a continuing, multidisciplinary process which begins in the preliminary design phase and extends through manufacturing and into the operational planning of the space structure. Figure 5.1 illustrates the main elements of this process. The purpose of this section is to provide guidance for implementing a general approach, defined in a Fracture Control Plan (FCP), to control and prevent damage that could cause a catastrophic failure due to the preexisting cracks in the structure. The general concepts used in this chapter are also used in other industries, such as nuclear, pressure vessels, and shipbuilding, where safety is of great concern.

The design philosophy, material selection, analysis approach, testing, quality control, inspection, and manufacturing are elements of the fracture control program that contribute to building and maintaining a trouble-free space structure.

The first part of this chapter (Sections 5.1 and 5.2) discusses the elements that are essential in implementing an effective fracture control plan. In the latter part of this chapter (Section 5.3), different nondestructive inspection techniques that are currently used for inspecting high risk critical space hardware are discussed. Section 5.1.1 briefly describes the two types of design concepts (namely slow crack growth and fail-safe) presently used in the design of aerospace and aircraft structures.

5.1.1 Design Philosophy

To provide adequate tolerance and safety, a space structure must be designed to withstand the environmental load throughout its service life, even when the struc-

254 Chap. 5 Fracture Control Program and Nondestructive Inspection

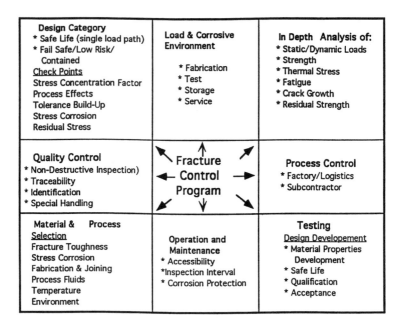

Figure 5.1 Main elements of a fracture control program

ture has preexisting flaws or when part of the structure has already failed. To provide adequate safety and to ensure that unstable crack growth is not reached, two types of damage tolerant design concepts are presented in this section: single load path (slow crack growth) and fail-safe design (also called multiple load path design). Figures 5.2 and 5.3 illustrate the slow crack growth and fail-safe design concepts, respectively.

5.1.1.1 Slow Crack Growth Design

The slow crack growth design concept simply states that single load path structures must be designed such that any preexisting cracks will grow in a slow and stable manner during their service life, and thus the growing crack will not achieve a critical size that could cause unstable propagation. In this design philosophy, the damage tolerance is ensured by maintaining a slow crack growth rate during the life of the structure. Moreover, the residual strength capability of the structural part must always be kept above the applied stress. The keel structure shown in Fig. 5.2 is used to attach a segment of a space structure to the shuttle during the flight. The pin is made of 455 custom stainless steel and was classified as a single load path structure. The structural integrity requirements of this part are to show that the pin has sufficient margin of safety based on static analysis, as well as a minimum of four service lives using a fracture mechanics approach.

5.1 Introduction

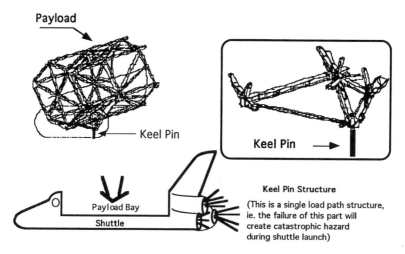

Figure 5.2 Illustration of a single load path structure. The payload is attached to the space shuttle payload bay by a keel pin structure that has a single load path design

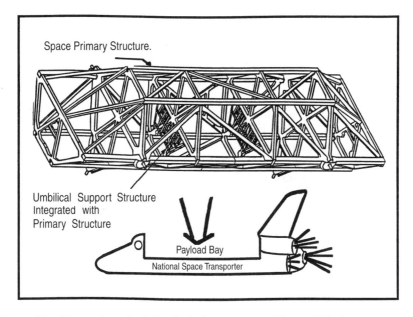

Figure 5.3a Illustration of a fail-safe design structure. The umbilical support structure carries avionics and fluid lines for space components.

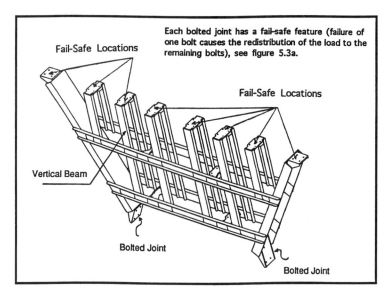

Figure 5.3b Fail-safe design of the umbilical structure and upper and lower bolted joints

5.1.1.2 Fail-Safe Design and Low Released Mass

In the fail-safe design philosophy, the structure is designed such that the failure of a structural component due to crack propagation is localized and safely contained or arrested. The damage tolerance in the fail-safe design category must be assured by allowing the partial structural failure, while the remaining undamaged structural components maintain their integrity and are able to operate safely under the remaining distributed load. In addition, it must be shown that all released fragments due to partial failure have a total mass of less than 0.25 lb (called low released mass) during the launch or else they are contained and will not pose a catastrophic hazard to the surrounding structural parts or to the shuttle. A fail-safe structure has a multiple load path feature (see Figs. 5.3a and b) in its design. Figure 5.3a illustrates a fail-safe designed space structural component called the umbilical support structure (situated in one of the sections of the main structure). It functions as a support structure to the avionics lines, which carry power, data, optical signals, instrumental wiring, and the fluid lines that carry ammonia (a coolant) to different regions of the space structure (avionics and fluid lines are not shown here). The failure of one beam (for example, one of the vertical beams shown in Fig. 5.3b) produces load redistribution to other members of the structure. By analysis, it can be shown that the remaining redistributed load will not cause damage to the umbilical structure or to the surrounding hardware during the launch and on-orbit environment.

Another example that can be useful to the reader for understanding the fail-safe design philosophy is the upper and lower joints that connect the umbilical support structure to the primary structure (8 bolted joints on the top and two on the bottom). Failure of any one of the bolted joints will redistribute the load to the remaining joints. In addition, each bolted joint has a fail-safe feature by having the four-bolt pattern shown in Fig. 5.3b. It can be shown by analysis that the failure of one bolt can cause the redistribution of the load in the joint and will prevent catastrophic damage to the main structure.

The residual strength capability of a fail-safe structure must be evaluated to ensure the structural integrity of the part. In other words, the redistributed load must fall below the residual strength capability of the structural part. To assess the residual strength of a fail-safe structure (1) the stress level at which partial damage occurs must be known and (2) the load-carrying capability of the remaining structure that must withstand the redistributed load must be evaluated.

5.1.1.3 Material Selection, Testing, Manufacture, and Inspection

Material selection is a critical element of the fracture control process. Tradeoff studies in the early stage of design are conducted between materials and the use of comparative property data is necessary in the selection process. Ultimate strength, yield strength, plane strain, and plane stress fracture toughness (including the fracture toughness for surface cracks) and stress corrosion property data must be considered in the material selection process for the expected environment. The crack growth rate as a function of stress intensity factor (da/dN, ΔK data) is also required. If the environment is corrosive, the crack growth rate data (da/dt versus ΔK) should be available to the analyst. Mechanical and fracture properties with respect to the anisotropic nature of certain materials must also be considered. A comprehensive list of materials that are currently recommended by NASA for space application (which may also be utilized for other industries) together with their static and fracture properties is available in Appendix A. In this appendix, the ultimate and yield strength as well as plane strain, K_{Ic}, and plane stress, K_c, part through fracture toughness, K_{Ie}, and crack growth properties for the listed material are available. Table 5.1 shows an example of static and fracture properties of 7075-T851 aluminum alloy that must be available for conducting a meaningful fracture analysis.

Testing methods for each critical part and assembly (an assembly is a piece of hardware that is made of several parts) are also developed and incorporated as part of the fracture control process.

Manufacturing processes must be selected for the high risk or critical parts such that they do not reduce the damage tolerance level required by the design. Gouges and surface scratches due to improper handling and machining must be avoided throughout the manufacturing process since their presence (if undetected) can reduce the life of the part. Control of processes and selection of inspection

Table 5.1 Fracture properties of 7075-T851 aluminum alloy (NASA/FLAGRO material library)

Yield Stress	Ult. Stress	K_{Ic}	K_{Ie}	A	B	C	n	p	q	ΔK_0	R_{cl}	α	SR
76	85	26	38	1.0	1.0	.2E-7	2.88	0.5	1.0	2.0	0.7	1.9	0.3
Strength Properties		Fracture Toughness Properties		Adjustement to Plane stress		Fatigue Crack Growth Properties							

procedures to maintain process quality are prime considerations of the fracture control plan.

This introduction to the fracture control plan illustrates the interconnection between design, testing, manufacture, and inspection and their roles in maintaining the desired damage tolerant structure and to reduce the incidence of fracture-related failures and loss.

The structural integrity of a hardware that contains several parts (such as an aircraft, a space structure, or a pressure vessel) can be maintained by ensuring the implementation of an effective fracture control plan. The implementation of a sound fracture control plan for space structures is required by the regulating agency (i.e., NASA). Throughout this chapter, the emphasis will be given to the fracture control policies and requirements for space structures. Section 5.2 describes the detailed fracture control policies and requirements that must be implemented for any space structure that will be used as a payload on the space shuttle. The fracture control plan described in Section 5.2 can be similarly applied to other structures where safety is a great concern, such as nuclear, military, or commercial aircraft. Prior to implementation of the fracture control program, a fracture control plan must be submitted to the customer for review. Implementation of the plan and its related activities are monitored by representatives of the procuring agency after its approval.

5.2 Fracture Control Plan

5.2.1 Purpose

The purpose of implementing a fracture control plan is to prevent unstable crack growth in structures of all elements and systems of flight vehicle structure during the design service life. Catastrophic failures as the result of unstable crack growth can cause the total loss of the structural hardware, and more importantly, the loss of human life. The required fracture control procedures are intended as an augmentation to, and are based heavily upon, the good engineering and manufacturing practices already embedded in the hardware development process. Fracture control imposes additional engineering and product assurance requirements needed to ensure the structural integrity of fracture critical structures throughout all phases of the component's lifetime.

5.2.2 Causes of Unstable Crack Growth

The basic assumptions that underlie the need for fracture control are based on the knowledge that all structures contain preexisting flaws that could be located in the most unfavorable location and orientation with respect to the applied stresses. Furthermore, it is essential that these cracks do not grow undetected to a critical length during the usage period.

Under sustained tensile stress in the presence of an aggressive environment, some materials exhibit a tendency to initiate cracks. The circumstances under which this stress corrosive cracking phenomenon occurs are sufficiently well documented to provide appropriate avoidance measures and criteria (MSFC-HDBK, Material Selection Guide for MSFC SpaceLab Payloads criteria or MSFC-SPC-522B called "Design Criteria for Controlling Stress Corrosion Cracking"). In noncorrosive environments, cracks will initiate owing to a sufficient number of high tensile stress cycles. However, since the assumption of preexisting cracks in all structural elements is more severe than the assumption of crack initiation (see Chapter 2 on crack initiation concept), the requirements for preventing failure are completely enveloped and satisfied by the Nondestructive Inspection (NDI) requirements (see Section 5.3 for a comprehensive review of NDI methods and their application to aerospace hardware). The reader should understand that the actual initial crack size in the material could be much smaller than the crack size assumption based upon the NDI method. Therefore, the analytical life evaluation of a part using the standard NDI initial flaw size yields conservative results.

The fracture control plan specifies the type of NDI and minimum initial crack sizes (a_i) that may be used for single load path versus fail-safe design. The initial crack size associated with any given NDI method can be reduced to extend the analytical life of the component if the contractor can demonstrate that the reduced initial flaw size is reliably detectable (see Section 5.2.8).

5.2.3 Fracture Control in the Stages of Hardware Development

Fracture control for flight hardware is implemented starting at conceptual design and continues through all phases of engineering, manufacture, and in-service usage. Therefore, the implementation of the program should be organized according to the logical sequence of a given structural component's life history. In the engineering phase, the part is designed, analyzed, and qualification tested. At this stage, the structural integrity and safe-life analysis are fully assessed. The end products of the engineering phase include the detailed production drawings used by manufacturing to produce the part.

The production phase occurs when the part is produced according to the production drawing, inspected, and/or acceptance tested. Satisfaction of NDI, proof testing, traceability, and special handling requirements are essential elements of this phase. During the operational phase, the part may be subjected to interval inspection, maintenance, and possibly repaired or replaced.

5.2.4 Typical Fracture Control Activities

All activities that influence the control of structural crack initiation and propagation in flight hardware elements and systems and their test articles are subject to the requirements and procedures defined in a Fracture Control Plan. These activities generally fall into the following categories:

1. Structural design
2. Structural analysis
3. Materials selection, procurement, and storage
4. Fabrication process control
5. Quality assurance and nondestructive inspection (NDI)
6. Test
7. In-service operations, maintenance, inspection, and repair

Examples of typical fracture control activities performed by most hardware developers include:

1. Minimization of flaw sizes through proper inspection, processing, and fabrication procedures
2. Selection of materials having adequate fracture toughness, resistance to stress corrosion cracking (SCC), and crack growth rate properties *(da/dN, ΔK* and possibly *da/dt, ΔK* data)
3. Minimization of residual stresses during processing and fabrication
4. Complete identification and documentation of design loads and environments
5. Design for sufficiently benign internal stresses, including minimization of stress concentrations
6. Use of proper analytical tools and tests to verify that the design fulfills fracture control requirements
7. If, owing to conflicting requirements, all fracture control requirements cannot be completely met, additional controls are placed on the design.

5.2.5 Data Required for Fracture Control

In support of fracture control activities, considerable data are developed and maintained by the contractor. Engineering data necessary for adequate assessment and classification of hardware include the following:

1. Definition of environments and load spectrum history. The environment could be inertia, thermal, pressure, sonic, corrosive, or a combination of several environments.
2. Detailed design and assembly drawings. All critical dimensions and characteristic must be denoted on the drawing.
3. Mechanical and fracture properties of materials in the appropriate environments
4. Stress and fracture analysis results
5. Level of NDI performed and the corresponding flaw size for through and part through cracks.

A large portion of these data is contractually required and generated for purposes other than fracture control (e.g., objective evidence of design and qualification documentation).

5.2.6 Contents of a Fracture Control Plan

Each major element of a fracture critical hardware system is governed by a Fracture Control Plan prepared by the contractor or system developer and approved by the responsible procuring agency. The Fracture Control Plan defines the elements of the Fracture Control Program and the responsibilities for managing and accomplishing them. The Plan must describe in detail the approach to be used to implement the following:

1. Fracture control classification of components
2. Analysis and/or testing to determine fracture control acceptability of hardware
3. Control of materials, manufacturing processes, nondestructive inspections, testing, design changes, and part handling transportation
4. Overall review and assessment of fracture control activities and results

Implementation approaches for the above four major elements of the fracture control program are described in Sections 5.2.7 through 5.2.10.

5.2.7 Fracture Control Classification of Components

Fracture control classification of all hardware components is typically determined as shown in the logic diagram of Fig. 5.4. Components that are classified as non-hazardous released parts, contained, fail-safe, or low risk must meet the specific requirements associated with at least one of these categories. These components

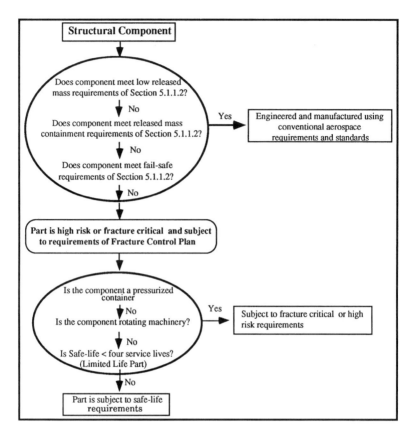

Figure 5.4 Evaluation of structural components for fracture control classification

are then considered nonfracture critical and are not required to be further evaluated for fracture control. However, a life expectancy analysis of nonfracture critical parts, either through a conventional fatigue (stress to life approach using the S–N curve data) or a damage tolerance approach that assumes a smaller initial flaw size, must be performed. Components classified as fracture critical or high-risk parts will have their damage tolerance and/or safe-life (one safe-life is equal to four service lives) verified by test and/or analysis, in addition to meeting the standard design, control, and verification requirements for aerospace structures. Note that for other structures, such as aircraft, one safe-life may be defined as two service lives. As fracture control classification and analysis of parts are performed, an up-to-date listing is maintained of all fracture critical parts.

5.2.8 Analysis and/or Testing to Determine Fracture Control Acceptability of Hardware

A fracture critical component is acceptable if it can be shown by analysis or test that the largest undetected flaw that could exist in the component will not grow to failure when subjected to the cyclic and sustained loads encountered in four service lifetimes. One service lifetime includes all significant loading cycles occurring after flaw screening (to establish minimum initial flaw size) and will include testing, transportation, flight, operations, or any other loading cycles considered relevant. Figure 5.5 illustrates a simplified version of a complicated load spectrum that an aircraft structural part can be exposed to. The number of cycles associated with each event must be considered in evaluating the life of the part.

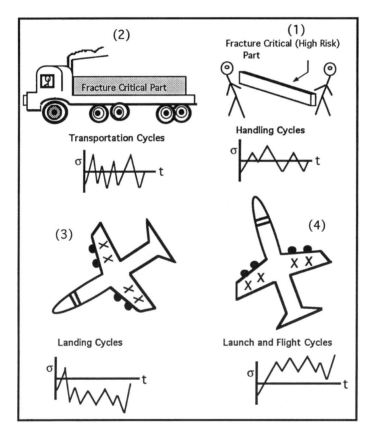

Figure 5.5 Illustrating a simplified version of an aircraft load environments. Load cycles for each environment contributes to accumulation of failure on aircraft structural parts

264 Chap. 5 Fracture Control Program and Nondestructive Inspection

Either of two analysis approaches may be used to show that a part subjected to NDI meets safe-life requirements. The first or direct approach is to select the appropriate NDI technique and to use the flaw sizes representative of the capability of that technique to show that the part will survive at least four lifetimes. The second approach is to calculate the critical (i.e., maximum) initial crack size for which the part can survive the required lifetimes and to verify by inspection that there are no cracks greater than or equal to this size. In the second approach, the initial crack length is estimated and by analysis the final crack length, a_f, corresponding to four service life is calculated. Figure 5.6 illustrates the two above approaches for a beam subjected to a fluctuating load environment.

In addition to the safe-life analysis or testing, all fracture critical parts must be subjected to NDI inspection to screen flaws. The selection of NDI methods and level of inspection is based primarily on the safe-life acceptance requirements of the part. The initial crack sizes correspond to a 90% probability with a 95% confidence level of inspection reliability. Minimum detectable initial crack sizes for specific NDI methods are given in Table 5.1 for the crack geometries shown in Figure 5.16. These are the minimum sizes that must be used in the safe-life analysis. Use of initial crack sizes for other geometries or NDI techniques requires the approval of the procuring agency. Where adequate NDI of finished parts cannot

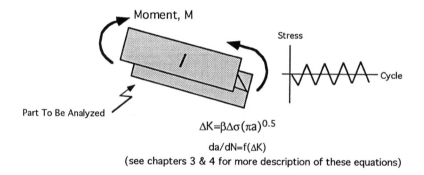

First Approach

Select a_i from appropriate NDI method and compare the calculated K (after four service life) with the critical K value (Kc).

Second Approach

Evaluate the critical crack length, a_f, for four service life and compare it's value with the initial crack length, a_i ($a_i < a_f$).

Figure 5.6 Illustrating the two approaches used for safe-life analysis

be accomplished, NDI may be required on the raw material and/or on the part itself at the most suitable step of fabrication.

Pressure systems (pressure vessels), rotating machinery, and fasteners that are classified high risk or as fracture critical have additional analysis and/or test requirements imposed upon them. For example, all pressure systems (such as a pressure vessel with stored energy of 14,240 ft-lb or greater) and high-energy rotating machinery components (a rotating mechanical assembly that has a kinetic energy of more than 14,240 ft-lb based on $K_{Energy} = I\omega^2/2$ where the two quantities I and ω are the mass moment of inertia and ω is the rotational frequency in radians per second respectively) require qualification proof testing. The purpose of this proof test is to verify the integrity of the structural part. The stress associated with proof testing for pressure vessels and rotating machinery is selected as 1.5 × limit load (see Figure 5.7).

5.2.9 Traceability and Design Annotation for Drawings

Traceability is defined as the ability to relate designated product data to specific products by retrieving information logged through each level of hardware records to the configuration item level. For example, the material mill heat number should be traceable from the part at any stage of manufacture and assembly. In addition, one should also be able to trace the NDI performed on the hardware and other critical characteristic records of fracture critical hardware at any stage of its development. Traceability is required to be maintained on all fracture critical parts

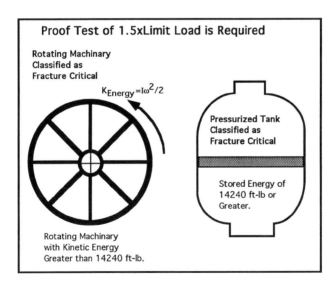

Figure 5.7 High energy (high risk) systems. A pressurized tank and A rotating wheel

throughout their development, manufacture, testing, and flight. Specific procedures for traceability are established at the discretion of the procuring agency, and may include serialization or other means of configuration control. For manned-rated flight hardware, serialization is typically required only for pressure vessels and components of high-energy rotating machinery. All other fracture critical hardware must maintain traceability to the raw material heat treat lot as a minimum. For each pressure vessel, a log is maintained to record all pressure cycles and associated environmental conditions occurring during the time period from fabrication to the end of the service life of the vessel.

Engineering drawings and equipment specifications for fracture critical parts contain notes that identify the part as fracture critical and specify the appropriate inspection or flaw screening method to be used on the part. Critical characteristics are also noted on the drawings to identify mandatory inspection points. Some examples of critical characteristics are critical dimensions, traceability, proof test, and NDI results. An example of fracture control program notes on the fracture critical part drawings (designated by the symbol [FCP]) are shown in Fig. 5.8.

[FCP] note number 1 indicates that this component is a high risk fracture critical part and must meet fracture critical requirements per XXX document. [FCP] note number 2 is related to the material serialization. [FCP] note number 3 is related to NDI penetrant inspection of the part after machining per XXX document. Finally, [FCP] note number 4 describes the identification method using electrochemical etch masking. Here the XXX refers to special documents that contain the procedures to be used to satisfy fracture control requirements.

Changes in design or process specifications, manufacturing discrepancies, repairs, and finished part modifications for all fracture critical parts are reviewed according to criteria established by the responsible fracture control authority to make certain that the parts still meet fracture control requirements. Changes to the Fracture Control Plan must be incorporated into a revised Plan, which is then resubmitted to the procuring agency for approval.

[FCP] 1: THIS PART HAS BEEN IDENTIFIED AS A FRACTURE CRITICAL PART. SHOULD MEET FRACTURE CRITICAL HARDWARE REQUIREMENTS PER XXX DOCUMENT.

[FCP] 2: SERIALIZE PER YYY DOCUMENT. EACH ITEM SHALL HAVE A DIFFERENT SERIAL NUMBER. GAPS IN SERIAL SEQUENCE ARE PERMISSIBLE.

[FCP] 3: PENETRANT INSPECT PER XXX DUCUMENT. ETCH SURFACES PRIOR TO PENETRANT INSPECTION.

[FCP] 4: IDENTIFY USING ELECTROCHEMICAL ETCH METHOD PER XXX DOCUMENT. DIRECT MARKING LOCATION APPROXIMATELY IN AREA SHOWN ON THE DRAWING

Figure 5.8 Related fracture control notes on drawings for high-risk or fracture critical parts

Special handling requirements are applied to fracture critical hardware to avoid material damage that could act as a crack initiator. Fracture critical parts and their assemblies must be protected at all times during the manufacturing process and installation operations from damage caused by environmental deterioration or rough handling. The procedures include requirements to ensure individual handling in protective packaging labeled with "FRACTURE CRITICAL—HANDLE WITH CARE." The XXX document mentioned in the drawing notes (see Fig. 5.5) contains program derived requirements for traceability, NDE inspection, and the special handling of fracture critical parts.

5.2.10 Overall Review and Assessment of Fracture Control Activities and Results

Each contractor or hardware system developer designates a specific individual or group to be the responsible fracture control authority. This authority is responsible for direction and implementation of its Fracture Control Program and for assuring its effectiveness. The designee is responsible for monitoring, reviewing, and approving fracture control activities performed both internally and by subcontractors. These activities are coordinated with other key organizations, including engineering, manufacturing, safety, reliability, and quality assurance as required. The total Fracture Control Program is subject to oversight and review by the procuring agency.

To certify compliance with fracture control requirements, the prime contractor or subcontractor responsible for the system development prepares a fracture control summary report on the total system for review and approval at a Design Certification Review by the procuring agency. The documents required to support the acceptability of a fracture critical part include a crack-growth analysis (or safe-life test) report and a Nondestructive Inspection (NDI) or proof test report. A documented description of the load spectrum and material crack growth properties used in the analysis is included in the safe-life analysis report. The NDI report includes the date of inspection, serial number or identification of the part inspected, and name of the inspector. If special NDI was used (to obtain a smaller flaw size), additional data are required in the inspection report to assure acceptability and traceability of the process. Documents or analysis supporting the fracture control summary report and the fracture control classification of components are normally kept for the life of the system and made available for audit.

5.2.11 Fracture Control Board

The purpose of the Fracture Control Board (FCB) is to direct and oversee the Fracture Control Program. The FCB is responsible for:

1. Ensuring preparation, maintenance, review, and approval of the FCP
2. Monitoring, coordinating, reviewing, and approving fracture control activities performed both internally and by subcontractors and vendors
3. Ensuring effective implementation of the fracture control program through activities in all affected organizations, and for maintaining awareness of fracture control progress or problems arising within these organizations.
4. Ensuring that each subcontractor engaged in the design and manufacture of fracture critical hardware establishes a Fracture Control organization that assures conformance with the requirements of the FCP. The typical FCB organization consists of a permanently assigned member from each of the following disciplines:
   Program Management (Chairman)
   Strength Analysis
   Materials and Processes (M&P)
   Quality Assurance (QA)
   Safety
   Structures Design
   Mechanical Systems Design
   Flight Crew Systems Design
   Fluid Systems Design
   Subcontract Technical Manager (STM) for subcontractor hardware

At FCB meetings concerning a particular system or element, the FCB Chairman and the assigned members representing Strength Analysis, M&P, QA, Safety, and the cognizant design authority or their delegates will be present. Figure 5.9 shows the organization of a typical FCB.

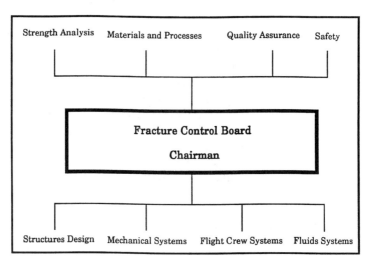

Figure 5.9 Organization of a typical Fracture Control Board

5.3 Nondestructive Inspection Techniques

5.3.1 Introduction

Nondestructive Inspection (NDI) can be defined as the use of nonintrusive methods to ascertain the integrity of a material or structure. Many nonintrusive methods have been developed to evaluate materials for property determination, verify quality of workmanship, and evaluate a component for the existence of flaws. A flaw, in this sense, can be considered as any nonconformity that exceeds an established size criteria.

Flaw detection is by far the most important aspect of NDI in regards to safe-life assessment of fracture critical parts. Fracture mechanics analysis assumes the existence of a maximum flaw size in the part that grows in a stable manner during its service life. NDI provides the assurance that a flaw larger than the identified maximum size does not exist in the part. From a safety perspective, the initial assumed crack lengths provided by the NDI methods are longer than any preexisting flaw that could be present in the structure after inspection. However, the degree of conservatism as a result of the longer initial crack size assumption (used to evaluate the life) must be realistic enough not to impact the structural weight or cause unnecessary rejection of parts.

There are numerous NDI methods utilized for flaw detection in structural components. Although many specialized methods are developed for specific materials and configuration, most techniques are variations on several general methods that use visual enhancement of defects or measure some form of energy transmission through materials and its interaction with defects. The most prevalent of these NDI techniques commonly used in the detection of flaws in aerospace components are: liquid penetrant, magnetic particle, eddy current, ultrasonic, and radiography. The purpose of this section is to briefly describe and compare the common NDI methods and to discuss the variables affecting them.

5.3.2 Liquid Penetrant Inspection

Liquid penetrant inspection is the most widely applied NDI method for the detection of all types of surface flaws, porosity, shrinkage area, and similar types of discontinuities in metals. Essentially an enhanced visual evaluation, penetrant inspection relies on capillary action to penetrate into flaws open to the surface. Excess dye is removed and a white powder developer is applied to the surface to act as a blotter to the trapped dye, thus enhancing flaw detection sensitivity. Fluorescent dyes are most commonly used in penetrant fluids, providing high sensitivity for critical part inspections. A schematic depicting the mechanism of crack detection by the penetrant process is shown in Fig. 5.10 [1]. Penetrant inspection is fast, portable, and easy to interpret, but does require that the flaw be open to the surface for detection. For metal components subjected to mechanical operations

270 Chap. 5 Fracture Control Program and Nondestructive Inspection

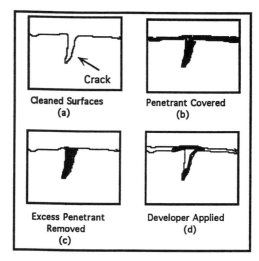

Figure 5.10 Mechanism of penetrant flaw detection

that may mask existing flaws, such as grinding or machining, etching may be required to remove the smeared metal prior to inspection. Variables affecting the penetrant inspection are discussed in Section 5.3.9.1.

5.3.3 Magnetic Particle Inspection

Magnetic particle inspection is another NDI method which uses a fluorescent media to reveal flaws. This approach is applicable only to ferromagnetic materials, such as plain carbon or alloy steels and ferritic stainless steels. A magnetic field is generated on the part surface by flowing electric current through the part or adjacent to it. Surface and near surface discontinuities cause variations in the magnetic field. The fluorescent magnetic oxide particles are then flowed over the part surface and are attracted to the magnetic field variations, where they can be detected under ultraviolet light. Figure 5.11 illustrates two types of magnetic particle inspection setups. Figure 5.8a shows an inspection of a tube segment for longitudinal cracks, while Figure 5.8b depicts a weld inspection utilizing contact prods [2]. The magnetic particle inspection method is quick and simple in principle, however, several different magnetic field orientations and magnetization strengths may be required to thoroughly inspect the suspected flaw orientations. Demagnetization of the part is also required after the inspection is completed. Variables affecting the magnetic particle inspection are discussed in Section 5.3.9.2.

5.3.4 Eddy Current Inspection

Another nondestructive technique used to detect surface and near surface flaws is eddy current inspection. Eddy current is an electromagnetic technique that utilizes

5.3 Nondestructive Inspection Techniques

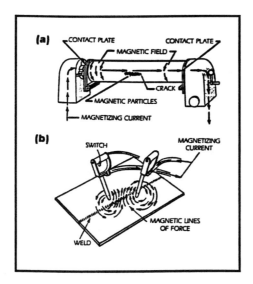

Figure 5.11 Magnetic particle circular magnetization setups

small diameter coils to induce electric currents in conductive materials. The eddy currents induced in the material, in turn, generate their own magnetic field. The magnitudes, time lags, phase angles, and flow patterns of the eddy currents within the test material are detected as changes in the electrical characteristics of the inducing coil. The presence of defects causes significant disruptions in the eddy current flow, and results in impedance and voltage variations. Calibration is performed on reference defects of known size.

Eddy current inspection methods have been used extensively in the aircraft industry for in-service inspections of components subjected to a fatigue environment. The high sensitivity of eddy current methods for crack detection provides a notable advantage over other NDI methods for this application. Conventional eddy current equipment designed for crack detection utilizes a CRT screen that displays the phase and amplitude of the received signal on an impedance plane [Inductive Reactance (X) versus Resistance (R)]. A typical signal response from a crack in aluminum is shown in Figure 5.12a. Figure 5.12b illustrates the change in signal response at various gain settings [3]. Variables affecting the Eddy current inspection are discussed in Section 5.3.9.3.

5.3.5 Ultrasonic Inspection

Ultrasonic evaluation methods are most prominently used to detect subsurface flaws in components. This capability permits ultrasonic evaluation of material in its raw stock form, thereby screening material containing rejectable flaws before a large investment in material processing is made. Ultrasonic evaluations require

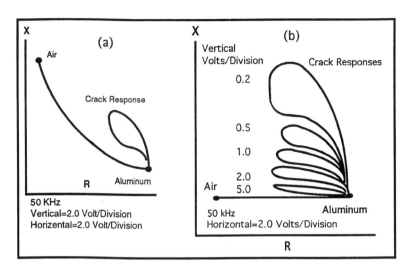

Figure 5.12 Eddy current CRT response from a crack in aluminum (a) Normal impedance plane response, (b) response with phase rotated to horizontal and at various gain settings

generating high-frequency (1 MHz to 50 MHz) acoustic waves with a vibrating crystal transducer and then introducing them into a component. A liquid or gel couplant is used to transmit the ultrasonic waves between the transducer and the material. As the ultrasonic wave encounters an interface, such as at the couplant/test part surface, part of the wave is reflected back to the transducer while part is transmitted into the part. The amplitude of the signal reflected from and transmitted by the liquid/solid interface is dependent upon the ratio of the acoustic impedances of the part and the coupling media. This reflectance and transmittal of the ultrasonic wave occurs at every subsequent interface (e.g., defects as well as the front and back surface of the test part). The amplitude and position in time of the reflected or transmitted signal are monitored producing a marked signal change when encountering material flaws. The signal can be represented as an analog trace on an oscilloscope (A-scan), through-thickness depth position (B-scan), or a plane view (C-scan). Ultrasonic signal displays for each of these data representations are shown in Figs 5.13 a, b, and c [4]. Variables affecting the ultrasonic inspection are discussed in Section 5.3.9.5.

5.3.6 Radiographic Inspection

Radiography, or x-ray imaging, is another NDI method employed in the detection of subsurface flaws. This method relies on the differential absorption of x-ray photons as they propagate through a material. In conventional film radiography, x-rays are generated via electrons striking a tungsten target and exit the x-ray tube

5.3 Nondestructive Inspection Techniques

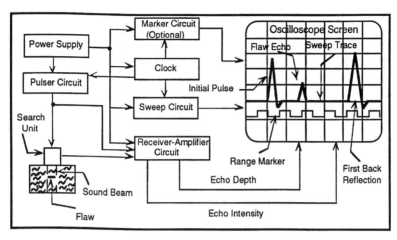

Figure 5.13a A-Scan display

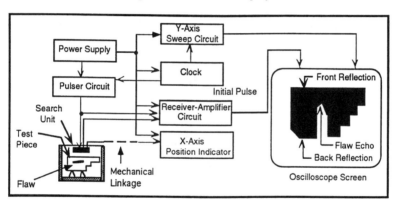

Figure 5.13b B-Scan display

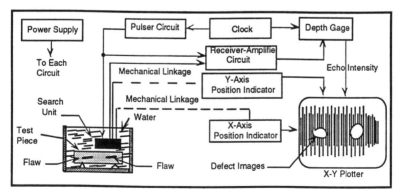

Figure 5.13c C-Scan display

in a conical beam. The photons that penetrate the material are imaged on the underlying photographic film. Areas of greater thickness or density will absorb more of the penetrating photons. If the difference in density between a flaw and the surrounding material is significant this difference will appear on the x-ray film as a difference in film density. A typical film setup is shown in Fig. 5.14 [5]. X-ray systems are usually sensitive to changes that result in an apparent change of at least 1% to 2% of the material thickness or density. This limits the flaw sensitivity to defects oriented parallel to the x-ray beam such as cracks, voids, and inclusions. X-ray inspections are of particular value to evaluations of welds and castings where defect type and orientation is favorable. Variables affecting the radiographic inspection are discussed in Section 5.3.9.5.

5.3.7 Flaw Size Verification

Although great efforts are expended to produce critical components free from defects, material processing operations have the potential for creating flaws. Castings are susceptible to voids, cold shuts, differential cooling cracks, and inclusions. Mill operations such as rolling of bar, plate, sheet, and extrusions can create

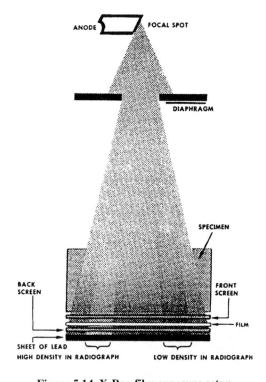

Figure 5.14 X-Ray film exposure setup

laminar defects such as seams, laps, stringers, and cracks. Machining with improper feeds and speeds can cause tears, laps, and cracks. Heat treating to obtain desired mechanical properties can also cause various types of cracking upon quenching or cooling. Joining operations such as welding or brazing can result in casting defects in the weld area as well as cracks in the adjacent heat-affected zone.

Once a component has been placed in service, flaws can initiate and propagate from a wide variety of factors. The most prevalent of these include: fatigue, corrosion, overload, creep, stress corrosion cracking, and hydrogen embrittlement, to name a few. Regularly scheduled in-service inspections using non-destructive evaluation techniques can allow components to be utilized far beyond what could safely be achieved without inspection. The importance of NDI for in-service inspections is illustrated in the crack growth rate curve and corresponding residual strength curve shown in Figure 5.15. The crack growth rate curve (top) plots crack size as a function of time based on anticipated service load profile. Underneath this curve is the corresponding residual strength curve which shows the reduction in strength with time and consequently crack instability. Point A corresponds to the crack size at which the defect becomes reliably detectable for the specific NDI method. Point B is the point at which residual strength has decreased to the failure load. The time between Point A and Point B is then the period available for the detection of cracks that may propagate in service before attaining a critical size [6].

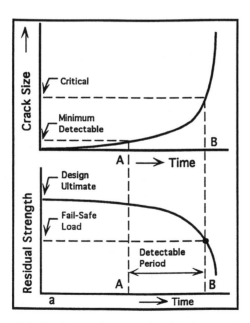

Figure 5.15 Crack growth rate and residual strength curves

Computational approaches to fracture mechanics require the assumption of a maximum initial flaw size. The true value of NDI for stress analysis lies in its application to screen critical components for flaws, assuring the validity of the critical maximum flaw size assumption. For NDI flaw screening to be effective, the detection capability of a particular NDI technique for a given application must be established. The task of assessing NDI capability is complex and is under constant refinement. The basis for all quantitative NDI capability assessments is to subject test specimens containing known flaws of varying sizes and locations to specific NDI inspection processes and measuring capability as a function of the detected flaw size. This is not a trivial task.

Extensive testing programs have been undertaken by NASA as well as several aerospace contractors to characterize the flaw detection capability of the predominant NDI methods. In typical NDI reliability testing, specimens containing fatigue cracks are utilized to provide uniform controlled flaws for meaningful comparisons. Starting with an electrodischarge machined (EDM) notch in the test specimen, a crack is initiated and grown until the desired size is reached. The EDM starter notch is then machined off, leaving only the fatigue crack. Several flaws of various orientations can be grown in a single specimen. The use of fatigue cracks allows tight control over flaw length, depth, and aspect ratio. However, it also adds some conservatism in the flaw size detection capability as tight fatigue cracks represent one of the most difficult flaws to detect. For the inspection of new production hardware, fatigue cracks are not an expected type of defect unless the hardware has already been subjected to a fatigue environment.

5.3.8 Probability of Detection Statistics

Nondestructive evaluation involves the measurement of complex physical parameters with inherent variations in the measurement technique as well as in the test specimens [6]. From a statistical point of view, the inspection outcome is not a simple accept/reject decision but rather a conditional acceptance with four possible outcomes: true positive (flaw exists and is identified), false positive (flaw does not exist but is identified), false negative (flaw exists but is not identified), and true negative (flaw does not exist and is not identified). Two independent probabilities can be considered to describe the inspection reliability—the probability of detection (POD) and the probability of false alarms (POFA). These are determined as follows:

POD = Total True Positive Calls/Total Number of Defects

POFA = Total False Alarms/Opportunities for False Alarms

For a large flaw, the signal and noise distributions will be well separated, and the POD will be high and the POFA low as shown in Fig.5.13a. If the inspection is repeated on a smaller flaw, signal and noise probability density distributions will

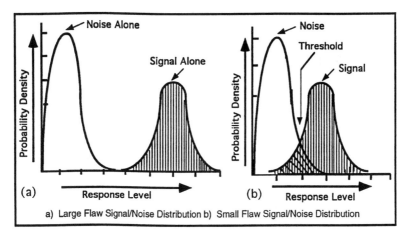

Figure 5.16 Signal/noise density distributions for large and medium size flaws

be closer together as depicted in Fig. 5.16b. For this flaw size, the signal and noise distributions overlap to a degree and the POD will decrease while the POFA increases [6].

NDI capability and reliability, however, are also affected by variations in the inspection process. Small flaws may be easily detected in the laboratory by a given technique, while large flaws may be missed by the same technique under less than ideal conditions. The factor of greatest importance in determining capability and reliability of an NDI technique is not the smallest flaw detected, but is defined by the largest flaw that could go undetected. Under this premise, NDI technique reliability is best determined by statistical methods. A valid statistical method of defining inspection performance is to specify percent probability (POD) at a given confidence level (C) for a crack of a given size. The values for POD and C commonly used for NDI reliability demonstrations have been 90% POD with 95% confidence level in order to meet MIL Handbook 5B requirements. The significance of establishing a confidence level for a given POD lies in the specification of a sample size to achieve statistical significance. The confidence level using a binomial distribution analysis is defined by the following [6]:

$$C = 1 - \sum_{X=S}^{N} (X^N)(\text{POD})^x(1 - \text{POD})^{N-x}$$

where

C = confidence level
N = sample size
S = the number of successes (flaws detected)
POD = the required probability of detection.

By specifying the confidence level (C) as 95% and the lower bound probability (POD) as 90%, a set of values for sample size (N) and number of successes (S) can be calculated. Each combination of N and S that satisfies the above equation indicates the number of inspections and the number of detections required to achieve the specified 90% POD at the 95% confidence level for a given flaw size. Table 5.2 lists the required number of successes for a given sample size that meets the 90% POD/95% confidence level criteria.

The binomial distribution analysis is an effective means for qualifying inspection personnel to a specific flaw size detection requirement. A more descriptive NDI capability is determined from the relationship between POD and crack size. This POD/crack size relationship is shown in Fig. 5.17. Various analytical techniques have been applied to generate POD/crack size curves. The two most widely applied statistical methods are the moving average method and the maximum likelihood method [6].

5.3.9 Variables Affecting NDI Flaw Detectability Testing

The results of reliability testing for a given NDI method are influenced by many factors. The tests are a measure not only of an inspector's ability to detect and characterize flaws but also of the capability of the equipment and processing

Table 5.2a Initial flaw size detectability assumptions for standard penetrant NDI processing

NDI method	Area	Part thickness (in.)	Crack type	Aspect ratio ($a/2c$)	Crack depth (in.)	Crack length (in.)
PT standard	Open surface	$t < 0.075$	TC		t	0.200
			PTC	0.1		0.250
				0.2		0.200
				0.5	(0.075 max.)	0.175
		$t \geqslant 0.075$	TC		t	0.150
			PTC	0.1		0.250
				0.2		0.200
				0.5		0.150
	Surface edges	$t < 0.075$	TC		t	0.100
		$t \geqslant 0.075$	CC		0.075	0.100

PT = Penetrant inspection
PTC = Part through crack
CC = Corner crack
TC = Through crack

Table 5.2b Initial flaw size detectability assumptions for standard eddy current NDI processing

NDI method	Area	Part thickness (in.)	Crack type	Aspect ratio ($a/2c$)	Crack depth (in.)	Crack length (in.)
ET standard	Open surface	$t \geq 0.020$	TC		t	0.100
			PTC	0.1		0.100
				0.2		0.100
				0.5		0.100
	Surface edges	$t \geq 0.020$	TC		t	0.050
			CC		0.020	0.050

ET = Eddy current inspection
PTC = Part through crack
CC = Corner crack
TC = Through crack

Table 5.2c Initial flaw size detectability assumptions for standard magnetic particle NDI processing

NDI method	Area	Part thickness (in.)	Crack type	Aspect ratio ($a/2c$)	Crack depth (in.)	Crack length (in.)
MT standard	Open surface	$t \geq 0.070$	TC		t	0.250
			PTC	0.1		0.375
				0.2		0.275
				0.5		0.250
	Surface edges	$t \geq 0.070$	TC		t	0.250
			CC		0.070	0.250

MT = Magnetic particle inspection
PTC = Part through crack
CC = Corner crack
TC = Through crack

steps. The following section briefly discusses the variables affecting each particular NDI method.

5.3.9.1 Variables Affecting the Liquid Penetrant Inspection

Liquid penetrant inspection is the simplest NDI method in principle. It is, however, a multistep process is which improper performance of any single step can

Table 5.2d Initial flaw size detectability assumptions for standard ultrasonic NDI processing

NDI method	Area	Part thickness (in.)	Crack type	Aspect ratio ($a/2c$)	Crack depth (in.)	Crack length (in.)
UT Standard (*L*-wave or *S*-wave)	Raw stock or machined	$t \geq 0.300$	Embedded or PTC, TC, CC	Class B	0.200 dia. (Equivalent Area)	
				Class A	0.130 dia. (Equivalent Area)	

UT = Ultrasonic inspection
PTC = Part through crack
CC = Corner crack
TC = Through crack

Table 5.2e Initial flaw size detectability assumptions for standard radiographic NDI processing

NDI method	Area	Part thickness (in.)	Crack type	Aspect ratio ($a/2c$)	Crack depth (in.)	Crack length (in.)
RT standard	Raw stock or machined	$t \geq 0.050$	Embedded	Ellipse	0.7t	t (0.150 min.)
			PTC, TC, CC	1/2 Ellipse	0.7t	t (0.150 min.)
		$t < 0.050$	Embedded	Ellipse	0.7t (0.025 min.)	0.150
			PTC, TC, CC	1/2 Ellipse	0.7t (0.025 min.)	0.150

RT = Radiographic inspection
PTC = Part through crack
CC = Corner crack
TC = Through crack

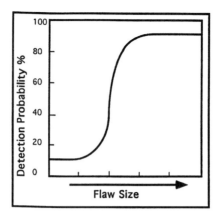

Figure 5.17 Typical probability of detection curve

seriously affect flaw detection. Surface preparation is the first area of concern, since flaws must be open to the surface for detection. If grinding, machining, or even polishing operations have been performed, flaws may be smeared over, reducing or precluding detectability. Acid etching a light layer (0.0002 in. typically) to remove the smeared metal may be required. Surfaces must also be clean and free from contaminants that may fill flaws preventing the penetrant dye from entering. Rough surfaces can also reduce detectability by holding penetrant dye in scratches and grooves, making the detection of defects difficult.

The selection and application of penetrant dyes also plays a serious role in defect detection. Penetrant dyes are classified as fluorescent (type I) or visible (type II) as well as by sensitivity level (1, 2, 3, or 4). Type I fluorescent dyes are easier to see and thus are more sensitive to defect detection. Sensitivity level refers to the brightness of the dye under ultraviolet light. The brighter the dye, the easier it is to detect. The brightest dyes are sensitivity level 4, specified mainly for highly stressed rotating engine component inspections. Penetrant dyes must also be allowed to dwell on the part long enough to penetrate into the defects. Removal of the penetrant must also be performed properly. Overzealous removal can wash out dye from shallow defects while underwashing will leave a significant background, adversely affecting detectability. Proper application of a light coating of developer is also important. Proper lighting during evaluation is also a concern. Excessive white light during fluorescent penetrant evaluations can reduce defect detection, as can using ultraviolet lights of inadequate intensity.

Figure 5.18a shows a POD curve developed using the maximum likelihood analysis method for a type I, sensitivity level 3 nonaqueous developer inspection of Haynes 188 alloy test specimens. The lower curve represents the 95% confidence level and indicates a flaw size of approximately 0.080 in. in length. In comparison, Figure 5.18b shows a POD curve for a type I sensitivity level 2 penetrant

5.3 Nondestructive Inspection Techniques

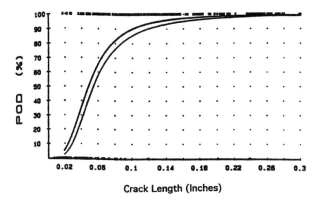

Figure 5.18a POD curve examples for fluorescent penetrant inspection (sensitivity level 2 penetrant). NASA/Inconel 718. Inspection #28. Type I. Method A. Form d. (Reproduced From Ref. [6])

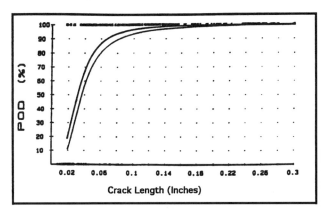

Figure 5.18b POD curve examples for fluorescent penetrant inspection (sensitivity level 3 penetrant). NASA/Inconel 718. Inspection #36. Type I. Method A. Form d. (Reproduced From Ref. [6])

with nonaqueous developer inspection of similar specimens. The 95% confidence level indicates a flaw size of about 0.120 in. in length [6].

5.3.9.2 Variables Affecting the Magnetic Particle Inspection

In reliability testing using magnetic particle methods the results are greatly affected by the proper development and application of the inspection technique. The proper selection of magnetizing currents is important to assure attraction of the ferromagnetic particles to the defect without obscuring its detection by too much current flow. As with penetrant inspection methods, magnetic particle detectability performance is also influenced by the type of materials used. Fluorescent particles provide better sensitivity; however, too much white light or inadequate ultraviolet light intensity will reduce sensitivity to defect detection. The orientation of the flaws away from 90° to the applied magnetic field will also decrease detectability, many times requiring multiple orientations and field strengths for complete inspection coverage. Magnetizing shots should also be performed in increasing field strength, as residual magnetism could obscure subsequent evaluations. Proper flow of the ferromagnetic particle medium during or immediately following magnetization is also important. Flaw depth can also affect inspection sensitivity. Subsurface flaw indications are usually less distinct than surface flaws. Complex part geometries and sharp edges also can produce false indications.

5.3.9.3 Variables Affecting the Eddy Current Inspection

Eddy current inspections are most seriously influenced by the equipment selection and scanning techniques. The pattern of eddy current flow is greatly affected by probe size, selected frequency, and the conductivity/magnetic permeability of the material being inspected. Higher frequencies and smaller probe coils provide a smaller eddy current field and thus increase the sensitivity to smaller surface defects. This, however, reduces the current flow into the material depth and reduces subsurface flaw detection capabilities. Other equipment settings such as gain and signal/noise ratio influence flaw detection performance.

Inspection of an entire part surface with eddy current requires scanning the probe over the surface. Defect detection can be affected by the speed of scanning, indexing increments, and the distance from the probe to the part, called lift-off. The flow of eddy current drops exponentially with distance from the probe coil. Rough surfaces and inconsistent probe pressure can adversely affect flaw detection capabilities of an eddy current inspection. Automated scanning methods with uniform speed, indexing, and probe pressure have been found to improve flaw detection capabilities. Figure 5.19 shows eddy current POD curves determined by the maximum likelihood analysis method on Haynes 188 test specimens using manual (Fig. 5.19a) and automated (Fig. 5.19b) inspection techniques [6].

5.3.9.4 Variables Affecting the Ultrasonic Inspections

Ultrasonic inspection flaw detectability is affected by many of the same variables as eddy current methods. Frequency selection and transducer design are significant factors. High-frequency scans offer greater sensitivity but are more easily attenuated, thus increasing signal noise. Smaller probe diameters or focused probes can also provide increased defect detectability. Complete part inspections with ultrasonics requires scanning the transducer over the surface using a liquid

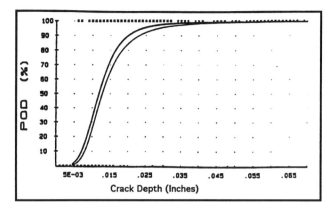

Figure 5.19a POD curve examples for eddy current inspections (manual inspection, differential probe). NASA/Haynes 188. Inspection #79. Eddy Current. 1 MHz. (Reproduced From Ref. [6])

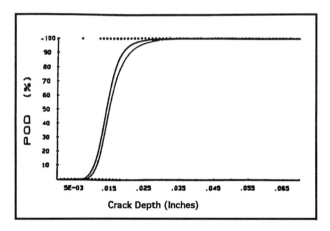

Figure 5.19b POD curve examples for eddy current inspections (automated inspection, absolute probe). Haynes 188. Inspection #82. Eddy Current. 1 MHz. (Reproduced From Ref. [6])

couplant such as water or oil. Insufficient couplant can cause signal loss while excess couplant on shear wave inspections can cause spurious reflections.

Flaw orientation and configuration also have a great effect on detectability performance. For optimum scan sensitivity, the ultrasonic beam must be oriented perpendicular to the flaw, thus providing maximum signal reflection back to the transducer. Also, as the ultrasonic signal propagates through a material it loses energy. Thus, a flaw located deep within a material produces an ultrasonic signal of less amplitude than an identical flaw located close to the surface. Figure 5.20

286   Chap. 5 Fracture Control Program and Nondestructive Inspection

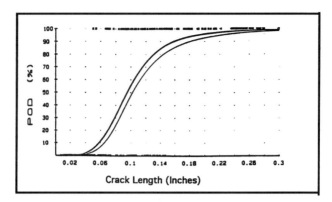

Figure 5.20 POD curve example for shear wave ultrasonic inspections. NASA/Iconel 718. Inspection #84. Hand Scan Ultrasonic (Reproduced From Ref. [6])

shows a POD curve determined by the maximum likelihood analysis method for a contact shear wave ultrasonic inspection of Inconel 718 specimens.

Similar to eddy current inspections, ultrasonic scans are also subject to scan speed and indexing inconsistencies. Automated scanning can be used to improve consistency, but may not always provide proper orientation for optimum flaw detection.

5.3.9.5 Variables Affecting the Radiographic Inspection

Radiographic defect detectability testing typically produces poor performance and relatively large maximum initial flaw sizes compared to other NDI methods. This can be attributed to the nature of the detectability test flaws which are exclusively fatigue cracks. Radiography is a volumetric inspection where density variations provide image contrast on photographic film for defect detection. While sensitive to changes of 1% to 2%, tight fatigue cracks do not provide significant volume unless the x-ray beam is oriented perfectly parallel with the crack. Other variables affecting detectability performance are x-ray tube kilovoltage and current, length of exposure time, type of x-ray film used, distance from the x-ray tube to the film, and distance from the flaw to the film. The farther a feature is from the film, the less distinct it appears.

X-ray images are essentially a 2-D density representation of a 3-D object. Complex or multilayer parts result in x-ray images with superimposed features that may be difficult to interpret. Interpretation is also adversely affected by excessive light during evaluation as well as defects in the x-ray film itself such as scratches and blemishes.

5.3.10 Critical Initial Flaw Sizes for Standard NDI

Conducting NDI performance testing for every flaw screening application would prove cost prohibitive for most manufacturing budgets. The need for standardized critical initial flaw sizes for fracture mechanics computations was evident. During the space shuttle development program in the 1970s, NASA addressed this issue and performed comprehensive NDI reliability testing in the five major NDI disciplines using a wide variety of inspection labs. The results were then compiled and minimum critical flaw sizes were established at a 90% POD and a 95% confidence level using standard NDI processing techniques. These are listed in Table 5.2 while the accompanying crack geometries are depicted in Figure 5.21 [7]. These flaw sizes represent the smallest flaws that can be detected using a particular NDI method at the 90% POD and 95% confidence level employing NDI industry standard processing methods. These results have been utilized on numerous military and commercial programs throughout the aerospace industry over the past 20 years. Recently, NASA has undertaken a program to reexamine the NDI POD flaw sizes in light of advanced processing techniques. Initial results indicate that although NDI evaluations are now performed in less time and at lower costs, the critical initial flaw detectability limit using standard processing methods is relatively unchanged from the initial study. The initial flaw sizes listed in Table 5.2 can be used by an analyst to conduct safe-life analysis of fracture critical parts. If the results of the analysis indicated that the part does not have sufficient life, a smaller initial flaw size must be assumed. Section 5.3.11 addresses the special NDI inspection method for obtaining the desire initial flaw size (smaller than the standard initial flaw sizes listed in Table 5.2) to ensure safe-life.

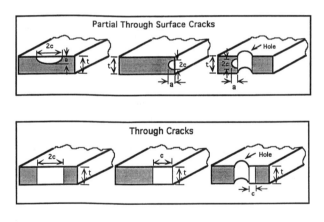

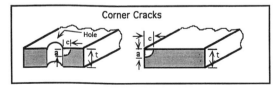

Figure 5.21 Initial crack geometries

5.3.11 Special Level Inspection

For some critical design applications, the maximum initial flaw sizes verifiable using standard NDI processing techniques are still too large to assure safe life. For these applications, NDI methods may still be used to screen for the desired flaw sizes by employing special level NDI. To perform special level NDI an inspector must first undergo NDI reliability testing in the applicable discipline using test materials and processing techniques similar to those of the components of interest. Results of this testing must verify that the inspector can detect flaws as small as the maximum initial design flaw at a 90% POD and 95% confidence level. Recertification is then required a minimum of every two years.

References

1. R. C., McMaster, (Ed.), Liquid Penetrant Tests, Nondestructive Testing Handbook, Second Edition, Vol. 2, American Society for Nondestructive Testing, 1982.
2. J. T., Schmidt and K. Skeie, (Technical Eds.), and P. MacIntire Ed., Magnetic Particle Testing, Nondestructive Testing Handbook, Second Edition, Vol. 6, American Society for Nondestructive Testing, 1989.
3. L. E. Bryant, (Technical Ed.) and P. MacIntire (Ed.), Radiography and Radiation Testing, Nondestructive Testing Handbook, Second Edition, Vol. 3, American Society for Nondestructive Testing, 1985.
5. A. S. Birks and R. E. Green, Jr. (Technical Eds.), and P. MacIntire (Ed.), Ultrasonic Testing, Nondestructive Testing Handbook, Second Edition, Vol. 7, American Society for Nondestructive Testing, 1991.
5. M. L. Mester, (Technical Ed.), and P. MacIntire, (Ed.), Electromagnetic Testing, Nondestructive Testing Handbook, Second Edition, Vol. 4, American Society for Nondestructive Testing, 1986.
6. B. K. Christner, D. L. Long, and W. D. Rummel, NDI Detectability of Fatigue-Type Cracks in High-Strength Alloys, NDI Reliability Assessments, MCR-88-1044, Martin Marietta Astronautics, 1988.
7. Standard NDE Guidelines and Requirements for Fracture Control Programs, MSFC-STD-1249, NDI Branch, Materials and Processes Laboratory, NASA-MSFC, 1985.

Chapter 6

The Fracture Mechanics of Ductile Metals Theory

6.1 Introduction

Almost all metallic materials manifest some plastic deformation in the region at the crack tip before catastrophic crack propagation. The general principle from which the Griffith theory [1] (see Section 3.1 of Chapter 3) is derived is not limited to materials that obey Hooke's law. The principle applies as well when dissipative mechanisms, such as plastic deformation, are present. Irwin and Orowan [2, 3] showed that Griffith's principle can also be applied to materials that manifest ductile behavior, that is:

$$\frac{\partial}{\partial c}[U_E - U_S - U_P] = 0 \qquad (6.1)$$

where U_E is the stored energy, and U_P is the energy consumed per unit thickness in plastic straining in the region at the crack tip. For ductile metals, where $U_P \gg U_S$, the expression for surface energy, U_S, was omitted from Eq. (6.1) to obtain:

$$\frac{\partial}{\partial c}[U_E - U_P] = 0 \qquad (6.2)$$

From Eq. (6.2), the following well-known Irwin–Orowan equivalent of the Griffith energy balance equation is obtained:

$$G_c = \frac{\pi \sigma^2 c}{E} = \frac{K_c^2}{E} \qquad (6.3)$$

290   Chap. 6 The Fracture Mechanics of Ductile Metals Theory

The quantity to the left is the energy release rate or the crack extension force at the instability, G_c, and has dimensions of force per unit of extension. The crack extension force, G, is the same for the cracked plate under fixed grip condition (where the external load cannot do work) and free ends where the work is done by the external load. In the former case the available energy is delivered by the elastic energy and in the latter case by the load. The quantity to the right of Eq. (6.3) is the critical value of the stress intensity factor, K_c, called fracture toughness which, for the thick sections when plane strain condition prevails, is designated by K_{1c}. Fracture toughness values for thicknesses smaller than the plane strain case are referred to as mixed mode where the maximum value is associated with the plane stress condition labeled as K_c. In Section 6.2, the fracture mechanics of ductile metals (FMDM) are discussed. A relationship between fracture stress and the corresponding critical half crack length is developed that enables the analyst to estimate the residual strength capability of the structural part without relying on test data generated in the laboratory. To verify the results and accuracy of the FMDM theory, several comparisons were made with the experimental data provided from different reliable sources. The correlation between the FMDM theory and the experimental data with various alloys is presented in Section 6.9. In addition, the fracture toughness values for several alloys were calculated by the FMDM approach and the results compared with ASTM fracture toughness testing (Section 6.10).

To appreciate the simplicity of calculating the fracture toughness with the FMDM theory, the reader may refer to Chapter 3 pertaining to the complexities associated with the ASTM specimen preparation procedures, including the machine starter slot, fatigue cracking the specimens prior to testing, method of applying load to the specimen, load recording, and crack measurements.

6.2 Fracture Mechanics of Ductile Metals

The theory of ductile fracture assumes the fracture characteristics of a metal, local to the crack tip, are directly related to strainability of metals which is of two kinds: (1) local strainability (necking) at the crack tip, the region of highly plastic deformation, and (2) uniform strainability near the crack tip. Thus, fracture is characterized by two parameters that represent the absorbed energy associated with two plastic deformation regions which can be shown to be determinable from the uniaxial stress–strain curve. The two aforementioned strained regions are illustrated in Figs. 6.1 and 6.2 for the crack tip and uniaxial stress–strain curves, respectively.

The total energy per unit thickness absorbed in plastic straining of the material around the crack tip, U_P, from Eq. (6.1), can be written as:

$$U_P = U_F + U_U \tag{6.4}$$

6.3 Determination of the $\partial U_F/\partial c$ Term

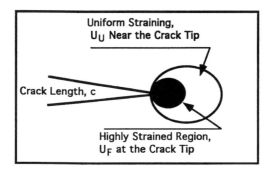

Figure 6.1 The crack tip plastic zones

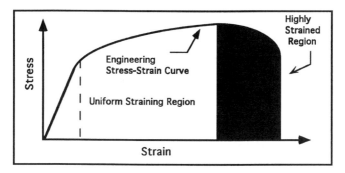

Figure 6.2 Different regions of the stress–strain curve

where U_F and U_U are the energy absorbed per unit thickness in plastic straining of the material beyond the ultimate at the crack tip and below the ultimate stress near the crack tip, respectively. Equation (6.1), in terms of U_F and U_U, described by Eq. (6.4), can be rewritten as:

$$\frac{\partial}{\partial c}[U_E - U_S - U_F - U_U] = 0 \qquad (6.5)$$

where $\frac{\partial U_F}{\partial c}$ and $\frac{\partial U_U}{\partial c}$ are the rates at which energy is absorbed in plastic straining beyond the ultimate stress at the crack tip and below the ultimate stress near the crack tip, respectively.

To obtain the residual strength capability curve of a cracked plate subjected to tension (the plot of the applied stress, σ_c, versus the critical half crack length, c) it is therefore necessary to determine the energy absorption rates for the two plastic regions around the crack tip. Sections 6.3 and 6.4 discuss the theoretical approach and assumptions used in obtaining the $\frac{\partial U_F}{\partial c}$ and $\frac{\partial U_U}{\partial c}$ terms for the plane stress condition. The plane strain condition is also discussed in Section 6.5.

6.3 Determination of the $\frac{\partial U_F}{\partial c}$ Term

To obtain the energy absorption rate for the highly strained region at the crack tip, $\frac{\partial U_F}{\partial c}$, let us assume that the remaining bulk of the material (except at the crack tip) is uniformly strained up to the ultimate of the material. This can be done by applying to the cracked plate an applied stress equal to the ultimate strength of the material, $\sigma_{\text{Applied}} = \sigma_{U\ell}$. In this case, the term U_U of Eq. (6.5) can be eliminated, since there is no energy absorption contribution at the crack tip from the U_U term (see Fig. 6.1). For $\sigma_{\text{Applied}} = \sigma_{U\ell}$, Eq. (6.5) becomes:

$$\frac{\pi \sigma_U^2 c_U}{E} = \frac{\partial}{\partial c}[U_S + U_F]$$

or

$$\frac{\pi \sigma_U^2 c_U}{E} = 2T + \frac{\partial U_F}{\partial c} \qquad (6.6)$$

where c_U is the crack length associated with $\sigma_{\text{Applied}} = \sigma_{U\ell}$. This value is important in the field of fracture mechanics when initial flaw size for fatigue crack growth analysis is needed (the quantity c_U can be viewed as the largest flaw size preexisting in the material). At any other applied stress below the ultimate, $\sigma_{\text{Applied}} < \sigma_{U\ell}$, Eq. (6.5) becomes:

$$\frac{\pi \sigma^2 c}{E} = 2T + \frac{\partial U_F}{\partial c} + \frac{\partial U_U}{\partial c} \qquad (6.7)$$

where T is the surface tension of the material, that is, the work done in breaking the atomic bonds and its value was given by Eq. (3.4) of Chapter 3. The quantity $\frac{\partial U_F}{\partial c}$ in Eq. (6.6) is set equal to $W_F h_F$, where W_F is equal to the unrecoverable energy density (energy per unit volume) represented by the area under the plastic uniaxial engineering stress–strain curve from the stress at which necking begins to the stress at fracture for an uncracked tensile specimen (see Fig. 6.3 for different types of stress–strain curves). It is assumed that essentially all of the energy represented by W_F is absorbed at the crack tip in a single dominant coarse slip band. The quantity h_F is the effective height of the highly strained portion of the small region at the crack tip, and its minimum value is considered to be equal to the effective height of a coarse slip band oriented to make an angle of 45° with the plane of the crack. The movement of the two slip surfaces by the amount of h_F due to movement of dislocations results in release of energy in the locality of the crack tip. The size of coarse slip bands at high stress and at room temperature is approximately 10 micrometers [4, 5] and therefore:

$$h_F = 0.00057 \text{ in.} \qquad (6.8a)$$

6.3 Determination of the $\partial U_F/\partial c$ Term

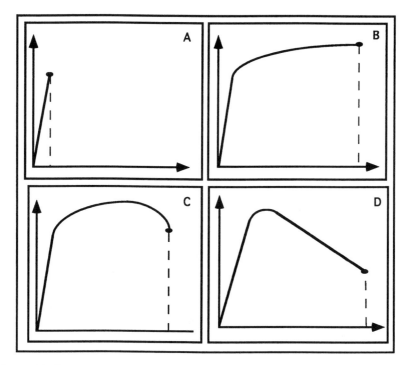

Figure 6.3 Typical engineering stress–strain curves for: (A) brittle materials, (B) metals that do not neck, (C) many structural metals, (D) some precipitation-hardening steel

An empirical relationship based on test data [6] has been developed for tough metals that have large "neck" strains (see for example case D of Fig. 6.3) which gives a higher h_F value and better correlation with the test data than indicated by Eq. (6.8a):

$$h_F = \left(\frac{8W_F E^2}{\pi \sigma_{U\ell}^3}\right)\alpha \tag{6.8b}$$

where α is equal to 6.31×10^{-5} and 5.68×10^{-5} for iron nickel and beryllium alloys, respectively. The area under the stress–strain curve from necking up to fracture, W_F, is equal to:

$$W_F = \overline{\sigma}_{UF}\varepsilon_{PN} \tag{6.9}$$

where the quantity $\overline{\sigma}_{UF}$ is the neck stress and its value is at the centroid of the plastic energy bounded on the top by the engineering stress–strain curve from the beginning of necking to fracture and on the bottom by a straight line from the beginning of necking to fracture. For material with a negligible amount of necking the neck stress $\overline{\sigma}_{UF} = (\sigma_U + \sigma_F)/2$. Moreover, ε_{PN} is the plastic straining at the onset of necking and its derivation is shown in Appendix B. Equation (6.7) in terms of h_F, $\overline{\sigma}_{UF}$, and ε_{PN} can be expressed as:

$$\frac{\pi\sigma^2 c}{E} = 2T + h_F \overline{\sigma}_{UF} \varepsilon_{PN} + \frac{\partial U_U}{\partial c} \qquad (6.7a)$$

The critical flaw size associated with applied stress equal to the ultimate stress of the material, c_U, described by Eq. (6.6), can be written as:

$$c_U = (2T + h_F \overline{\sigma}_{UF} \varepsilon_{PN}) \frac{E}{\pi\sigma_U^2} \qquad (6.10)$$

As discussed in Chapter 4, for structural life evaluation, when the load environment is cyclic, the initial crack size assumption (the original flaw that preexisted in the material prior to its usage) must be available in order to conduct a meaningful fracture mechanics analysis. From equation 6.10 it can be seen that for material that do not exhibit necking (where $h_F \sigma_{UF} \varepsilon_{PN} = 0$) the critical flaw size at ultimate, c_U, becomes small. This is in accordance with the experimental observation which shows that a brittle material can tolerate a much smaller initial flaw at failure as compared with ductile materials with an appreciable amount of necking prior to their final failure.

6.4 Determination of the $\frac{\partial U_U}{\partial c}$ Term

The quantity $\frac{\partial U_U}{\partial c}$ can be expressed as $W_U h_U$, where W_U is the unrecoverable energy density represented by the area under the plastic uniaxial stress–strain curve from the elastic limit stress, σ_L, to the ultimate stress, $\sigma_{U\ell}$, for an uncracked specimen. Equation (6.7a) becomes:

$$\frac{\pi\sigma^2 c}{E} = 2T + h_F \overline{\sigma}_{UF} \varepsilon_{PN} + W_U h_U \qquad (6.7b)$$

The effective height of the volume in which W_U is absorbed is h_U. The expression for $W_U h_U$ is derived in Sections 6.4.1 and 6.5 for the two cases of plane stress and plane strain conditions, respectively. In Section 6.4.1, the octahedral shear stress theory is used to develop an expression for the total true strain near the crack tip from elastic limit load up to the ultimate of the material. This theory is used here

6.4 Determination of the $\partial U_U/\partial c$ Term

to describe the crack tip multiaxial stress state in terms of the uniaxial stress obtainable through laboratory testing.

6.4.1 Octahedral Shear Stress Theory (Plane Stress Conditions)

From linear elastic fracture mechanics (LEFM), the crack tip true stresses for plane stress conditions (when $r/c \ll 1.0$) were derived in Section 3.2 of Chapter 3. In polar coordinates the crack tip true stress can be written as:

$$\sigma_{Tr}(r, \theta) = \frac{\sigma}{\sqrt{2}} \sqrt{\frac{c}{r}} \left(\frac{5}{4} \cos \frac{\theta}{2} - \frac{1}{4} \cos 3 \frac{\theta}{2} \right)$$

$$\sigma_{T\theta}(r, \theta) = \frac{\sigma}{\sqrt{2}} \sqrt{\frac{c}{r}} \left(\frac{3}{4} \cos \frac{\theta}{2} + \frac{1}{4} \cos 3 \frac{\theta}{2} \right)$$

$$\sigma_{Tr\theta}(r, \theta) = \frac{\sigma}{\sqrt{2}} \sqrt{\frac{c}{r}} \left(\frac{1}{4} \sin \frac{\theta}{2} + \frac{1}{4} \sin 3 \frac{\theta}{2} \right) \quad (6.11)$$

In Eq. (6.11), it is assumed that the true stresses are approximately equal to the engineering stresses within the elastic range. For simplicity let the quantities inside the parentheses [shown by Eq. (6.11)] be represented by:

$$\psi_r(\theta) \equiv \frac{5}{4} \cos \frac{\theta}{2} - \frac{1}{4} \cos 3 \frac{\theta}{2}$$

$$\psi_\theta(\theta) \equiv \frac{3}{4} \cos \frac{\theta}{2} + \frac{1}{4} \cos 3 \frac{\theta}{2}$$

$$\psi_{r\theta}(\theta) \equiv \frac{1}{4} \sin \frac{\theta}{2} + \frac{1}{4} \sin 3 \frac{\theta}{2} \quad (6.12)$$

Substituting Eq. (6.12) in Eq. (6.11) the crack tip true stress can be rewritten as:

$$\sigma_{Tr}(r, \theta) = \frac{\sigma}{\sqrt{2}} \sqrt{\frac{c}{r}} \psi_r(\theta) \quad (6.13a)$$

$$\sigma_{T\theta}(r, \theta) = \frac{\sigma}{\sqrt{2}} \sqrt{\frac{c}{r}} \psi_\theta(\theta) \quad (6.13b)$$

$$\sigma_{Tr\theta}(r, \theta) = \frac{\sigma}{\sqrt{2}} \sqrt{\frac{c}{r}} \psi_{r\theta}(\theta) \quad (6.13c)$$

The equation for octahedral shear stress, τ_{OCT} (similar to the strain energy of distortion) satisfactorily correlates states of biaxial tension stress (σ_{Tr} and $\sigma_{T\theta}$) with uniaxial tension stress, σ_T, and is used here to relate the stresses in the zone just ahead of the crack tip to the uniaxial tension stress, σ_T [7]. The octahedral shear stress, τ_{OCT}, measures the intensity of stress that is responsible for bringing a solid substance into the plastic state. The equation for octahedral shear in terms of crack tip stresses is:

$$\tau_{OCT} = \frac{1}{3}[(\sigma_x - \sigma_y)^2 + (\sigma_y - \sigma_z)^2 + (\sigma_z - \sigma_x)^2 + 6(\tau_{xy}^2 + \tau_{yz}^2 + \tau_{zx}^2)]^{1/2}$$

or in terms of r and θ:

$$\tau_{OCT} = \frac{1}{3}[(\sigma_{Tr} - \sigma_{T\theta})^2 + (\sigma_{T\theta} - \sigma_{Tz})^2 + (\sigma_{Tz} - \sigma_{Tr})^2 + 6(\tau_{Tr\theta}^2 + \tau_{T\theta z}^2 + \tau_{Tzr}^2)]^{1/2} \quad (6.14)$$

For plane stress, $\sigma_{Tz} = 0$, $\tau_{T\theta z} = 0$, and $\tau_{Tzr} = 0$ and Eq. (6.14) becomes:

$$\tau_{OCT} = \frac{1}{3}[(\sigma_{Tr} - \sigma_{T\theta})^2 + \sigma_{T\theta}^2 + \sigma_{Tr}^2 + 6\sigma_{Tr\theta}^2]^{1/2} \quad (6.15)$$

$$= \frac{\sqrt{2}}{3}\sigma_{Tr}\left[1 - \frac{\sigma_{T\theta}}{\sigma_{Tr}} + \left(\frac{\sigma_{T\theta}}{\sigma_{Tr}}\right)^2 + 3\left(\frac{\sigma_{Tr\theta}}{\sigma_{Tr}}\right)^2\right]^{1/2} \quad (6.16)$$

For simple uniaxial stress, σ_T, where $\sigma_{T\theta} = 0$, and $\sigma_{Tr\theta} = 0$:

$$\tau_{OCT} = \frac{\sqrt{2}}{3}\sigma_T \quad (6.17)$$

Equating τ_{OCT} [Eq. (6.16)] for uniaxial stress to τ_{OCT} [Eq. (6.17)] for plane stress:

$$\sigma_T = \sigma_{Tr}\left[1 - \frac{\sigma_{T\theta}}{\sigma_{Tr}} + \left(\frac{\sigma_{T\theta}}{\sigma_{Tr}}\right)^2 + 3\left(\frac{\sigma_{Tr\theta}}{\sigma_{Tr}}\right)^2\right]^{1/2} \quad (6.18)$$

or,

$$\sigma_T = \sigma_{Tr}\left[1 - \frac{\psi_\theta(\theta)}{\psi_r(\theta)} + \left(\frac{\psi_\theta(\theta)}{\psi_r(\theta)}\right)^2 + 3\left(\frac{\psi_{r\theta}(\theta)}{\psi_r(\theta)}\right)^2\right]^{1/2} \quad (6.19)$$

6.4 Determination of the $\partial U_U/\partial c$ Term

and,

$$\sigma_T = \frac{\sigma_{Tr}}{\psi_r(\theta)} [\psi_r^2(\theta) - \psi_r(\theta)\psi_\theta(\theta) + \psi_\theta^2(\theta) + 3\psi_{r\theta}^2(\theta)]^{1/2} \qquad (6.20)$$

Let the quantity inside the bracket be equal to $\psi_{oc}(\theta)$:

$$\psi_{oc}(\theta) = [\psi_r^2(\theta) - \psi_r(\theta)\psi_\theta(\theta) + \psi_\theta^2(\theta) + 3\psi_{r\theta}^2(\theta)]^{1/2} \qquad (6.21)$$

Substituting Eq. (6.21) in Eq. (6.20):

$$\sigma_T = \frac{\sigma_{Tr}}{\psi_r(\theta)} \psi_{oc}(\theta) \qquad (6.22)$$

The true crack tip uniaxial stress, σ_T, in terms of applied stress, σ, crack length, c, and the crack tip distance, r, using Eq. (6.13a) becomes:

$$\sigma_T = \frac{\sigma}{\sqrt{2}} \sqrt{\frac{c}{r}} \psi_{oc}(\theta) \qquad (6.23)$$

and in terms of strain, ε_T, Eq. (6.23) becomes:

$$\varepsilon_T = \frac{\sigma}{\sqrt{2}E} \sqrt{\frac{c}{r}} \psi_{oc}(\theta) \qquad (6.24)$$

Equations (6.23) and (6.24) define the variation of true stress and strains at the crack tip up to the elastic limit of the material, ε_{TL}. For small plastic deformation (when the crack tip plastic deformation is small and the bulk of the material is elastic) the quantity, σ_T, can still be determined by Eq. (6.23) with reasonable accuracy. However, for ductile materials where an appreciable amount of straining occurs at the crack tip, equation 6.23 must be avoided when assessing the stresses in that locality. Eq. (6.24) can be extended to describe the variation of uniform true straining, ε_T, from the elastic limit to the ultimate of the material. Figure 6.4 illustrates the variations of true elastic stress, σ_T, and strain, ε_T, as described by LEFM, and plastic strains from limit up to ultimate strain with respect to the crack tip distance, r, for a typical metal. The quantity ε_T as described by Eq. (6.24) can be used to determine the rate at which plastic energy is absorbed in the larger zone (uniform plastic region). The sum of all the elements acting on the area inside the uniform plastic zone that contribute to energy absorption rate as the crack extends, can be expressed as (Fig. 6.5):

298   Chap. 6 The Fracture Mechanics of Ductile Metals Theory

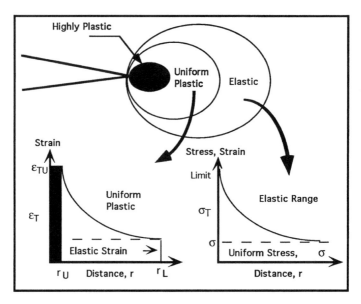

Figure 6.4 (1) Stress and strain distribution in the elastic region up to the limit and (2) uniform strain distribution near the crack tip up to the ultimate

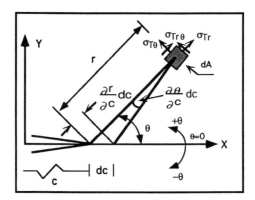

Figure 6.5 Illustration of an element (*dA*) chosen in the plastic zone for dU_U/d_c analysis

$$\frac{\partial U_U}{\partial c} = \int_{(A)} \sigma_T \frac{\partial \varepsilon_{TP}}{\partial c} dA = 2 \int_0^{\theta_c} \int_{r_U}^{r_L} \sigma_T \frac{\partial \varepsilon_{TP}}{\partial c} r \, dr \, d\theta \qquad (6.25)$$

where θ_c is the angle at which the plastic energy absorption rate, $\frac{\partial \varepsilon_{TP}}{\partial c}$, vanishes, i.e., the angle at which no contribution to plastic deformation is made at the crack

6.4 Determination of the $\partial U_U/\partial c$ Term

tip. Replacing the total strain expressed via Eq. (6.24) into Eq. (6.25) (by assuming the total strain is approximately equal to the plastic strain, $\varepsilon_T \approx \varepsilon_{TP}$):

$$\frac{\partial U_U}{\partial c} = 2 \int_0^{\theta_c} \int_{r_u}^{r_L} \sigma_T \frac{\partial \varepsilon_T}{\partial c} r \, dr \, d\theta \quad (6.26)$$

When the magnitude of the applied stress is in such a way that the net section yielding is avoided except in the localized region at the crack tip, the limits of the integral, described by Eq. (6.26), are correct. However, when the applied stress is above the elastic limit of the material or when the bulk of the structure due to the applied stress is plastic, Eq. (6.2) can be rewritten as:

$$\frac{\partial U_U}{\partial c} = 2 \int_0^{\theta_c} \int_{r_u}^{r} \sigma_T \frac{\partial \varepsilon_T}{\partial c} r \, dr \, d\theta \quad (6.26a)$$

Note that when the applied stress becomes larger than the elastic limit stress of the material, the plastic deformation contribution at the crack tip due to the energy absorption rate term associated with the uniform plastic region (shown in Fig. 6.4) is less and only a portion of it is included in Eq. (6.26). Later in Section 6.5, the FMDM results will be given in a general form, where both cases of applied stress (above or below the elastic limit of the material) can be used for obtaining the corresponding half critical crack length, c. The reader should note that one of the limitations associated with the LEFM, is that, the formation of the crack tip plasticity must be limited to the small region at the crack tip and the bulk of material must be elastic. Otherwise, the LEFM cannot be utilized to assess the residual strength capability determination of a part when the applied stress is above the yield of the material or the bulk of the structure is plastic.

From Eq. (6.24), the quantity $\frac{\partial \varepsilon_T}{\partial c}$ can be simplified as follows:

$$\frac{\partial \varepsilon_T}{\partial c} = \frac{\sigma \varphi_{oc}(\theta)}{2\sqrt{2} E \sqrt{c_r}} - \frac{\sigma \varphi_{oc}(\theta)}{2\sqrt{2} Er \sqrt{r}} \frac{\partial r}{\partial c} \quad (6.24a)$$

$$+ \frac{\sigma}{\sqrt{2} E} \sqrt{\frac{c}{r}} \frac{\partial \varphi_{oc}(\theta)}{\partial \theta} \frac{\partial \theta}{\partial c} + \frac{\partial \sigma}{\partial c} \frac{\varepsilon_T}{\sigma}$$

Under constant load conditions, the stress would be constant as the crack begins to grow catastrophically. In this case, $\frac{\partial \sigma}{\partial c}$ would be zero and the last term in Eq. (6.24a), $\frac{\partial \sigma}{\partial c} \frac{\varepsilon_T}{\sigma}$, would vanish. From simplification of Eq. (6.24a):

$$\frac{\partial \varepsilon_T}{\partial c} = \varepsilon_T \left[\frac{1}{2c} - \frac{1}{2r} \frac{\partial r}{\partial c} + \frac{\frac{\partial}{\partial \theta} \varphi_{oc}(\theta)}{\varphi_{oc}(\theta)} \frac{\partial \theta}{\partial c} \right] \quad (6.27)$$

Chap. 6 The Fracture Mechanics of Ductile Metals Theory

From Fig. 6.5, the quantities $\frac{\partial r}{\partial c}$ and $\frac{\partial \theta}{\partial c}$ can be replaced by:

$$\frac{\partial r}{\partial c} = -\cos\theta, \text{ and } \frac{\partial \theta}{\partial c} = \frac{\sin(\theta)}{r} \tag{6.28}$$

Substituting Eq. (6.28) in Eq. (6.27):

$$\frac{\partial \varepsilon_T}{\partial c} = \frac{\varepsilon_T}{2r}\left[\frac{r}{c} + \cos\theta + 2\sin\theta \frac{\frac{\partial}{\partial \theta}\varphi_{oc}(\theta)}{\varphi_{oc}(\theta)}\right] \tag{6.29}$$

But, $\frac{r}{c} \ll 1$; therefore:

$$\frac{\partial \varepsilon_T}{\partial c} = \frac{\varepsilon_T}{2r}\left[\cos\theta + 2\sin\theta \frac{\frac{\partial}{\partial \theta}\varphi_{oc}(\theta)}{\varphi_{oc}(\theta)}\right] \tag{6.30}$$

Let

$$f(\theta) = \left[\cos\theta + 2\sin\theta \frac{\frac{\partial}{\partial \theta}\varphi_{oc}(\theta)}{\varphi_{oc}(\theta)}\right] \tag{6.31}$$

Substituting Eq. (6.31) in Eq. (6.30):

$$\frac{\partial \varepsilon_T}{\partial c} = \frac{\varepsilon_T}{2r} f(\theta) \tag{6.32}$$

Substituting Eq. (6.32) in Eq. (6.26a), the energy absorption rate can be written as:

$$\frac{\partial U_U}{\partial c} = 2\int_0^{\theta_c}\int_{r_U}^{r_L} \sigma_T \frac{\varepsilon_T}{2r} f(\theta) r\, dr\, d\theta \tag{6.33}$$

$$= \int_0^{\theta_c}\int_{r_U}^{r_L} \sigma_T \varepsilon_T f(\theta)\, dr\, d\theta \tag{6.33a}$$

For metals whose true stress versus true plastic strain curve can be approximated by the Ramberg–Osgood Equation (when fitted at σ_{TL} and σ_{TU}) the true stress in terms of true plastic strain can be written as:

6.4 Determination of the $\partial U_U/\partial c$ Term

$$\sigma_T = \sigma_{TU}\left[\frac{\varepsilon_{TP}}{\varepsilon_{TPU}}\right]^{1/n} \quad (6.34)$$

The quantity ε_{TPU} is the true plastic strain at ultimate and n is the strain hardening coefficient (for perfectly plastic material, the value of $n = \infty$) and can be described by:

$$n = \frac{\ln\left(\dfrac{\varepsilon_{TPU}}{\varepsilon_{TPL}}\right)}{\ln\left(\dfrac{\sigma_{TU}}{\sigma_{TL}}\right)} \quad (6.35)$$

where ε_{TPL} is the true plastic strain at the elastic limit. For material with an appreciable amount of plastic strain where $\varepsilon_{TP} \approx \varepsilon_T$ and $\varepsilon_{TPN} \approx \varepsilon_{TN}$

$$\sigma_T = \sigma_{TU}\left[\frac{\varepsilon_T}{\varepsilon_{TU}}\right]^{1/n} \quad (6.36)$$

Substituting Eq. (6.36) in Eq. (6.33):

$$\frac{\partial U_U}{\partial c} = \sigma_{TU}\varepsilon_{TU}\int_0^{\theta_c}\int_{r_U}^{r_L}\left[\frac{\varepsilon_T}{\varepsilon_{TU}}\right]^{\frac{n+1}{n}} f(\theta)\,dr\,d\theta \quad (6.37)$$

From Eq. (6.24), the ratio of the quantity $\varepsilon_T/\varepsilon_{TU}$ can be replaced by its equivalent:

$$\frac{\varepsilon_T}{\varepsilon_{TU}} = \sqrt{\frac{r_U}{r}} \quad (6.38)$$

Substituting Eq. (6.38) in Eq. (6.37)

$$\frac{\partial U_U}{\partial c} = \sigma_{TU}\varepsilon_{TU}\int_0^{\theta_c}\int_{r_U}^{r_L}\left[\frac{r_U}{r}\right]^{\frac{n+1}{2n}} f(\theta)\,dr\,d\theta \quad (6.39)$$

Integrating and simplifying equation 6.39:

$$\frac{\partial U_U}{\partial c} = \sigma_{TU}\varepsilon_{TU}\int_o^{\theta_c}\frac{2n}{n-1}r_U\left[\left(\frac{r}{r_U}\right)^{\frac{n-1}{2n}}\right]_{r_U}^{r_L} f(\theta)\,d\theta$$

$$\frac{\partial U_U}{\partial c} = \frac{n}{n-1}\sigma_{TU}\varepsilon_{TU}\left[\left(\frac{r_L}{r_U}\right)^{\frac{n-1}{2n}} - 1\right]\int_0^{\theta_c} 2r_U f(\theta)\,d\theta \quad (6.40)$$

From test data for plane stress and infinite width, an empirical relationship has been established that can describe the r_U value for ductile metals [8]:

$$2r_U = Q\left(\frac{1}{\varepsilon_T}\right)\left[1 - \left(\frac{\sigma_T}{\sigma_{TU}}\right)^{n+1}\right]\psi_{oc}^2(\theta) \qquad (6.41)$$

where $Q = h(\varepsilon_{TF}\varepsilon_{TL})/\varepsilon_{TU}$ [the value of $h = 0.00057$ in.; see Eq. (6.8)]. For finite width, w:

$$2r_U = \frac{Q}{Y^6}\left(\frac{1}{\varepsilon_T}\right)\left[1 - \left(\frac{\sigma_T}{\sigma_{TU}}\right)^{n+1}\right]\psi_{oc}^2(\theta) \qquad (6.41a)$$

For center cracked panels, the width correction, Y, is expressed as:

$$Y = \frac{1}{\left(\cos\dfrac{\pi}{2}\dfrac{2c}{W}\right)^{1/2}}$$

The quantity r_U represents the radius of uniform plastic zone which extends from $r = r_L$ to $r = r_U$. This quantity is a function of ε_{TL}, n, σ_{TU}, and ε_T which in turn is a function of applied stress and the crack length via Eq. (6.24). The simplified presentations of plastic zone size through Irwin's and Von Mises' yield criteria was discussed in Chapter 3 [see Eq. (3.42)].

Substituting Eq. (6.41) in Eq. (6.40) and noting that the ratio of $(r_L/r_U)^{1/2} = \varepsilon_{TU}/\varepsilon_{TL}$ [see Eq. 6.24]:

$$\frac{\partial U_U}{\partial c} = \frac{n}{n-1}\sigma_{TU}\varepsilon_{TU}\left[1 - \left(\frac{\sigma_T}{\sigma_{TU}}\right)^{n+1}\right]h\left[\frac{\varepsilon_{TF}\varepsilon_{TL}}{\varepsilon_{TU}\varepsilon_T}\right]$$

$$\left[\left(\frac{\varepsilon_{TU}}{\varepsilon_{TL}}\right)^{\frac{n-1}{n}} - 1\right]\underbrace{\int_0^{\theta_c}\psi_{oc}^2(\theta)f(\theta)d\theta}_{\beta} \qquad (6.42)$$

By numerical evaluation the thickness parameter, β, for plane stress is:

$$\beta = \int_0^{\theta_c}\psi_{oc}^2(\theta)f(\theta)d\theta = 1.30 \qquad (6.43)$$

The term thickness parameter, β, is used here to show the distinction between the energy absorption rate, $\frac{\partial U_U}{\partial c}$, defined by Eqs. (6.42) and (6.49) (defined in Section

6.5) for the two cases of plane strain and plane stress conditions, respectively; see Eqs. (6.43) and (6.50a) which describe the β values.

For the general case of loadings where the applied stress could be either below or above the elastic limit stress of the material, Eq. (6.42) can be rewritten as:

$$\frac{\partial U_U}{\partial c} = \frac{n}{n-1} \sigma_{TU} \varepsilon_{TU} \left[1 - \left(\frac{\sigma_T}{\sigma_{TU}}\right)^{n+1}\right] h \left[\frac{\varepsilon_{TF} \varepsilon_{TL}}{\varepsilon_{TU} \varepsilon_T}\right]$$

$$\left[\left(\frac{\varepsilon_{TU}}{\varepsilon_T}\right)^{\frac{n-1}{n}} - 1\right]\beta \qquad (6.43a)$$

6.5 Octahedral Shear Stress Theory (Plane Strain Conditions)

Using the crack tip stresses described by Irwin for the plane strain condition [see Eq. (6.11)] where:

$$\sigma_{TZ} = v(\sigma_{Tr} + \sigma_{T\theta}), \quad \tau_{T\theta Z} = 0 \text{ and } \tau_{TZr} = 0$$

The octahedral shear stress from Eq. (6.14) becomes:

$$\tau_{OCT} = \frac{1}{3}[(\sigma_{Tr} - \sigma_{T\theta})^2 + (\sigma_{T\theta} - v\sigma_{Tr} - v\sigma_{T\theta})^2$$
$$+ (\sigma_{Tr} - v\sigma_{Tr} - v\sigma_{T\theta})^2 + 6\tau_{Tr\theta}^2] \qquad (6.44)$$

or:

$$\tau_{OCT} = \frac{\sqrt{2}}{3}[(1 - v + v^2)(\sigma_{Tr}^2 + \sigma_{T\theta}^2)$$
$$- (1 + 2v - 2v^2)\sigma_{Tr}\sigma_{T\theta} + 3\tau_{Tr\theta}^2]^{1/2} \qquad (6.44a)$$

For plastic straining, $v = 0.5$
Then:

$$\tau_{OCT} = \frac{\sqrt{2}}{3}\left[\frac{3}{4}(\sigma_{Tr}^2 + \sigma_{T\theta}^2) - \frac{3}{2}\sigma_{Tr}\sigma_{T\theta} + 3\tau_{Tr\theta}^2\right]^{1/2}$$

or:

$$\tau_{OCT} = \frac{\sqrt{2}}{3}\sigma_{Tr}\left[\frac{3}{4}\left(1 + \frac{\sigma_{T\theta}^2}{\sigma_{Tr}^2}\right) - \frac{3}{2}\frac{\sigma_{T\theta}}{\sigma_{Tr}} + 3\left(\frac{\tau_{Tr\theta}}{\sigma_{Tr}}\right)^2\right]^{1/2} \qquad (6.44b)$$

As before, equating Eq. (6.44) to simple uniaxial stress, where $\sigma_{T\theta} = 0$ and $\sigma_{Tr\theta} = 0$:

$$\sigma_T = \sigma_{Tr} \left[\frac{3}{4}\left(1 + \frac{\sigma_{T\theta}^2}{\sigma_{Tr}^2}\right) - \frac{3}{2}\frac{\sigma_{T\theta}}{\sigma_{Tr}} + 3\left(\frac{\tau_{Tr\theta}}{\sigma_{Tr}}\right)^2 \right]^{1/2} \quad (6.45)$$

or:

$$\sigma_T = \sigma_{Tr} \left[\frac{3}{4}\left(1 + \frac{\psi_\theta^2(\theta)}{\psi_r^2(\theta)}\right) - \frac{3}{2}\frac{\psi_\theta(\theta)}{\psi_r(\theta)} + 3\left(\frac{\psi_{r\theta}(\theta)}{\psi_r(\theta)}\right)^2 \right]^{1/2} \quad (6.45a)$$

and:

$$\sigma_T = \frac{\sigma_{Tr}}{\psi_r(\theta)} \left[\frac{3}{4}(\psi_r^2(\theta) + \psi_\theta^2(\theta)) - \frac{3}{2}\psi_r(\theta)\psi_\theta(\theta) + 3\psi_{r\theta}^2(\theta) \right]^{1/2} \quad (6.46)$$

Let:

$$\psi_{oc}(\theta) \equiv \left[\frac{3}{4}(\psi_r^2(\theta) + \psi_\theta^2(\theta)) - \frac{3}{2}\psi_r(\theta)\psi_\theta(\theta) + 3\psi_{r\theta}^2(\theta) \right]^{1/2} \quad (6.47)$$

For the plane strain condition, the energy absorption rate in the larger zone can be simplified the same way as was done for the plane stress condition; see Eq. (6.25). The quantity r_U from test data for plane strain and infinite width:

$$2r_U = hQ\left(\frac{1}{\varepsilon_T}\right)\left[1 - \left(\frac{\sigma_T}{\sigma_{TU}}\right)^{n+1}\right]\frac{2}{\pi^2}\psi_{oc}^2(\theta) \quad (6.48a)$$

For finite width:

$$2r_U = \frac{h}{Y^6}Q\left(\frac{1}{\varepsilon_T}\right)\left[1 - \left(\frac{\sigma_T}{\sigma_{TU}}\right)^{n+1}\right]\frac{2}{\pi^2}\psi_{oc}^2(\theta) \quad (6.48b)$$

The energy absorption rate in the larger zone, $\frac{\partial U_U}{\partial c}$, in its final form for infinite width is given by:

$$\frac{\partial U_U}{\partial c} = \frac{n}{n+1}\sigma_{TU}\varepsilon_{TU}\left[1 - \left(\frac{\sigma_T}{\sigma_{TU}}\right)^{n+1}\right]h\left(\frac{\varepsilon_{TF}\varepsilon_{TL}}{\varepsilon_{TU}\varepsilon_T}\right)$$

$$\left[\left(\frac{\varepsilon_{TU}}{\varepsilon_T}\right)^{\frac{n-1}{n}} - 1\right]\frac{2}{\pi^2}\int_0^{\theta_c}\psi_{oc}^2(\theta)f(\theta)d\theta \quad (6.49)$$

By numerical evaluation, the thickness parameter, β, for plane strain is:

$$\beta = \frac{2}{\pi^2} \int_0^{\theta_c} \psi_{oc}^2(\theta) f(\theta) d\theta = 0.127 \tag{6.50a}$$

Note that the value of β for a thickness, t, greater than the maximum thickness (t_o) for plane stress and less than the thickness for plane strain (this condition is called mixed mode where the value of β falls between plane stress and plane strain condition) is given by:

$$\beta = 1.30(t_o/t) + 0.127[(t-t_o)/t] \tag{6.50b}$$

6.6 Applied Stress, σ, and Half Crack Length, c, Relationship

The quantity W_U of Eq. (6.10a) can be evaluated from the area under the stress–strain curve taken from the stress equal to the applied stress to the ultimate of the material:

$$W_U = \int_{\sigma_T}^{\sigma_{TU}} \sigma_T \, d\varepsilon_{TP} \tag{6.51}$$

From Eq. (6.36):

$$\varepsilon_{TP} = \varepsilon_{TPU} \left[\frac{\sigma_T}{\sigma_{TU}} \right]^n \tag{6.52}$$

Taking the derivative:

$$d\varepsilon_{TP} = n\varepsilon_{TPU} \left(\frac{\sigma_T}{\sigma_{TU}} \right)^{n-1} \frac{d\sigma_T}{\sigma_{TU}} \tag{6.53}$$

Substituting Eq. (6.53) in Eq. (6.51):

$$W_U = \int_{\sigma_T}^{\sigma_{TU}} n\varepsilon_{TPU} \left(\frac{\sigma_T}{\sigma_{TU}} \right)^n d\sigma_T \tag{6.54}$$

Integrating:

$$W_U = \frac{n}{n+1} \varepsilon_{TPU} \left(\frac{\sigma_T}{\sigma_{TU}}\right)^{n+1} \sigma_{TU} \int_{\sigma_T}^{\sigma_{TU}} \quad (6.55)$$

$$W_U = \frac{n}{n+1} \varepsilon_{TPU} \sigma_{TU} \left[1 - \left(\frac{\sigma_T}{\sigma_{TU}}\right)^{n+1}\right] \quad (6.56)$$

Substituting Eqs. (6.56) and (6.50) in Eq. (6.49):

$$\frac{\partial U_U}{\partial c} = \frac{n+1}{n-1} W_U h \frac{\varepsilon_{TF} \varepsilon_{TL}}{\varepsilon_{TU} \varepsilon_T} \left[\left(\frac{\varepsilon_{TU}}{\varepsilon_T}\right)^{\frac{n-1}{n}} - 1\right] \beta \quad (6.57)$$

Define h_U by:

$$\frac{\partial U_U}{\partial c} = W_U h_U \quad (6.58)$$

From Eq. (6.58) and Eq. (6.57):

$$h_U = \frac{n+1}{n-1} h \frac{\varepsilon_{TF} \varepsilon_{TL}}{\varepsilon_{TU} \varepsilon_T} \left[\left(\frac{\varepsilon_{TU}}{\varepsilon_T}\right)^{\frac{n-1}{n}} - 1\right] \beta \quad (6.59)$$

From Eq. (6.10) and Eq. (6.5):

$$\frac{\pi \sigma^2 c}{E} = 2T + W_F h_F + W_U h_U$$

$$c = \frac{E}{\pi \sigma^2} [2T + W_F h_F + W_U h_U] \quad (6.60)$$

For a thin plate of infinite width (plane stress conditions, $\beta = 1.3$):

$$W_F = \overline{\sigma}_{TUF} [\varepsilon_{TPF} - \varepsilon_{TPU}] = \overline{\sigma}_{UF} \varepsilon_{PN} \quad (6.61)$$

Substituting Eqs. (6.56), (6.59), and (6.61) into Eq. (6.60) gives the critical half crack length in a plate of infinite width for plane stress fracture at applied stress σ:

$$c = \frac{E}{\pi \sigma^2} \left\{ 2T + \overline{\sigma}_{UF} \varepsilon_{PN} h_F + \frac{n}{n+1} \varepsilon_{TPU} \sigma_{TU} \left[1 - \left(\frac{\sigma_T}{\sigma_{TU}}\right)^{n+1}\right] \right.$$

$$\left. * \left(\frac{n+1}{n-1}\right) h \frac{\varepsilon_{TF} \varepsilon_{TL}}{\varepsilon_{TU} \varepsilon_T} \left[\left(\frac{\varepsilon_{TU}}{\varepsilon_T}\right)^{\frac{n-1}{n}} - 1\right] \beta \right\} \quad (6.62)$$

For mixed mode fracture, Eq. (6.62) becomes:

$$c = \frac{E}{\pi\sigma^2\mu}\left\{2T + \bar{\sigma}_{UF}\varepsilon_{PN}h_F k + \frac{n}{n-1}\varepsilon_{TPU}\sigma_{TU}\left[1 - \left(\frac{\sigma_{TL}}{\sigma_{TU}}\right)^{n+1}\right]\right.$$

$$\left. * h\frac{\varepsilon_{TF}\varepsilon_{TL}}{\varepsilon_{TU}\varepsilon_T}\left[\left(\frac{\varepsilon_{TU}}{\varepsilon_T}\right)^{\frac{n-1}{n}} - 1\right]\beta\right\} \qquad (6.63)$$

and for the finite width condition:

$$c = \frac{E}{\pi\sigma^2\mu Y^2}\left\{2T + \bar{\sigma}_{UF}\varepsilon_{PN}h_F k + \frac{n}{n-1}\frac{\varepsilon_{TPU}\sigma_{TU}}{Y^6}\left[1 - \left(\frac{\sigma_{TL}}{\sigma_{TU}}\right)^{n+1}\right]\right.$$

$$\left. * h\frac{\varepsilon_{TF}\varepsilon_{TL}}{\varepsilon_{TU}\varepsilon_T}\left[\left(\frac{\varepsilon_{TU}}{\varepsilon_T}\right)^{\frac{n-1}{n}} - 1\right]\beta\right\} \qquad (6.63a)$$

(See Section 6.7 for a detailed description of the thickness parameters β, k, and μ).

Equations (6.62) (plane stress) and (6.63) (mixed mode) simply state that, for a given crack length $2c$, the corresponding fracture stress, σ, can be easily computed provided that the full stress–strain curve for the material under study is available. The plot of fracture stress versus half crack length, c, is called the residual strength capability diagram and is extremely important in the field of fracture mechanics. From the residual strength data obtained from Eq. (6.63) the material resistance to fracture for different thicknesses and crack length can be computed.

6.7 Mixed Mode Fracture and Thickness Parameters

Expressions for k, β, and μ for mixed mode fracture [shown in Eq. (6.63)] are described as follows:

A shear lip is formed in plane stress fracture, whereas plane strain fracture is characterized by a flat fracture surface. It is assumed that the energies released and absorbed in plane stress and plane strain are in proportion to these thicknesses, as shown in Fig. 6.6.

Let t_o be the maximum thickness for plane stress fracture. The thickness corrections for mixed mode are applied to two strained regions at the crack tip and near the crack tip (Figure 6.1). The thickness parameter, k, for the highly strained region at the crack tip [see Eq. (6.6)] can be written as:

$$\frac{\pi\sigma_U^2 c_U}{E}\left[\frac{t_o}{t} + (1-v^2)\left(\frac{t-t_o}{t}\right)\right] = 2T + W_F h_F\left[\frac{t_o}{t} + 2\left(\frac{t-t_o}{t}\right)\right] \qquad (6.64)$$

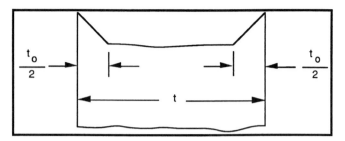

Figure 6.6 Shear lip and flat fracture surfaces

where the two quantities k and μ are defined as:

$$k = \frac{t_o}{t} + 2\left(\frac{t - t_o}{t}\right) \qquad (6.65a)$$

and

$$\mu = \left[\frac{t_o}{t} + (1 - v^2)\left(\frac{t - t_o}{t}\right)\right] \qquad (6.65b)$$

The minimum values of k and μ are equal to unity when $t = t_o$. This is associated with the maximum plane stress condition where complete shear lip prevails. Note that, for the case of maximum plane stress condition where $t = t_o$, equation 6.64 can be simplified to equation 6.6. Furthermore, the thickess parameter, β, for the mixed mode condition in the uniform strained region near the crack tip [see Eq. (6.63)] can be described as:

$$= 2T + W_F h_F k + W_U h_U \left[1.30 \frac{t_o}{t} + 0.127\left(\frac{t - t_o}{t}\right)\right] \qquad (6.66)$$

In terms of β and k Eq. (6.66) can be rewritten as:

$$= 2T + W_F hk + (W_U h_U)\beta \qquad (6.66a)$$

where

$$\beta = \left[1.30 \frac{t_o}{t} + 0.127\left(\frac{t - t_o}{t}\right)\right] \qquad (6.66b)$$

The maximum and minimum values of β are 1.3 and 0.127 [see Eq. (6.66)] for the plane stress (when $t = t_o$) and plane strain (when $t_o = 0$) conditions, respectively [see also Eq. (6.43) and (6.50a)].

The thickness parameters κ, μ, and β, described by equations 6.65a, 6.65b, and 6.66b, each contain two terms which are normalized to the total plate thickness, t.

The first term (t_0/t) represents the adjustment to the plane stress condition. The latter one describes the amount of flat fracture, $\frac{t-t_0}{t}$, that differs for the two absorbed energy terms (quantities to the right of equation 6.66) as well as for the energy released term to the left of equation 6.64.

6.8 The Stress–Strain Curve

The short-time uniaxial full-range tension test is used to determine the mechanical properties of structural metals. Information obtained from this test can be used to calculate the fracture properties of structural metals. As shown in Fig. 6.3, a typical stress–strain curve for a ductile metal shows that the metal undergoes initial elastic straining, then uniform plastic straining and finally local plastic straining (necking). The stress–strain curve in Fig. 6.3A is typical of brittle material, such as glass. The curve in Fig. 6.3B is representative of metals that do not "neck," such as cross-rolled beryllium. The curve in Fig. 6.3C is typical of many structural metals, and the curve in Fig. 6.3D, which shows very little uniform plastic straining, is typical of some precipitation hardening steels.

In a uniaxial tension test, strains are obtained from deflections measured by a gage that is usually longer than the necked portion of the test specimen. The typical gage length employed for most uniaxial tension tests has a 2 in. length. This average strain must be corrected to obtain the actual local strain in that part of the test specimen. The correction can be made by multiplying the local average strain by the length of the gage and dividing by the thickness, t_s, of the tensile test specimen. The correction for the actual true "neck" strain and the procedure to obtain the true "neck" strain, ε_{pn}, is discussed in Appendix B.

Typical full-range uniaxial tensile stress–strain curves should be obtained from reputable and reliable sources. If such curves are not available, five or six full-range tensile stress–strain curves should be obtained by tests. Either flat or round test specimens are suitable. A typical stress–strain curve should be developed from the test data by the strain departure method, which is outlined in MIL-HDBK-5, Metallic Materials and Elements for Aerospace Vehicle structures.

6.9 Verification of FMDM Results with the Experimental Data

The following are the calculated residual strength capability data provided by the FMDM theory for several aerospace alloys compared with test data gathered from several reliable sources. Both cases of finite and infinite cracked plates were used to check the analysis. In Chapter 3, the ASTM procedures for determination of fracture toughness, including standard specimen preparation, surface finish, and prefatigue cracking of the specimen, as well as data gathering, are briefly discussed. In Section 6.10 the FMDM approach for determination of fracture toughness for 2219-T87 and 7075-T73 is discussed and the calculated results compared with the test data obtained through the ASTM procedure.

To employ the FMDM theory for constructing the residual strength diagram (fracture stress as a function of half critical crack length), it is necessary to have the full range stress–strain curve for the material under study. The full range stress–strain curve must be obtained from reliable sources. The results of the analysis conducted by the FMDM theory indicated exceptionally good agreement with experimental data. In Fig. 6.7, the stress–strain curve for 2219-T87 aluminum alloy (with plate thickness $t_s = 0.125$ in.) is plotted from MIL-HDBK-5. The pertinent information is extracted as input to the FMDM equation to obtain the variation of fracture stress as a function of half critical crack length, as shown in Fig. 6.8. The experimental data plotted in Fig. 6.8 were extracted from Reference [6]. An excellent correlation between the test data and the FMDM results was observed. As another example, the HP-9NI-4CO-0.2C Steel [8] was selected and proper information that is needed for generating a residual strength diagram was taken from the stress–strain curve shown in Fig. 6.9. Figure 6.10 is the constructed residual strength capability curve which describes the variation of the fracture stress as a function of half critical crack length. Good agreement between

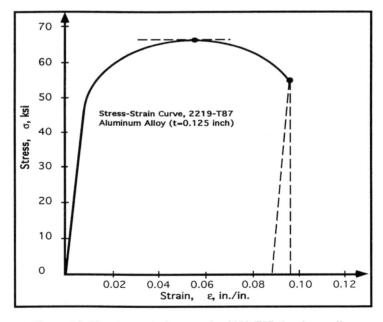

Figure 6.7 The stress–strain curve for 2219-T87 aluminum alloy

6.9 Verification of FMDM Results with the Experimental Data

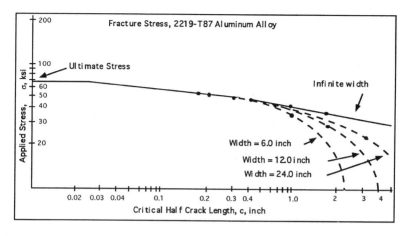

Figure 6.8 Residual strength diagram for 2219-T87 aluminum alloy

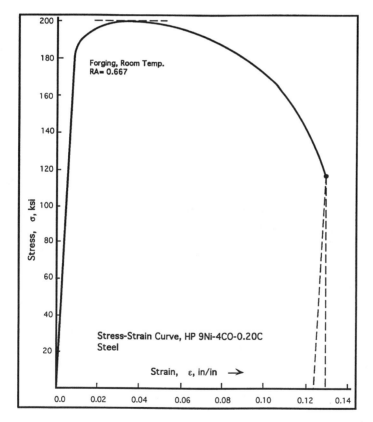

Figure 6.9 The stress–strain curve for HP-9NI-4CO-0.2C steel

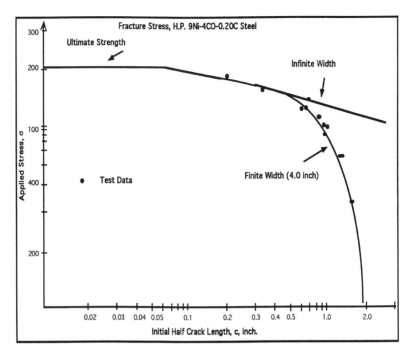

Figure 6.10 Residual strength diagram for HP-9NI-4CO-0.2C steel

the test data and the FMDM result can be seen. Additional data that can validate the results of the FMDM analysis with the test data are available in Reference [6].

6.10 Fracture Toughness Computation by the FMDM Theory

6.10.1 Introduction

As mentioned previously, plane stress fracture toughness testing is costly and time consuming because of its dependency on the material thickness and the crack length. This dependency necessitates having standard specimens with different thicknesses that must be prepared based on the ASTM procedures. The specimen preparation for fracture toughness testing includes the machine starter slot, fatigue cracking of the specimens prior to testing, method of applying load to the specimen, load recording, and crack length measurements, all of which are elements that can contribute to a costly and time-consuming process. The following scenario, which can often happen in the aircraft and aerospace industries, clearly illustrates the need for knowing the variation in fracture toughness as a function of

6.10 Fracture Toughness Computation by the FMDM Theory

material thickness. For example, consider a situation in which the fracture mechanics analysis of a component of an aircraft structure shows that a part cannot survive the load varying environment. Thus, there is a need to increase the part thickness, t_1, in order to reduce the applied stress, σ. A new fracture mechanics analysis of the part with increased thickness ($t_2 > t_1$) requires a corrected fracture toughness value that corresponds to the new assigned thickness, t_2. With the FMDM theory, the K_c value can easily be computed for the material, provided that the material stress–strain curve from reliable sources, such as MIL-HDBK-5, is available to the analyst.

Equations (6.62) and (6.63) provide the variation in fracture stress, σ, with respect to the crack length, c. Incorporating the fracture stress, σ, and half critical crack length, c, into the crack tip stress intensity factor equation (for a center crack panel in an infinite plate or plate of finite width, W), the obtained K value will correspond to the apparent fracture toughness. The stress intensity factor equation for a cracked plate having thickness t and width W can be written as:

$$K_c = \sigma(\pi c \sec(\pi c/W))^{1/2} \text{ where } \sigma < \sigma_{\text{Yield}} \quad (6.67)$$

The analyst must remember that the computed value of the fracture stress from the FMDM theory should not fall above the material yield value nor is net-section yielding allowed when using Eq. (6.67) for the fracture toughness computation. That is:

$$(\sigma_c)\text{net-section} = \sigma_c[W/(W - 2c)] < \sigma_{\text{Yield}} \quad (6.68)$$

Sections 6.10.2 and 6.10.3 contain example problems pertaining to computation of the fracture toughness by the FMDM theory [9, 10] for 2219-T87 and 7075-T73 aluminum alloys which have stress–strain curves taken directly from MIL-HDBK-5.

6.10.2 Fracture Toughness, K_c, Evaluation for 2219-T87 Aluminum Alloy

The stress–strain curve for 2219-T87 aluminum alloy (sheet and plate, longitudinal direction, room temperature) is shown in Fig. 6.7. The FMDM residual strength capability data generated by the FMDM Computer code [see Eqs. (6.62) and (6.63)] is plotted in Figs. 6.11 through 6.14 for plate thicknesses of $t = 0.1$, 0.2, 0.4, and 0.8 in., respectively. Using Eq. (6.67) the variation of the calculated fracture toughness, K_c, with respect to material thickness ($t = 0.1, 0.2, 0.4, 0.8$, 1.0, 1.4, and 2.0 in.) can be plotted (Fig. 6.15). The FMDM residual strength capability diagrams for other thicknesses ($t = 1.0, 1.4,$ and 2.0 in.) are not shown here.

Because fracture toughness is a function of crack length, two arbitrary values of crack length were chosen. One is associated with a fracture stress $\sigma_c = 90\%$ to

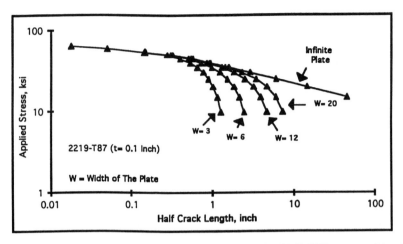

Figure 6.11 The residual strength capability diagram for 2219-T87 generated by the FMDM computer code (thickness $t = 0.1$ in.)

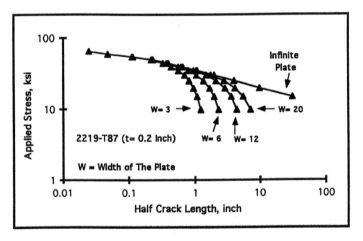

Figure 6.12 The residual strength capability diagram for 2219-T87 generated by the FMDM computer code (thickness $t = 0.2$ in.)

95% of σ_{Yield}, and the other one with half crack length $c = 1.5$ in. Two other intermediate crack lengths were also used and the corresponding fracture toughnesses are shown in Fig. 6.15. The reader should remember that the larger the original crack length, the higher is the computed fracture toughness value, K_c. Therefore, the minimum fracture toughness is expected to be associated with a half critical crack length having fracture stress of $\sigma_c = 90\%$ to 95% of σ_{Yield}. A minimum fracture toughness value, corresponding to the minimum crack length, where

6.10 Fracture Toughness Computation by the FMDM Theory 315

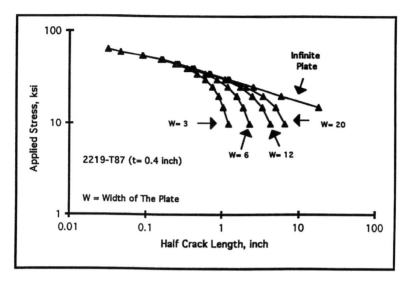

Figure 6.13 The residual strength capability diagram for 2219-T87 generated by the FMDM computer code (thickness $t = 0.4$ in.)

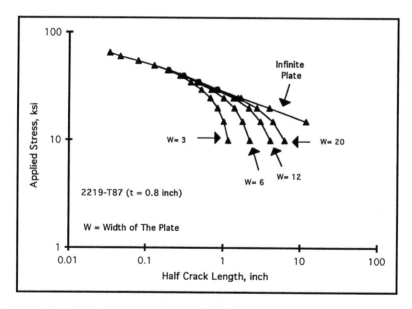

Figure 6.14 The residual strength capability diagram for 2219-T87 generated by the FMDM computer code (thickness $t = 0.8$ in.)

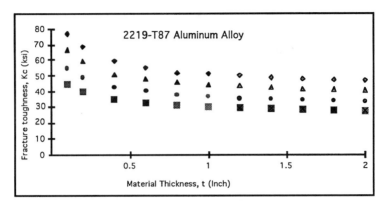

■ → Minimum Crack Length is associated with applied stress of 90 to 95% of the material yield.
●, ▲ → Intermediate Crack Length   ◆ → Max Crack Length
(2c = 3.0 inches)

Figure 6.15 Variation of material fracture toughness vs. thickness for 2219-T87 aluminum alloy

$\sigma_c = 0.9$ to $.95 \times \sigma_{\text{Yield}}$, gives a conservative result when conducting crack growth rate analysis of a structural component. In most real cases, the initial crack length assumption in the structure is small, so that the minimum fracture toughness is an acceptable value to use in the analysis. Also note that the fracture toughness values shown in Appendix A [11] are the lower bound K_c values based on minimum initial crack size. The fracture stress corresponding to the minimum crack size should not cause plastic deformation in the bulk of the test specimen. Figure 6.16 shows the variation of the experimentally obtained fracture toughness, K_c, versus material thickness for 2219-T87 aluminum alloy taken from Reference [11]. Fracture toughness curves shown in Fig. 6.16 correspond to half crack lengths $c = 0.24$ to 0.38, 0.45 to 0.68, 1.0 to 1.2, and 1.5 to 1.8 in. Both figures (6.15 and 6.16) clearly illustrate that the fracture toughness value increases with decreasing thickness and, moreover, that K_c is a function of crack length, c.

6.10.3 Fracture Toughness, K_c, Evaluation for 7075-T73 Aluminum Alloy

The pertinent information extracted from the 7075-T73 aluminum alloy (sheet and plate, longitudinal direction at room temperature) stress–strain curve and it is shown in Fig. 6.17. Figure 6.17 is a portion of the FMDM computer output that contains information which can be used to construct the FMDM residual strength capability curve. Using the FMDM computer code the residual strength diagrams were plotted in Figs. 6.18 through 6.21 for different plate thicknesses $t = 0.1, 0.2,$

6.10 Fracture Toughness Computation by the FMDM Theory

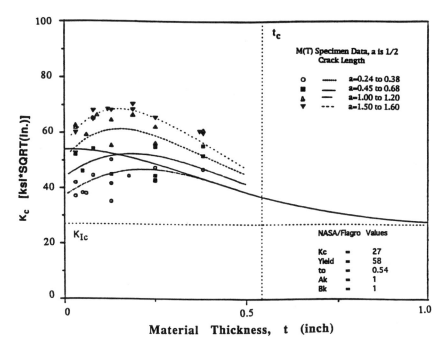

Figure 6.16 Variation of K_C as a function of thickness, t, for 2219-T87 aluminum alloy [10]

0.4, and 0.8 in., respectively. Equation (6.67) was utilized to calculate the material fracture toughness and to establish the variation of K_c versus material thicknesses ($t = 0.1, 0.2, 0.4, 0.8, 1.0, 1.4$, and 2.0 in.) for different crack lengths, $2c$ (Fig. 6.22). No attempt was made to include all the residual strength capability diagrams for the thicknesses that were used to generate Fig. 6.22.

A minimum fracture toughness value corresponding to the minimum crack length is assigned to a fracture stress equal to 90% to 95% of the material yield strength. Also, a half crack length $c = 1.5$ in. was selected as the upper limit of the fracture toughness (two other intermediate crack lengths were also used and the corresponding fracture toughnesses are shown in Fig. 6.22). Note that the upper limit of the fracture toughness was arbitrarily chosen to be associated with a half critical crack length, $c = 1.5$ in. A higher fracture toughness value for this material can be obtained by using a larger crack length. The fracture toughness dependency of 7075-T73 aluminum alloy on the crack length and the material thickness is also shown in Fig. 6.23, in which data points were obtained through laboratory tests conducted in accordance with the ASTM standards. Both figures (6.22 and 6.23) clearly illustrate that the fracture toughness is dependent on the material thickness (decreases as thickness increases) and the crack length, c (increases as crack length increases).

> **Aluminum Alloy, 7075-T73, Plate, L-T Dir., Room Temp.**
> The following mechanical properties were obtained from flat tensile test bars.
>
> | Atom. Spacing. | Ult. Str. | Yld. Str. | Rup. Str. | Neck Str. | E |
> |---|---|---|---|---|---|
> | 2.860 | 83.000 | 73.000 | 75.500 | 80.900 | 10300.00 |
>
> | Poisn. R. | Ult. Strn. | Rup. Strn. | Thk. Tst. Sp. | Gage L. |
> |---|---|---|---|---|
> | 0.3300 | 0.0880 | 0.12 | 0.0625 | 2.000 |
>
> Computed Basic Data
>
> | t_0 | WF | EPN | STU | STF | n |
> |---|---|---|---|---|---|
> | 0.146 | 68.82 | 0.850 | 90.037 | 152.18 | 17.89 |
>
> Atomic spacing is in Angstroms; stresses, modulus and WF are in ksi; t_0, thicknesses, gage length, widths and crack sizes are in inches.
>
> Neck Str. is the average stress from the beginning of necking to rupture of the tensile test specimen.
>
> Ult. Strn. is the strain at the beginning of necking.
>
> t_0 is the maximum thickness of a through cracked test specimen for plane-stress fracture.
>
> WF is the density of plastic energy under the stress-strain curve from strain at the beginning of necking to strain at rupture.
>
> EPN is the corrected neck uniaxial plastic tensile strain. (corrected from gage length).
>
> STU is the uniaxial true ultimate tensile stress.
>
> STF is the uniaxial true rupture tensile stress.
>
> n is the exponent in the Ramberg-Osgood relation for uniaxial true plastic tensile strain.
>
> Str. is the computed gross area fracture stress.

Figure 6.17 Portion of the FMDM computer output for 7073-T73

6.10 Fracture Toughness Computation by the FMDM Theory 319

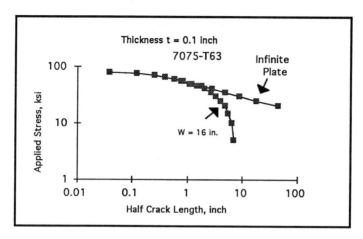

Figure 6.18 The residual strength capability diagram for 7075-T63 generated by the FMDM computer code (thickness $t = 0.1$ in.)

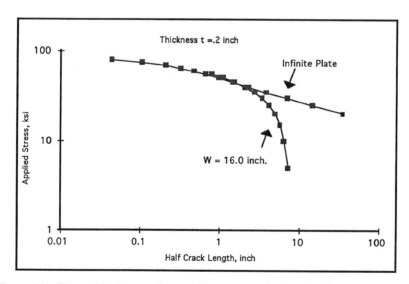

Figure 6.19 The residual strength capability diagram for 7075-T63 generated by the FMDM computer code (thickness $t = 0.2$ in.)

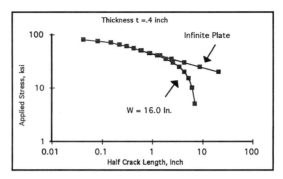

Figure 6.20 The residual strength capability diagram for 7075-T63 generated by the FMDM computer code (thickness $t = 0.4$ in.)

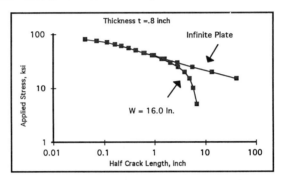

Figure 6.21 The residual strength capability diagram for 7075-T63 generated by the FMDM computer code (thickness $t = 0.8$ in.)

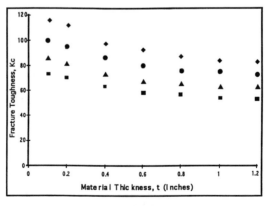

Figure 6.22 Variation of material fracture toughness versus thickness for 7075-T63 aluminum alloy

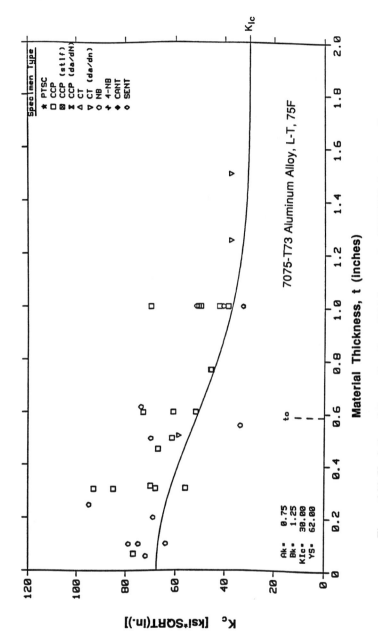

Figure 6.23 Variation of K_C as a function of thickness, t, for 7075-T73 aluminum alloy [10]

References

1. A. A. Griffith, "The Phenomena of Rupture and Flow in Solids," Philos. Trans., R. Soc. Lond., Ser. A., Vol. 221, 1920, p. 163.
2. G. R. Irwin, "Fracture Dynamics," Fracture of Metals, ASM, 1948, pp. 147–166.
3. E. Orowan, "Fracture and Strength of Solids," Rep. Prog. Physics, Vol. 12, 1949, pp. 185–232.
4. A. H. Cottrell, Dislocations and Plastic Flow in Crystals, The Claredon Press, 1958, p. 3.
5. H. Conrad, "Experimental Evaluation of Creep and Stress Rupture," Chapter 8, in Mechanical Behavior of Materials at Elevated Temperature, edited by John E. Dorn, McGraw-Hill, 1961.
6. G. E. Bockrath and J. B. Glassco, "A Theory of Ductile Fracture," McDonnell Douglas Astronautics Company, Report MDC G2895, August 1972, revised April 1974.
7. A. Nadai, Plasticity. McGraw-Hill, 1960, pp. 103, 260.
8. J. B. Glassco, B-1 Fracture Analysis by Ductile Fracture Method, Rockwell International Report TFD-76-1368-L, 1977.
9. B. Farahmand and G. E. Bockrath, "A Theoretical Approach for Evaluating the Plane Strain Fracture Toughness," Engin. J. Fract. Mech., March 1996.
10. B. Farahmand, "A New Analytical Approach to Obtain the Plane Stress Fracture Toughness Value by the FMDM Theory," The Fourth Pan American Congress of Applied Mechanics, Buenos Aires, Argentina, January 5-9, 1994.
11. Fatigue Crack Growth Computer Program "NASA/FLAGRO," developed by R. G. Forman, V. Shivakumar, and J. C. Newman. JSC-22267A, January 1993.

Appendix A

NASA/FLAGRO 2.0 Materials Constants U. S. Customary Units (ksi, ksi $\sqrt{\text{in}}$)

Material; Condition; Environment*	Code	YS	UTS

[A] Iron, alloy or cast

ASTM Specification
A536 Grd 80-55-06

As cast	A1AC50AB1	58	80

[B] ASTM spec. grd. steel

A10 Series
A36

Plt (Dyn Klc, < 500Hz); LA, HHA, 3% NaCl	B0CB10AB1	44	78
ES Weld & HAZ (Dyn Klc, < 500Hz); LA, HHA, 3% NaCl	B0CZK1AB1	44	78

A200 Series
A203 Grd E (3.5% N1)

Plt	B2CE12AB1	71	80
Plt; $-100F$	B2CE12LA7	82	96

A216 Grd WCC

Casting	B2GC51AB1	48	80

A300 Series
A302 Grd B

Plt	B3AB12AB1	55	90

A372 Type IV

Forg	B3GD21AB1	65	105

A387 Grd 22, Cl 2

Plt	B3IQ10AB1	50	75

A400 Series
A469 Cl 4

Forg	B4JD26AB1	85	105

A469 Cl 5

Forg	B4JE20AB1	95	115

A500 Series
A508 Cl2 & Cl3

Forg	B5AC21AB1	65	100

A514 Typ F

Plt	B5BF10AB1	105	120
GMA SR Weld	B5BFC2AB1	100	115

A517 Grd F (T1 Steel)

Plt	B5DF12AB1	100	125

A533-B, Cl1 & Cl2

Plt	B5HD10AB1	70	100
SMA Weld	B5HDF1AB1	60	90

A553 Typ I

Plt	B5QA12AB1	95	110
Plt; $-320F$	B5QA12LA4	150	175

* Unless noted, assume Lab Air (LA) environment and any orientation except S-T, S-L, C-R, C-L, and R-L.

(continued)

K_{Ie}	K_{Ic}	A_k	B_k	C	n	p	q	DK_0	R_{cl}	a	S.R.
45	32	0.75	0.50	.700E-9	2.900	0.5	0.5	8.0	0.7	2.5	0.3
100	70	0.75	0.5	.100E-8	3.000	0.5	0.5	7.0	0.7	2.0	0.3
100	70	0.75	0.5	.100E-8	3.000	0.5	0.5	7.0	0.7	2.0	0.3
230	170	0.75	0.5	.915E-9	2.688	0.5	0.5	7.0	0.7	2.5	0.3
280	200	0.75	0.5	.915E-9	2.688	0.5	0.7	7.0	0.7	2.5	0.3
200	150	0.75	0.5	.300E-9	3.000	0.5	0.5	7.0	0.7	2.5	0.3
120	100	0.75	0.5	.600E-10	3.480	0.5	0.5	7.0	0.7	2.5	0.3
120	100	0.75	0.5	.600E-10	3.480	0.5	0.5	6.0	0.7	2.5	0.3
120	100	0.75	0.5	.400E-9	3.000	0.5	0.5	7.0	0.7	2.5	0.3
230	170	0.75	0.5	.900E-9	2.800	1.0	0.5	5.5	0.7	2.5	0.3
230	170	0.75	0.5	.100E-8	2.800	0.5	0.5	4.3	0.7	2.5	0.3
140	100	0.75	0.5	.100E-8	2.800	0.5	0.5	6.0	0.7	2.5	0.3
115	85	0.75	0.5	.200E-8	2.570	0.25	0.25	4.0	0.7	2.5	0.3
115	85	0.75	0.5	.150E-8	2.570	0.25	0.25	4.0	0.7	2.5	0.3
140	100	0.75	0.5	.100E-9	3.500	0.25	0.25	6.0	0.7	2.5	0.3
200	150	0.75	0.5	.100E-8	2.700	0.5	0.5	6.5	0.7	2.5	0.3
140	100	0.75	0.5	.100E-8	2.700	0.5	0.5	8.0	0.7	2.5	0.3
250	180	0.75	0.5	.150E-7	2.000	0.25	0.25	4.5	0.7	2.5	0.3
140	100	0.75	0.5	.300E-9	3.200	0.25	0.25	6.0	0.7	2.5	0.3

(continued)

Material; Condition; Environment*	Code	YS	UTS
[B] ASTM spec. grade steel			
A500 Series			
A579 Grd 75 (12% Nl)			
Forg	B5VW20AB1	180	190
A588 Grd A & Grd B			
Plt	B5XA11AB1	55	80
Plt; 3% NaCl, > 0.2 Hz	B5XA11WB1	55	80
A645 (5% Nl)			
Plt	B6GA12AB1	75	105
Plt; -320F	B6GA12LA4	110	165
[C] AISI - SAE Steel			
AISI 10xx - 12xx Steel			
Low Carbon 1005 - 1012			
Hot rolled plt	C1AB11AB1	25	45
Low Carbon 1015 - 1025			
Hot rolled plt	C1BB11AB1	30	58
AISI 43xx - 48xx Steel			
4330V MOD			
180-200 UTS; Plt & Forg	C4BS10AB1	175	190
200-220 UTS; Plt & Forg	C4BT10AB1	195	210
220-240 UTS; Plt & Forg	C4BU10AB1	215	230
4340			
160-180 UTS; Plt & Forg	C4DC21AB1	155	170
180-200 UTS; Plt & Forg; HHA	C4DD11AD1	175	190
180-200 UTS; Plt & Forg	C4DD21AB1	175	190
200-220 UTS; Plt & Forg	C4DE11AB1	195	210
200-220 UTS; Plt & Forg; $-50F$	C4DE11LB7	200	220
220-240 UTS; Plt & Forg	C4DF11AB1	215	230
220-240 UTS; Plt & Forg; $-50F$	C4DF11LB7	225	240
240-280 UTS; Plt & Forg	C4DG11AB1	240	260
240-280 UTS; Plt & Forg; $-50F$	C4DG11LB7	250	270
[D] Misc. U. S. Spec. Grade Steel			
SAE Spec. Steel			
0030 Cast	D5AC50AB1	44	72

* Unless noted, assume Lab Air (LA) environment and any orientation except S-T, S-L, C-R, C-L, and R-L.

(*continued*)

K_{Ie}	K_{Ic}	A_k	B_k	C	n	p	q	DK_0	R_{cI}	a	$S.R.$
125	90	0.75	0.5	.150E-7	2.000	0.25	0.25	3.0	0.7	2.5	0.3
140	100	0.75	0.5	.350E-9	3.300	0.5	0.5	8.5	0.7	2.5	0.3
140	100	0.75	0.5	.700E-9	3.300	0.5	0.5	8.5	0.7	2.5	0.3
250	180	0.75	0.5	.170E-8	2.500	0.25	0.25	5.0	0.7	2.5	0.3
110	80	0.75	0.5	.170E-9	3.500	0.25	0.25	7.0	0.7	2.5	0.3
100	70	0.75	0.5	.800E-10	3.600	0.5	0.5	8.0	0.7	2.5	0.5
100	70	0.75	0.5	.800E-10	3.600	0.5	0.5	8.0	0.7	2.5	0.5
150	110	0.75	0.5	.130E-8	2.700	0.25	0.25	5.5	0.7	2.5	0.3
110	80	0.75	0.5	.130E-8	2.700	0.25	0.25	4.5	0.7	2.5	0.3
85	65	0.75	0.75	.130E-8	2.700	0.25	0.25	4.0	0.7	2.5	0.3
190	135	0.75	0.5	.170E-8	2.700	0.25	0.25	6.0	0.7	2.5	0.3
155	110	0.75	0.5	.160E-8	2.700	0.25	0.25	5.5	0.7	2.5	0.3
155	110	0.75	0.5	.130E-8	2.700	0.25	0.25	5.5	0.7	2.5	0.3
110	80	0.75	0.5	.130E-8	2.700	0.25	0.25	4.5	0.7	2.5	0.3
70	55	0.75	0.5	.170E-8	2.700	0.25	0.25	4.5	0.7	2.5	0.3
85	65	0.75	0.75	.130E-8	2.700	0.25	0.25	4.0	0.7	2.5	0.3
55	45	0.75	0.75	.170E-8	2.700	0.25	0.25	4.0	0.7	2.5	0.3
65	55	0.75	1.0	.130E-8	2.700	0.25	0.25	3.5	0.7	2.8	0.3
45	40	0.75	1.0	.170E-8	2.700	0.25	0.25	3.5	0.7	2.8	0.3
95	70	0.75	0.5	.200E-10	4.000	0.25	0.25	8.0	0.7	2.5	0.3

(*continued*)

Material; Condition; Environment*	Code	YS	UTS
[E] Trade/common name steel			
Ultra High Strength Steel			
18 Nl Maraging			
250 Grd; Plt & Forg	E1AD10AB1	240	260
300 Grd; Plt & Forg	E1AE10AB1	280	290
300M			
270-300 UTS; Plt & Forg	E1BF21AB1	240	285
AF1410			
220-240 UTS; Plt & Forg; −65F	E1CC12AA7	240	250
220-240 UTS; Plt & Forg; LA, HHA/DW > 1Hz	E1CC12AB1	220	230
D6AC			
220-240 UTS; Plt & Forg; Nom. K_{Ic} (70); −40F	E1DD10AA8	225	240
220-240 UTS; Plt & Forg; Nom. K_{Ic} (70)	E1DD10AB1	215	230
220-240 UTS; Plt & Forg; Nom. K_{Ic} (70); HHA/DW > 0.1 Hz	E1DD10AD1	215	230
220-240 UTS; Plt & Forg; High K_{Ic} (90)	E1DJ10AB1	215	230
HP-9-4-20			
190-210 UTS; Plt & Forg; L-T, T-L; HHA, SW > 1 Hz	E1EB23AB1	190	200
190-210 UTS; Plt & Forg; L-T, T-L; −65F	E1EB23AC7	200	210
190-210 UTS; GTA Weld + SR; LA, HHA, SW > 1 Hz	E1ECB2WA1	185	195
190-210 UTS; GTA Weld + SR; −65F	E1ECB2AC7	195	205
HP-9-4-30			
220-240 UTS; Plt & Forg; L-T, T-L; LA, HHA, SW > 1 Hz	E1GC23AB1	205	230
220-240 UTS; Plt & Forg; L-T, T-L; -65F	E1GC23AA7	215	240
220-240 UTS; Plt & Forg; L-T, T-L; 600F	E1GC23AA14	165	195
HY-180 (10Nl)			
Plt & Forg	E1IB13AB1	180	200
Plt & Forg; DW, ASW > 0.1 Hz	E1IB13WA1	180	200
Plt & Forg; SW > 0.1 Hz	E1IB13WB1	180	200
HY-TUF			
220-240 UTS; VAR Forg	E1JB23AB1	200	230
H-11 MOD			
240-260 UTS; Plt & Forg	E1LE23AB1	215	250

* Unless noted, assume Lab Air (LA) environment and any orientation except S-T, S-L, C-R, C-L, and R-L.

(continued)

K_{Ie}	K_{Ic}	A_k	B_k	C	n	p	q	DK_0	R_{cI}	a	S.R.
90	75	0.75	0.75	.350E-8	2.600	0.25	0.25	3.5	0.7	2.5	0.3
85	70	0.75	0.75	.300E-8	2.600	0.25	0.25	3.0	0.7	2.8	0.3
65	55	0.75	1.0	.500E-8	2.460	0.25	0.25	3.0	0.7	2.8	0.3
150	110	0.75	.0.50	.600E-8	2.200	0.25	0.25	3.5	0.7	2.5	0.3
180	135	0.75	0.50	.100E-7	2.200	0.25	0.25	3.5	0.7	2.5	0.3
70	50	0.75	0.75	.150E-8	2.570	0.25	0.25	4.0	0.7	2.5	0.3
100	70	0.75	0.75	.320E-8	2.570	0.25	0.25	4.0	0.7	2.5	0.3
100	70	0.75	0.75	.150E-7	2.570	0.25	0.25	4.0	0.7	2.5	0.3
120	90	0.75	0.75	.320E-8	2.570	0.25	0.25	4.0	0.7	2.5	0.3
135	110	0.75	0.50	.820E-8	2.320	0.25	0.25	4.5	0.7	2.5	0.3
135	110	0.75	0.50	.500E-8	2.320	0.25	0.25	4.5	0.7	2.5	0.3
135	110	0.75	0.50	.820E-8	2.320	0.25	0.25	4.5	0.7	2.5	0.3
135	110	0.75	0.50	.500E-8	2.320	0.25	0.25	4.5	0.7	2.5	0.3
115	90	0.75	0.50	.820E-8	2.320	0.25	0.25	4.0	0.7	2.5	0.3
115	90	0.75	0.50	.820E-8	2.320	0.25	0.25	4.0	0.7	2.5	0.3
110	85	0.75	0.50	.820E-8	2.320	0.25	0.50	4.0	0.7	2.5	0.3
200	150	0.75	0.50	.600E-8	2.300	0.25	0.25	5.0	0.7	2.5	0.3
200	150	0.75	0.50	.120E-7	2.300	0.25	0.25	5.0	0.7	2.5	0.3
200	150	0.75	0.50	.600E-7	2.300	0.25	0.25	5.0	0.7	2.5	0.3
150	110	0.75	0.75	.350E-8	2.500	0.25	0.25	5.0	0.7	2.5	0.3
60	50	0.75	0.75	.150E-8	2.700	0.25	0.25	4.5	0.7	2.5	0.3

(*continued*)

Material; Condition; Environment*	Code	YS	UTS
[E] Trade/Common Name Steel			
Pressure Vessel/Piping			
HY 80			
Plt	E2AA13AB1	90	105
Plt; 3.5% NaCl/SW > 0.1 Hz	E2AA13WB1	90	105
HY 130			
Plt	E2CA13AB1	140	150
Plt; 3.5% NaCl/SW > 0.1 Hz	E2CA13WB1	140	150
GMA Weld	E2CAC1AB1	130	140
SMA Weld	E2CAF1AB1	130	140
Construction Grade			
HT-80			
Plt	E3BA13AB1	110	120
SA Weld	E3BAH1AB1	95	115

* Unless noted, assume Lab Air (LA) environment and any orientation except S-T, S-L, C-R, C-L, and R-L.

(*continued*)

K_{Ie}	K_{Ic}	A_k	B_k	C	n	p	q	DK_0	R_{cI}	a	S.R.
250	200	0.75	0.50	.150E-8	2.500	0.25	0.25	5.5	0.6	2.5	0.3
250	200	0.75	0.50	.300E-8	2.500	0.25	0.25	5.5	0.6	2.5	0.3
250	200	0.75	0.50	.300E-8	2.500	0.25	0.25	5.0	0.6	2.5	0.3
250	200	0.75	0.50	.700E-8	2.500	0.25	0.25	5.0	0.6	2.5	0.3
250	200	0.75	0.50	.150E-8	2.500	0.25	0.25	5.0	0.6	2.5	0.3
250	200	0.75	0.50	.150E-8	2.500	0.25	0.25	5.0	0.6	2.5	0.3
200	150	0.75	0.50	.700E-9	3.000	0.25	0.25	8.0	0.7	2.5	0.3
200	150	0.75	0.50	.250E-9	2.700	0.25	0.25	2.5	0.7	2.5	0.3

(*continued*)

Material; Condition; Environment*	Code	YS	UTS
[F] AISI type stainless steel			
AISI 300 Series			
AISI 301/302			
Ann Plt & Sht	F3AA13AB1	40	90
1/2 Hard sht	F3AC13AB1	125	165
Full Hard sht	F3AE13AB1	190	205
AISI 304/304L			
Ann Plt & Sht, Cast; 550F Air	F3DA13AA13	24	64
Ann Plt & Sht, Cast; 800F Air, > 1Hz	F3DA13AA16	20	63
Ann Plt & Sht, Cast	F3DA13AB1	40	90
Ann Plt & Sht, Cast; -320F LN2	F3DA13LA4	100	205
SA weld (308 filler) + SR; 800F Air, > 1Hz	F3FAH2AA16	49	69
SA weld (308 filler) + SR	F3FAH2AB1	66	107
AISI 316/316L			
Ann Plt & Sht, Cast; 600F Air	F3KA13AA14	32	60
Ann Plt & Sht, Cast; 800F Air	F3KA13AA16	20	60
Ann Plt & Sht, Cast	F3KA13AB1	36	90
Ann Plt & Sht, Cast; -452F LHe	F3KA13LA2	80	215
Ann Plt & Sht, Cast; -320F LN2	F3KA13LA4	70	185
SMA weld (316 filler) + SR; 800F Air, > 1Hz	F3KAH2AA16	49	69
20% CW Plt & Sht	F3KB13AB1	100	110
AISI 400 Series			
AISI 430 VAR			
Ann Rnd, C-R	F4LA16AB1	34	59

* Unless noted, assume Lab Air (LA) environment and any orientation except S-T, S-L, C-R, C-L, and R-L.

(*continued*)

K_{Ie}	K_{Ic}	A_k	B_k	C	n	p	q	DK_0	R_{cI}	a	S.R.
280	200	1.0	0.50	.600E-9	3.0	0.25	0.25	3.5	0.7	2.5	0.3
140	100	1.0	0.50	.130E-8	3.0	0.25	0.25	3.5	0.7	2.5	0.3
110	80	1.0	0.50	.550E-8	2.2	0.25	0.25	3.5	0.7	2.5	0.3
200	150	1.0	0.50	.400E-10	4.0	0.25	0.25	6.0	0.7	2.5	0.3
140	100	1.0	0.50	.120E-9	4.0	0.25	0.25	7.0	0.7	2.5	0.3
280	200	1.0	0.50	.600E-9	3.0	0.25	0.25	3.5	0.7	2.5	0.3
280	200	1.0	0.50	.120E-9	3.2	0.25	0.25	7.0	0.7	2.5	0.3
140	100	1.0	0.5	.100E-9	4.0	0.25	0.25	8.0	0.7	2.5	0.3
200	150	1.0	0.5	.580E-10	3.68	0.25	0.25	6.0	0.7	2.5	0.3
200	150	1.0	0.5	.750E-10	4.0	0.25	0.25	6.0	0.7	2.5	0.3
140	100	1.0	0.5	.100E-9	4.0	0.25	0.25	8.0	0.7	2.5	0.3
280	200	1.0	0.50	.800E-9	3.0	0.25	0.25	3.5	0.7	2.5	0.3
280	200	1.0	0.5	.220E-9	3.2	0.25	0.25	8.0	0.7	2.5	0.3
280	200	1.0	0.5	.220E-9	3.2	0.25	0.25	7.0	0.7	2.5	0.3
110	80	1.0	0.5	.260E-11	5.0	0.25	0.25	8.0	0.7	2.5	0.3
140	100	1.0	0.50	.400E-9	3.0	0.25	0.25	5.0	0.7	2.5	0.3
114	80	1.0	0.5	.350E-10	3.8	0.50	0.25	14.0	0.7	5.84	1.0

(*continued*)

Material; Condition; Environment*	Code	YS	UTS
[G] Misc. CRES/heat resistant steel			
PHxx-x Alloys			
PH13-8 Mo			
H1000; Plt, Forg, Extr	G1AD13AB1	200	208
H1000; Plt & Forg; DW & SW, > 1Hz	G1AD13WD1	200	208
H1050; Plt & Forg	G1AF13AB1	185	190
H1050; Plt & Forg; DW & SW, > 0.1Hz	G1AF13WD1	185	190
xx-xPH Alloys			
15-5PH			
H900; Rnd, C-R	G2AB16AB1	170	190
H1025; Rnd, C-R	G2AD16AB1	165	175
H1025; Forg	G2AD23AB1	155	165
H1100; Rnd, C-R	G2AF16AB1	145	155
17-4PH			
H900; Plt, L-T	G2CB11AB1	170	195
H900; Plt, T-L	G2CB12AB1	170	195
H1050; Plt	G2CE13AB1	155	160
H1025; Rnd, C-L	G2CE19AB1	160	163
H1025; Cast; HHA	G2CE50AD1	160	163
H1100; Plt; HHA	G2CH13AD1	145	150
17-7PH			
TH1050; Plt	G2EH13AB1	170	190
AMxxx Alloys			
AM 350			
CRT; Sht, L-T	G4AH11AB1	185	205
AM 367			
SCT (850); Sht	G4FC11AB1	240	243
Custom xxx Alloys			
Custom 455			
H1000; Plt & Forg	G5BD23AB1	195	205
H1025; Forg, C-R	G5BE26AB1	185	195
Nitronic xx Alloys			
Nitronic 33			
Ann; Plt	G7AA13AB1	64	115
Ann; Plt; −452F LHe	G7AA13LA2	220	260
Ann; Plt; −320F LN2	G7AA13LA4	165	220
Nitronic 50			
Ann; Plt	G7CA13AB1	77	120
Ann; Plt; −452F LHe	G7CA13LA2	210	275
Ann; Plt; −320F LN2	G7CA13LA4	170	230
Nitronic 60			
HR, CR; Rnd Rod	G7DC18AB1	137	192

* Unless noted, assume Lab Air (LA) environment and any orientation except S-T, S-L, C-R, C-L, and R-L.

(continued)

K_{Ie}	K_{Ic}	A_k	B_k	C	n	p	q	DK_0	R_{cI}	a	S.R.
140	100	0.75	0.75	.250E-8	2.620	0.25	0.25	5.0	0.6	2.5	0.3
140	100	0.75	0.75	.350E-8	2.800	0.50	0.50	6.0	0.6	2.5	0.3
160	115	0.75	0.75	.200E-8	2.620	0.25	0.25	5.0	0.6	2.5	0.3
160	115	0.75	0.75	.250E-9	4.100	0.50	0.50	7.0	0.4	2.5	0.3
65	50	1.0	0.5	.780E-10	4.000	0.25	0.25	5.0	0.7	2.5	0.3
80	60	1.0	0.5	.200E-9	3.600	0.25	0.25	5.0	0.7	2.5	0.3
150	110	1.0	0.5	.830E-9	2.870	0.25	0.25	5.0	0.7	2.5	0.3
110	80	1.0	0.5	.450E-9	3.100	0.25	0.25	5.0	0.7	2.5	0.3
60	50	1.0	0.5	.600E-9	3.110	0.25	0.25	4.0	0.7	2.5	0.3
55	45	1.0	0.5	.600E-9	3.400	0.25	0.25	4.0	0.7	2.5	0.3
80	60	1.0	0.5	.340E-9	3.420	0.25	0.25	4.0	0.7	2.5	0.3
70	55	1.0	0.5	.150E-9	3.500	0.25	0.25	4.0	0.7	2.5	0.3
70	55	1.0	0.5	.400E-9	3.500	0.25	0.25	4.0	0.7	2.5	0.3
120	90	1.0	0.5	.250E-8	2.470	0.25	0.25	4.0	0.7	2.5	0.3
60	50	1.0	1.0	.350E-9	3.400	0.25	0.25	3.0	0.7	2.5	0.3
120	90	1.0	1.5	.130E-8	2.500	0.25	0.25	5.0	0.7	2.5	0.3
80	65	1.0	1.0	.350E-8	2.150	0.25	0.25	4.0	0.7	0.7	0.3
140	100	1.0	0.5	.300E-8	2.440	0.25	0.25	4.0	0.7	2.5	0.3
150	110	1.0	0.5	.910E-9	2.760	0.25	0.25	4.2	0.7	2.5	0.3
280	200	1.0	0.5	.150E-10	3.850	0.25	0.25	9.0	0.7	2.5	0.3
80	65	1.0	0.5	.700E-11	5.000	0.25	0.25	9.0	0.7	2.5	0.3
150	110	1.0	0.5	.500E-10	4.000	0.25	0.25	9.0	0.7	2.5	0.3
250	180	1.0	0.5	.200E-10	3.800	0.25	0.25	9.0	0.7	2.5	0.3
120	90	1.0	0.5	.230E-11	4.300	0.25	0.25	9.0	0.7	2.5	0.3
150	110	1.0	0.5	.170E-10	3.650	0.25	0.25	9.0	0.7	2.5	0.3
84	60	0.75	0.5	.475E-8	2.100	0.50	0.50	3.1	0.5	5.85	1.0

(*continued*)

Material; Condition; Environment*	Code	YS	UTS
[H] High temperature steel			
Nickel Chromium			
A286 (140 ksi)			
Plt & Sht; 600-800F	H1AB13AA15	95	138
Plt & Sht	H1AB13AB1	100	140
Forg, L-T, T-L, L-R	H1AB23AB1	100	140
A286 (160 ksi)			
Plt & Sht; 600-800F	H1AC13AA15	95	138
Plt & Sht	H1AC13AB1	105	160
Forg. rod, L-R	H1AC28AB1	120	160
A286 (200 ksi Bolt Material)			
Forg. rod, L-R	H1AD28AB1	190	200
JBK-75			
ST-CR-A; Plt, T-L	H1CB12AB1	150	180
[J] Tool Steel			
AISI Tool Steel			
M-50			
61-63 Rc; Plt	J1IK10AB1	325	375
T1 (18-4-1)			
60-63 Rc; Plt	J1MA10AB1	325	350

* Unless noted, assume Lab Air (LA) environment and any orientation except S-T, S-L, C-R, C-L, and R-L.

(continued)

K_{Ie}	K_{Ic}	A_k	B_k	C	n	p	q	DK_0	R_{cI}	a	S.R.
110	80	1.0	0.5	.150E-8	2.930	0.25	0.25	6.0	0.7	2.5	0.3
140	100	1.0	0.5	.500E-9	3.000	0.25	0.25	6.0	0.7	2.5	0.3
140	100	1.0	0.5	.100E-9	3.300	0.25	0.25	6.0	0.7	2.5	0.3
110	80	1.0	0.5	.120E-8	3.000	0.25	0.25	6.0	0.7	2.5	0.3
140	100	1.0	0.5	.260E-9	3.200	0.25	0.25	6.0	0.7	2.5	0.3
140	100	1.0	0.5	.177E-8	2.450	0.25	0.25	4.5	0.5	3.0	0.3
140	100	1.0	0.5	.300E-8	2.100	0.25	0.25	3.5	0.2	3.0	0.3
120	90	1.0	0.5	.200E-8	2.400	0.25	0.25	8.5	0.7	2.5	0.3
13.5	13	1.0	0.5	.140E-8	3.180	0.1	0.1	2.4	0.7	3.0	0.3
15.5	15	1.0	0.5	.350E-9	3.800	0.1	0.1	2.4	0.7	3.0	0.3

(*continued*)

Material; Condition; Environment*	Code	YS	UTS
[M] 1000-9000 Series aluminum			
2000 Series			
2014-T6			
Plt & Sht; L-T	M2AD11AB1	65	74
Plt & Sht; T-L	M2AD12AB1	63	71
2014-T651			
Plt & Sht; L-T	M2AF11AB1	64	71
Plt & Sht; T-L	M2AF12AB1	64	71
Plt & Sht; GTA Weld	M2AFB1AB1	24	47
Plt & Sht; GTA Weld, SR	M2AFB2AB1	14	27
2020-T651			
Plt & Sht; L-T	M2CB11AB1	77	82
Plt & Sht; T-L	M2CB12AB1	78	82
2024-T3			
Clad, Plt & Sht; L-T; LA & HHA	M2EA11AB1	53	66
Clad, Plt & Sht; T-L; LA & HHA	M2EA12AB1	48	65
Clad, Plt & Sht; L-T; DW	M2EA11WA1	53	66
Clad, Plt & Sht; T-L; DW	M2EA12WA1	48	65
2024-T351			
Plt & Sht; L-T; 300F to 400F Air	M2EB11AA11	52	66
Plt & Sht; L-T; LA & HHA	M2EB11AB1	54	68
Plt & Sht; T-L; LA & HHA	M2EB12AB1	52	68
2024-T3511			
Extr; L-T; LA & HHA	M2EC31AB1	55	77
2024-T62			
Plt & Sht; L-T; LA, HHA & ASW	M2EG11AB1	58	66
Plt & Sht; T-L; LA, HHA & ASW	M2EG12AB1	57	66
2024-T81			
Plt & Sht; L-T; 350F Air	M2EI11AA11	52	61
Plt & Sht; L-T	M2EI11AB1	63	75
Plt & Sht; L-T; DA	M2EI11AC1	63	70
Plt & Sht; L-T; HHA	M2EI11AD1	63	70
Plt & Sht; T-L; 350F Air	M2EI12AA11	52	65
Plt & Sht; T-L	M2EI12AB1	62	68
2024-T851			
Plt & Sht; T-L; 300F to 350F Air	M2EJ12AA11	56	61
Plt & Sht; LA, DA, JP-4	M2EJ13AB1	64	74
Plt & Sht; 3.5% NaCl	M2EJ13WB1	64	70

* Unless noted, assume Lab Air (LA) environment and any orientation except S-T, S-L, C-R, C-L, and R-L.

Appendix A NASA/FLAGRO 2.0 Materials Constants

(*continued*)

K_{Ie}	K_{Ic}	A_k	B_k	C	n	p	q	DK_0	R_{cI}	a	S.R.
38	27	1.0	1.0	.350E-7	2.800	0.5	1.0	2.7	0.7	1.5	0.3
23	18	1.0	1.0	.350E-7	2.800	0.5	1.0	2.7	0.7	1.5	0.3
28	22	1.0	1.0	.150E-7	2.800	0.5	1.0	2.7	0.7	1.5	0.3
27	20	1.0	1.0	.150E-7	2.800	0.5	1.0	2.7	0.7	1.5	0.3
22.4	16	1.0	1.0	.646E-8	3.918	0.5	1.0	9.5	0.7	1.5	0.3
22.4	16	1.0	1.0	.115E-8	5.005	0.5	0.5	4.0	0.7	1.5	0.3
29	22.5	1.0	1.0	.310E-8	3.695	0.5	0.5	2.2	0.7	1.5	0.3
21	17	1.0	1.0	.135E-7	3.074	0.5	0.5	2.2	0.7	1.5	0.3
46	33	1.0	1.0	.829E-8	3.284	0.5	1.0	2.9	0.7	1.5	0.3
41	29	1.0	1.0	.244E-7	2.601	0.5	1.0	2.9	0.7	1.5	0.3
46	33	1.0	1.0	.169E-7	3.090	0.5	1.0	2.9	0.7	1.5	0.3
41	29	1.0	1.0	.892E-8	3.282	0.5	1.0	2.9	0.7	1.5	0.3
46	33	1.0	1.0	.334E-7	2.956	0.5	1.0	2.6	0.7	1.5	0.3
48	34	1.0	1.0	.922E-8	3.353	0.5	1.0	2.6	0.7	1.5	0.3
41	29	1.0	1.0	.922E-8	3.353	0.5	1.0	2.6	0.7	1.5	0.3
35	25	1.0	1.0	.200E-7	2.700	0.5	1.0	2.9	0.7	1.5	0.3
50	36	1.0	1.0	.100E-7	3.200	0.5	1.0	2.9	0.7	1.5	0.3
42	30	1.0	1.0	.100E-7	3.200	0.5	1.0	2.9	0.7	1.5	0.3
35	25	1.0	1.0	.105E-6	2.761	0.5	1.0	2.9	0.7	1.5	0.3
30	22	1.0	1.0	.804E-7	2.763	0.5	1.0	2.8	0.7	1.5	0.3
29.7	22	1.0	1.0	.217E-7	2.890	0.5	1.0	2.8	0.7	1.5	0.3
29.7	22	1.0	1.0	.863E-7	2.566	0.5	1.0	2.8	0.7	1.5	0.3
32	23	1.0	1.0	.340E-7	3.255	0.5	1.0	2.9	0.7	1.5	0.3
28.1	21	1.0	1.0	.219E-7	3.313	0.5	1.0	2.8	0.7	1.5	0.3
34	24	1.0	1.0	.243E-8	4.308	0.5	1.0	2.8	0.7	1.5	0.3
31	23	1.0	1.0	.150E-7	3.100	0.5	1.0	2.8	0.7	1.5	0.3
31.3	23	1.0	1.0	.566E-7	2.480	0.5	1.0	2.8	0.7	1.5	0.3

(continued)

Material; Condition; Environment*	Code	YS	UTS
[M] 1000-9000 Series aluminum			
2000 Series			
2024-T852			
Forg; LA & DA	M2EK23AB1	57	70
2024-T861			
Plt & Sht; L-T; 300F to 400F Air	M2EL11AA11	56	58
Plt & Sht; L-T; LA & HHA	M2EL11AB1	73	76
Plt & Sht; T-L	M2EL12AB1	72	76
2048-T851			
Plt & Sht; L-T; LA, DA	M2FC11AB1	63	69
Plt & Sht; T-L; LA, DA	M2FC12AB1	62	69
2124-T851			
Plt & Sht; L-T; 120F to 350F Air	M2GC11AA10	53	55
Plt & Sht; L-T; LA, DA, HHA	M2GC11AB1	63	71
Plt & Sht; T-L; 300F - 400F Air	M2GC12AA11	53	55
Plt & Sht; T-L; LA, HHA	M2GC12AB1	64	72
Plt & Sht; T-L; $-200F$ to $-150F$ GN2	M2GC12GB6	72	80
Plt & Sht; S-T, S-L; LA, HHA	M2GC15AB1	60	69
2219-T62			
Plt & Sht; L-T	M2IA11AB1	43	61
Plt & Sht; L-T & T-L; 350F Air	M2IA13AA11	37	46
Plt & Sht; T-L	M2IA12AB1	43	61
Plt & Sht; L-T & T-L; $-320F$ LN2	M2IA13LA4	51	76
2219-T851			
Plt & Sht; L-T, LA, DA	M2IC11AB1	53	65
Plt & Sht; T-L; LA, DA	M2IC12AB1	50	66
2219-T87			
Plt & Sht; L-T; 300F to 350F Air	M2IF11AA11	43	49
Plt & Sht; L-T	M2IF11AB1	57	68
Plt & Sht; L-T; $-320F$ LN2	M2IF11LA4	68	83
Plt & Sht; T-L; 300F to 350F Air	M2IF12AA11	44	49
Plt & Sht; T-L	M2IF12AB1	58	69
Plt & Sht; T-L; $-320F$ LN2	M2IF12LA4	66	85
Plt & Sht; GTA weld, PAR	M2IFB1AB1	20	42
Plt & Sht; GTA weld, PAR; $-320F$ LN2	M2IFB1LA4	27	46
2324-T39			
Plt & Sht; L-T	M2JA11AB1	65	72
2090-T8E41			
Plt & Sht; L-T	M2PA11AB1	80	85

* Unless noted, assume Lab Air (LA) environment and any orientation except S-T, S-L, C-R, C-L, and R-L.

(continued)

K_{Ie}	K_{Ic}	A_k	B_k	C	n	p	q	DK_0	R_{cl}	a	S.R.
39.2	28	1.0	1.0	.666E-7	2.060	0.5	1.0	2.9	0.7	1.5	0.3
50	36	1.0	1.0	.283E-7	3.181	0.5	1.0	2.8	0.7	1.5	0.3
30	23	1.0	1.0	.280E-7	2.920	0.5	1.0	2.2	0.7	1.5	0.3
24	19	1.0	1.0	.309E-7	2.895	0.5	1.0	2.2	0.7	1.5	0.3
49	35	1.0	1.0	.386E-8	3.460	0.5	1.0	2.7	0.7	1.5	0.3
42	30	1.0	1.0	.203E-8	3.951	0.5	1.0	2.7	0.7	1.5	0.3
43	31	1.0	1.0	.162E-7	3.233	0.5	0.5	2.9	0.7	1.5	0.3
42	30	1.0	1.0	.141E-7	3.146	0.5	0.5	2.7	0.7	1.5	0.3
36	26	1.0	1.0	.452E-8	3.715	0.5	1.0	2.9	0.7	1.5	0.3
31	23	1.0	1.0	.119E-7	3.061	0.5	1.0	2.7	0.7	1.5	0.3
35	26	1.0	1.0	.200E-8	3.400	0.5	1.0	2.2	0.7	1.5	0.3
28	21	1.0	1.0	.259E-8	3.867	0.5	1.0	2.7	0.7	1.5	0.3
43	31	1.0	1.0	.377E-7	2.657	0.5	1.0	3.6	0.7	1.5	0.3
42	30	1.0	1.0	.458E-7	2.611	0.5	1.0	3.8	0.7	1.5	0.3
41	29	1.0	1.0	.420E-7	2.648	0.5	1.0	3.6	0.7	1.5	0.3
42	30	1.0	1.0	.257E-8	3.178	0.5	1.0	2.9	0.7	1.5	0.3
46	33	1.0	1.0	.119E-7	3.156	0.5	1.0	3.0	0.7	1.5	0.3
43	31	1.0	1.0	.307E-7	2.863	0.5	1.0	2.1	0.7	1.5	0.3
39	28	1.0	1.0	.293E-7	2.533	0.5	1.0	3.5	0.7	1.5	0.3
42	30	1.0	1.0	.572E-7	2.487	0.5	1.0	2.9	0.7	1.5	0.3
57	41	1.0	1.0	.617E-8	3.151	0.5	1.0	2.3	0.7	1.5	0.3
38	27	1.0	1.0	.269E-8	3.903	0.5	1.0	3.5	0.7	1.5	0.3
38	27	1.0	1.0	.700E-8	3.400	0.5	1.0	2.9	0.7	1.5	0.3
46	33	1.0	1.0	.450E-10	5.279	0.5	1.0	2.3	0.7	1.5	0.3
50	28	1.0	1.0	.640E-8	4.060	0.5	1.0	4.0	0.7	1.5	0.3
38	27	1.0	1.0	.304E-10	6.140	0.5	1.0	4.0	0.7	1.5	0.3
55	39	1.0	1.0	.831E-8	3.072	0.5	1.0	2.9	0.7	1.5	0.3
46	33	1.0	1.0	.485E-8	3.267	0.5	1.0	4.0	0.7	1.5	0.3

(*continued*)

Material; Condition; Environment*	Code	YS	UTS
[M] 1000-9000 Series aluminum			
5000 Series			
5083-O			
Plt; T-L	M5BA12AB1	20	43
6000 Series			
6061-T6			
Plt; T-L	M6AB13AB1	41	45
Plt; GTA weld, PAR	M6ABA1AB1	23	26
6061-T651			
Plt; 300F Air	M6AC13AA10	36	37
Plt	M6AC13AB1	44	47
6063-T5			
Plt & Sht; T-L; LA	M6BA12AB1	21	27
7000 Series			
7005-T6 & T63			
Plt & Sht; L-T	M7BA11AB1	48	53
Plt & Sht; T-L	M7BA12AB1	48	53
7010-T73651			
Plt & Sht; L-T & L-S	M7DA11AB1	64	73
7050-T73511			
Extr; L-T; LA, HHA, DA	M7GE31AB1	72	80
7050-T736 & T74			
Forg; L-T	M7GI21AB1	65	72
Forg; T-L	M7GI22AB1	62	72
7050-T73651 & T7451			
Plt & Sht; L-T; LA & HHA	M7GJ11AB1	66	77
Plt & Sht; L-T; DA	M7GJ11AC1	66	77
Plt & Sht; T-L; LA & HHA	M7GJ12AB1	65	77
Plt & Sht; T-L; DA	M7GJ12AC1	65	77
Plt & Sht; S-T	M7GJ15AB1	61	75
7050-T74511			
Extr; L-T; LA & DW	M7GL31AB1	70	79
7050-T73652 & T7452			
Forg; L-T	M7GM21AB1	70	79
Forg; T-L	M7GM22AB1	70	79
7050-T7651			
Plt & Sht; L-T; LA & HHA	M7GQ11AB1	75	80
Plt & Sht; T-L	M7GQ12AB1	75	80
7050-T76511			
Extr; L-T; LA & HHA	M7GS31AB1	79	87
Extr; T-L	M7GS32AB1	79	87

* Unless noted, assume Lab Air (LA) environment and any orientation except S-T, S-L, C-R, C-L, and R-L.

Appendix A NASA/FLAGRO 2.0 Materials Constants

(continued)

K_{Ie}	K_{Ic}	A_k	B_k	C	n	p	q	DK_0	R_{cI}	a	S.R.
63	45	1.0	0.1	.249E-6	1.938	0.5	1.0	4.5	0.7	1.5	0.3
36	26	1.0	0.75	.900E-7	2.300	0.5	0.5	3.5	0.7	2.0	0.3
36	26	1.0	1.0	.113E-7	3.524	0.5	1.0	4.5	0.7	1.5	0.3
38	27	1.0	1.0	.288E-8	3.884	0.5	1.0	4.0	0.7	1.5	0.3
38	27	1.0	1.0	.133E-6	2.248	0.5	1.0	3.5	0.7	1.5	0.3
34	24	1.0	0.75	.946E-8	3.447	0.5	1.0	3.5	0.7	1.9	0.3
64	46	1.0	1.0	.990E-7	2.248	0.5	1.0	3.4	0.7	1.75	0.3
56	40	1.0	1.0	.839E-7	2.313	0.5	1.0	3.4	0.7	1.75	0.3
43	31	1.0	1.0	.929E-8	3.122	0.5	1.0	2.5	0.7	1.75	0.3
49	35	1.0	1.0	.133E-7	2.908	0.5	1.0	2.1	0.7	1.9	0.3
46	33	1.0	1.0	.615E-8	3.368	0.5	1.0	2.3	0.7	1.9	0.3
33	24	1.0	1.0	.213E-7	2.677	0.5	1.0	2.3	0.7	1.9	0.3
43	31	1.0	1.0	.462E-7	2.379	0.5	1.0	2.4	0.7	1.9	0.3
43	31	1.0	1.0	.363E-7	2.084	0.5	1.0	2.4	0.7	1.9	0.3
35	25	1.0	1.0	.246E-7	2.738	0.5	1.0	2.4	0.7	1.9	0.3
35	25	1.0	1.0	.660E-7	2.865	0.5	1.0	2.4	0.7	1.9	0.3
33	24	1.0	1.0	.132E-7	2.674	0.5	1.0	2.8	0.7	1.9	0.3
50	36	1.0	1.0	.279E-7	2.682	0.5	1.0	2.2	0.7	1.9	0.3
43	31	1.0	1.0	.266E-7	2.654	0.5	1.0	2.4	0.7	1.9	0.3
27	21	1.0	1.0	.266E-7	2.654	0.5	1.0	2.4	0.7	1.9	0.3
43	31	1.0	1.0	.446E-7	2.412	0.5	1.0	2.0	0.7	1.9	0.3
38	28	1.0	1.0	.446E-7	2.412	0.5	1.0	2.0	0.7	1.9	0.3
41	30	1.0	1.0	.502E-7	2.329	0.5	1.0	2.0	0.7	1.9	0.3
32	24	1.0	1.0	.502E-7	2.329	0.5	1.0	2.0	0.7	1.9	0.3

(*continued*)

Material; Condition; Environment*	Code	YS	UTS
[M] 1000-9000 Series aluminum			
7000 Series			
7075-t6			
Plt, Sht & Clad; LA	M7HA13AB1	75	84
Plt, Sht & Clad; HHA	M7HA13AD1	75	84
7075-T651			
Plt & Sht; L-T; LA, DA	M7HB11AB1	76	85
Plt & Sht; L-T; HHA	M7HB11AD1	76	85
Plt & Sht; L-T; 3.5% NaCl	M7HB11WB1	76	85
Plt & Sht; T-L; DW	M7HB12WA1	76	85
Plt & Sht; S-T	M7HB15AB1	66	75
7075-T6510			
Extr; L-T; LA & DA	M7HC31AB1	79	88
7075-T6511			
Extr; L-T; LA & HHA	M7HD31AB1	80	87
Extr; T-L	M7HD32AB1	80	87
7075-T73			
Plt & Sht; L-T; LA & HHA	M7HG11AB1	60	74
Plt & Sht; T-L	M7HG12AB1	61	71
7075-T7351			
Plt & Sht; L-T	M7HH11AB1	62	71
Plt & Sht; L-T; DA	M7HH11AC1	62	71
Plt & Sht; L-T; HHA	M7HH11AD1	62	71
Plt & Sht; T-L; LA & DA	M7HH12AB1	60	71
Plt & Sht; S-T; LA	M7HH15AB1	58	65
7075-T73510			
Extr; L-T; LA	M7HI31AB1	64	75
7075-T73511			
Extr; L-T; LA, DA, HHA	M7HJ31AB1	65	74
7075-T7352			
Plt, Sht & Forg; L-T; LA & DA	M7HK11AB1	59	68
Plt, Sht & Forg; T-L; LA & DA	M7HK12AB1	53	68
7075-T7651			
Plt & Sht; L-T; LA & DA	M7HM11AB1	66	76
Plt & Sht; T-L; LA & DA	M7HM12AB1	66	75
7079-T651			
Plt & Sht; L-T	M7IC11AB1	75	83

* Unless noted, assume Lab Air (LA) environment and any orientation except S-T, S-L, C-R, C-L, and R-L.

Appendix A NASA/FLAGRO 2.0 Materials Constants

(continued)

K_{Ie}	K_{Ic}	A_k	B_k	C	n	p	q	DK_0	R_{cI}	a	S.R.
37	27	1.0	1.0	.209E-7	2.947	0.5	1.0	2.0	0.7	1.9	0.3
37	27	1.0	1.0	.145E-6	2.497	0.5	1.0	2.0	0.7	1.9	0.3
38	28	1.0	1.0	.233E-7	2.885	0.5	1.0	2.0	0.7	1.9	0.3
38	28	1.0	1.0	.147E-7	3.985	0.5	1.0	2.5	0.7	1.9	0.3
38	28	1.0	1.0	.339E-6	2.135	0.5	1.0	2.0	0.7	1.9	0.3
32	24	1.0	1.0	.191E-6	1.917	0.5	1.0	2.0	0.7	1.9	0.3
23	18	1.0	1.0	.578E-7	2.435	0.5	1.0	2.3	0.7	1.9	0.3
38	28	1.0	1.0	.184E-6	1.869	0.5	1.0	1.8	0.7	1.9	0.3
38	28	1.0	1.0	.609E-7	2.324	0.5	1.0	1.8	0.7	1.9	0.3
32	24	1.0	1.0	.609E-7	2.324	0.5	1.0	1.8	0.7	1.9	0.3
50	36	1.0	1.0	.149E-7	3.321	0.5	1.0	2.7	0.7	1.9	0.3
32	23	1.0	1.0	.252E-7	2.908	0.5	1.0	2.7	0.7	1.9	0.3
41	29	1.0	1.0	.348E-7	2.529	0.5	1.0	2.6	0.7	1.9	0.3
41	29	1.0	1.0	.120E-7	2.994	0.5	1.0	2.6	0.7	1.9	0.3
41	29	1.0	1.0	.655E-8	3.696	0.5	1.0	2.6	0.7	1.9	0.3
35	25	1.0	1.0	.397E-8	3.652	0.5	1.0	2.6	0.7	1.9	0.3
25	19	1.0	1.0	.200E-7	2.300	0.5	1.0	2.6	0.7	1.9	0.3
43	31	1.0	1.0	.138E-6	1.918	0.5	1.0	2.4	0.7	1.9	0.3
46	33	1.0	1.0	.347E-7	2.508	0.5	1.0	2.4	0.7	1.9	0.3
46	33	1.0	1.0	.150E-7	3.100	0.5	1.0	2.7	0.7	1.9	0.3
35	25	1.0	1.0	.150E-7	3.100	0.5	1.0	2.7	0.7	1.9	0.3
45	32	1.0	1.0	.552E-7	2.305	0.5	1.0	2.4	0.7	1.9	0.3
32	23	1.0	1.0	.654E-8	3.338	0.5	1.0	2.4	0.7	1.9	0.3
36	26	1.0	1.0	.212E-6	2.101	0.5	1.0	2.0	0.7	1.9	0.3

(continued)

Material; Condition; Environment*	Code	YS	UTS
[M] 1000-9000 Series aluminum			
7000 Series			
7149-T73511			
Extr; L-T; LA	M7NA31AB1	66	76
Extr; T-L; LA	M7NA32AB1	63	74
7178-T6 & T651			
Plt & Sht; L-T; LA & HHA	M7RA11AB1	84	89
Plt & Sht; T-L; LA & HHA	M7RA12AB1	79	89
7178-T7651			
Plt & Sht; L-T	M7RF11AB1	72	80
Plt & Sht; T-L	M7RF12AB1	70	79
7475-T61			
Plt, Sht, Clad; L-T; LA, DA, HHA	M7TB11AB1	74	78
Plt, Sht, Clad; T-L; LA, DA, HHA	M7TB12AB1	71	78
7475-T651			
Plt & Sht; L-T; LA, DA, HHA	M7TD11AB1	75	85
7475-T7351			
Plt & Sht; L-T; LA, DA, HHA, DW	M7TF11AB1	63	73
Plt & Sht; T-L; LA, DA, HHA	M7TF12AB1	57	70
7475-T7651			
Plt & Sht; L-T; LA, DA, HHA, DW, 3.5% NaCl	M7TJ11AB1	69	78
[O] Misc. and cast aluminum			
300 Series cast			
A356-T60			
Cast	O3FB50AB1	31	40

* Unless noted, assume Lab Air (LA) environment and any orientation except S-T, S-L, C-R, C-L, and R-L.

(*continued*)

K_{Ie}	K_{Ic}	A_k	B_k	C	n	p	q	DK_0	R_{cI}	a	S.R.
43	31	1.0	1.0	.278E-6	1.544	0.5	1.0	2.4	0.7	1.9	0.3
33	24	1.0	1.0	.352E-6	1.435	0.5	1.0	2.4	0.7	1.9	0.3
31	24	1.0	1.0	.156E-6	2.180	0.5	1.0	2.3	0.7	1.9	0.3
27	21	1.0	1.0	.405E-7	2.736	0.5	1.0	1.8	0.7	1.9	0.3
39	28	1.0	1.0	.140E-6	1.800	0.5	1.0	2.1	0.7	1.9	0.3
31	23	1.0	1.0	.140E-6	1.800	0.5	1.0	2.1	0.7	1.9	0.3
43	31	1.0	1.0	.898E-7	2.211	0.5	1.0	2.1	0.7	1.9	0.3
36	26	1.0	1.0	.360E-6	1.496	0.5	1.0	2.1	0.7	1.9	0.3
52	37	1.0	1.0	.232E-6	2.328	0.5	1.0	3.0	0.7	1.9	0.3
60	43	1.0	1.0	.175E-7	2.877	0.5	1.0	2.4	0.7	1.9	0.3
49	35	1.0	1.0	.696E-7	2.212	0.5	1.0	2.8	0.7	1.9	0.3
56	40	1.0	1.0	.181E-7	2.808	0.5	1.0	2.2	0.7	1.9	0.3
22	16	1.0	1.0	.142E-9	4.980	0.5	1.0	6.3	0.7	1.9	0.3

(continued)

Material; Condition; Environment*	Code	YS	UTS
[P] Titanium alloys			
T1 Unalloyed			
T1-55			
Plt & Sht	P1AA13AB1	55	65
Plt & Sht; DW & SW	P1AA12WA1	55	65
T1-70			
Plt & Sht	P1CA13AB1	70	80
Plt & Sht; DW & SW	P1CA13WA1	70	80
Binary Alloys			
T1-2.5 Cu; STA			
Sht; LA, HHA, DW	P2AA13AB1	97	110
Ternary Alloys			
T1-5 Al-2.5Sn; Annealed			
Sht; LA, HHA, DW	P3CA13AB1	120	130
T1-5 Al-2.5Sn (ELI); Annealed			
Forg	P3CB23AB1	115	120
Forg; $-423F$ LH2	P3CB23LA3	200	210
T1-3 Al-2.5V; CW, SR (750F)			
Extr	P3DB33AB1	105	125
T1-6 Al-4V (MA)			
Plt & Sht, $-100F$	P3EA13AA7	165	170
Plt & Sht	P3EA13AB1	138	146
Forg	P3EA23AB1	135	145
Extr	P3EA33AB1	125	140
T1-6 Al-4V; BA (1900F/.5h + 1325F/2h)			
Plt & Sht; LA, DA, 3.5% NaCl	P3EB12AB1	120	135
Forg; LA, DA, HHA, 3.5% NaCl	P3EB23AB1	110	125
T1-6 Al-4V; RA			
Sht; L-T; LA, DA, HHA, DW, 3.5% NaCl	P3EC11AB1	140	150
Sht; T-L; LA, DA, HHA, DW, 3.5% NaCl	P3EC12AB1	140	150
Plt; $-100F$	P3EC13AA7	160	175
Plt; LA, DA, HHA, DW	P3EC13AB1	125	135
Forg; LA, DA, HHA, 3.5% NaCl	P3EC23AB1	115	130
T1-6 Al-4V; ST (1750F) + A (1000F/4h)			
Plt & Sht; SR (1000F/4h)	P3ED13AA1	155	167
Plt & Sht; SR (1000F/8h)	P3ED13AB1	140	150
Plt & Sht; SR (1000F/4h); $-320F$ LN2	P3ED13LA4	230	240
Forg; SR (1000F/4h)	P3ED20AB1	150	163
Forg; SR (1000F/4h); $-320F$ LN2	P3ED20LA4	225	235

* Unless noted, assume Lab Air (LA) environment and any orientation except S-T, S-L, C-R, C-L, and R-L.

Appendix A NASA/FLAGRO 2.0 Materials Constants

(continued)

K_{Ie}	K_{Ic}	A_k	B_k	C	n	p	q	DK_0	R_{cl}	a	S.R.
70	50	1.0	0.5	.542E-9	3.670	0.5	0.5	5.0	0.7	2.5	0.3
70	50	1.0	0.5	.100E-8	3.670	0.5	0.5	5.0	0.7	2.5	0.3
70	50	1.0	0.5	.210E-8	3.180	0.5	0.5	5.0	0.7	2.5	0.3
70	50	1.0	0.5	.300E-8	3.180	0.5	0.5	5.0	0.7	2.5	0.3
70	50	1.0	0.5	.750E-8	2.750	0.5	0.5	5.0	0.7	2.5	0.3
90	65	1.0	1.0	.120E-7	2.500	0.5	0.5	5.0	0.4	2.5	0.3
90	65	1.0	1.0	.900E-8	2.600	0.5	0.5	5.0	0.4	2.5	0.3
80	60	1.0	0.5	.150E-8	2.850	0.5	0.5	5.0	0.4	2.5	0.3
70	50	1.0	0.5	.300E-9	4.000	0.25	0.75	4.5	0.7	2.5	0.3
65	50	1.0	0.5	.100E-8	3.300	0.25	0.75	3.5	0.7	2.5	0.3
65	50	1.0	0.5	.252E-8	3.010	0.25	0.75	3.5	0.7	2.5	0.3
65	50	1.0	0.5	.311E-9	3.667	0.25	0.75	3.5	0.7	2.5	0.3
75	60	1.0	0.5	.147E-9	3.834	0.25	0.75	5.0	0.7	2.5	0.3
100	80	1.0	0.5	.900E-9	3.250	0.25	0.75	3.5	0.7	2.5	0.3
100	80	1.0	0.5	.600E-9	3.250	0.25	0.75	6.0	0.7	2.5	0.3
80	60	1.0	0.5	.150E-8	3.300	0.25	0.75	3.5	0.7	2.5	0.3
80	60	1.0	0.5	.250E-8	3.300	0.25	0.75	3.5	0.7	2.5	0.3
85	65	1.0	0.5	.120E-9	4.200	0.50	0.75	5.0	0.7	2.5	0.3
100	75	1.0	0.5	.250E-8	3.000	0.50	0.75	6.0	0.7	2.5	0.3
100	75	1.0	0.5	.500E-9	3.300	0.50	0.75	8.0	0.7	2.5	0.3
55	45	0.5	1.0	.500E-8	2.850	0.25	0.75	3.5	0.5	2.5	0.3
55	45	0.5	1.0	.200E-8	3.000	0.25	0.75	3.5	0.5	2.5	0.3
44	38	1.0	0.5	.400E-8	2.750	0.25	0.75	3.5	0.5	2.5	0.3
50	42	0.75	0.75	.150E-8	3.200	0.25	0.75	3.5	0.5	2.5	0.3
47	40	0.5	0.75	.220E-9	3.300	0.25	0.75	3.5	0.5	2.5	0.3

(continued)

Material; Condition; Environment*	Code	YS	UTS
[P] Titanium alloys			
GTA Weld; SR; thk < 0.2"	P3EDB2AB1A	125	135
GTA Weld; SR; thk >= 0.2"	P3EDB2AB1B	125	135
T1-6 Al-4V; ELI; BA (1900F/.5h) + (1325F/2h)			
Plt & sht; LA, 3.5% NaCl	P3EL12AB1	115	127
T1-6 Al-4V (ELI) RA			
Plt	P3EM13AB1	120	130
Forg; −100F	P3EM23AA7	145	155
Forg	P3EM23AB1	120	130
Forg; −452F LHe	P3EM23LA2	240	248
Forg; −320F LN2	P3EM23LA4	200	213
Forg; EB welded, SR; weldine	P3EMD2AB1	120	130
Forg; EB welded, SR; weldine; −320F LN2	P3EMD2LA4	200	213
Forg; EB welded, SR; HAZ	P3EMD8AB1	120	130
Forg; EB welded, SR; HAZ; −320F LN2	P3EMD8LA4	200	213
Quaternary Alloys			
T1-4.5 Al-5 Mo-1.5 Cr			
Plt; LA, 3.5% NaCl	P4BA10AB1	170	182
T1-8 Al-1 Mo-1V			
Sht	P4CB11AB1	138	150
T1-6 Al-6V-2Sn MA			
Plt, Forg, Extr; LA, DA, HHA, DW	P4DA33AB1	150	160
T1-6 Al-6V-2Sn RA			
Plt	P4DB12AB1	150	160
T1-6 Al-6V-2Sn BA			
Plt	P4DC12AB1	140	155
T1-6 Al-6V-2Sn ST (1600F); A (1000F/6h)			
Forg; −65F	P4DD21AA7	205	212
Forg; 300F	P4DD21AA10	165	172
Forg; LA, DA, HHA	P4DD21AB1	180	190
T1-10V-2Fe-3Al			
STA (140-160 UTS, 70Klc) Plt & Forg	P4MD23AB1	140	150
STA (160-180 UTS, 60Klc) Plt & Forg	P4MF23AB1	153	170
STA (180-200 UTS, 40Klc) Plt & Forg	P4MG13AB1	178	190
STA (180-200 UTS, 30Klc) Plt & Forg	P4MG20AB1	178	190
STA (180-200 UTS, 25Klc)Forg	P4MG23AB1	178	190
T1-6A; -2Zn-2Sn-2Mo-2Cr (ST or STA)			
Plt; HHA	P5FB11AD1	155	165

* Unless noted, assume Lab Air (LA) environment and any orientation except S-T, S-L, C-R, C-L, and R-L.

Appendix A NASA/FLAGRO 2.0 Materials Constants

(*continued*)

K_{Ie}	K_{Ic}	A_k	B_k	C	n	p	q	DK_0	R_{cl}	a	S.R.
50	40	1.0	0.5	.800E-8	2.580	0.25	0.25	3.5	0.7	2.5	0.3
70	55	1.0	0.5	.800E-8	2.580	0.25	0.25	3.5	0.7	2.5	0.3
100	80	1.0	0.5	.900E-9	3.250	0.25	0.75	3.5	0.7	2.5	0.3
95	75	1.0	0.5	.250E-8	3.000	0.50	0.75	6.0	0.7	2.5	0.3
95	75	1.0	0.5	.440E-9	3.300	0.50	0.75	6.0	0.7	2.5	0.3
95	75	1.0	0.5	.150E-8	3.000	0.50	0.75	8.0	0.7	2.5	0.3
60	50	1.0	0.5	.500E-9	3.800	0.50	0.50	5.0	0.7	2.5	0.3
65	55	1.0	0.5	.140E-8	3.000	0.50	0.50	6.0	0.7	2.5	0.3
90	65	1.0	0.5	.100E-7	2.000	0.50	0.75	5.0	0.7	2.5	0.3
65	55	1.0	0.5	.800E-8	2.000	0.50	0.50	5.0	0.7	2.5	0.3
90	65	1.0	0.5	.120E-7	2.000	0.50	0.75	5.0	0.7	2.5	0.3
65	55	1.0	0.5	.350E-8	2.500	0.50	0.50	5.0	0.7	2.5	0.3
57	45	1.0	0.5	.650E-8	2.650	0.25	0.75	3.5	0.7	2.5	0.3
77	55	1.0	0.5	.500E-8	2.500	0.25	0.75	3.5	0.7	2.5	0.3
65	50	1.0	0.5	.600E-8	2.700	0.25	0.75	3.5	0.7	2.5	0.3
91	65	1.0	0.5	.150E-7	2.400	0.25	0.75	3.5	0.7	2.5	0.3
84	60	1.0	0.5	.161E-7	2.207	0.25	0.75	3.5	0.7	2.5	0.3
32	28	1.0	0.5	.162E-8	3.021	0.25	0.75	3.5	0.7	2.5	0.3
70	53	1.0	0.5	.370E-8	2.635	0.25	0.75	3.5	0.7	2.5	0.3
35	30	1.0	0.5	.736E-8	2.279	0.25	0.75	3.5	0.7	2.5	0.3
95	70	1.0	0.5	.120E-7	2.500	0.25	0.25	2.0	0.7	2.5	0.3
80	60	1.0	0.5	.200E-7	2.500	0.25	0.25	2.0	0.7	2.5	0.3
48	40	1.0	0.5	.200E-7	2.500	0.25	0.25	2.0	0.7	2.5	0.3
35	30	1.0	0.5	.200E-7	2.500	0.25	0.25	2.0	0.7	2.5	0.3
28	25	1.0	0.5	.300E-7	2.500	0.25	0.25	2.0	0.7	2.5	0.3
75	55	1.0	0.5	.125E-7	2.500	0.25	0.75	3.5	0.7	2.5	0.3

(*continued*)

Material; Condition; Environment*	Code	YS	UTS
[Q] NI Alloys/superalloys			
Hastelloy Alloys			
Hastelloy B			
Rnd Rnd	Q1AA16AB1	60	127
Hastelloy X-280; ST (2150F)			
Plt; 600-800F Air	Q1QA10AA15	39	91
Plt; 1000-1200F Air; >.67Hz	Q1QA10AA19	33	84
Plt	Q1QA10AB1	53	109
Inconel Alloys			
Inconel 600			
Plt & Sht; 1000F	Q3AB10AA18	28	83
Plt & Sht; 75-800F	Q3AB10AB1	35	94
Inconel 625			
Plt & Sht; 600F	Q3EA10AA14	50	122
Plt & Sht; 800F	Q3EA10AA16	48	119
Plt & Sht; 1000F	Q3EA10AA18	48	119
Plt & Sht;	Q3EA10AB1	66	133
Inconel 706; ST (1800-1950F); A (1375F/8h; 1150F/5-8h)			
Forg & Extr	Q3JB33AB1	145	177
Forg & Extr; −452F LHe	Q3JB33LA2	177	230
ST Plt - GTA weld - STA	Q3JBB3AB1	145	164
ST Plt - GTA weld - STA; −452F LHe	Q3JBB3LA2	177	213
Inconel 718; ST (1700-1850F) + A (1325F/8h 1 1150F/10h)			
Plt; 600F air, >.3Hz	Q3LB11AA14	150	190
Plt; 800F air, >.3Hz	Q3LB11AA16	142	180
Plt; 1000F air, >.3Hz	Q3LB13AA18	135	178
Sht (t <.25")	Q3LB13AB1A	175	210
Plt	Q3LB13AB1B	170	200
Forg	Q3LB23AB1	165	190
Forg; 300F air, >.3Hz	Q3LB26AA10	160	187
Forg; 600F air, >.3Hz	Q3LB26AA14	155	185
Forg; 800F air, >.3Hz	Q3LB26AA16	147	173
Forg; 1000F air, >.3Hz	Q3LB26AA18	140	167
GTA weld-STA; 600F air, >.6Hz	Q3LBB3AA14	140	170
GTA weld-STA; 800F air, >.6Hz	Q3LBB3AA16	135	160
GTA weld-STA; 1000F air, >.6Hz	Q3LBB3AA18	132	160
ST plt-GTA weld-aged	Q3LBB3AB1	160	192
ST plt-GTA weld-aged; −320F LN2	Q3LBB3LA4	178	227
ST plt-EB weld-aged	Q3LBD3AB1	163	205
ST plt-EB weld-aged; −320F LN2	Q3LBD3LA4	193	224

* Unless noted, assume Lab Air (LA) environment and any orientation except S-T, S-L, C-R, C-L, and R-L.

Appendix A NASA/FLAGRO 2.0 Materials Constants

(*continued*)

K_{Ie}	K_{Ic}	A_k	B_k	C	n	p	q	DK_0	R_{cI}	a	S.R.
267	100	1.0	0.5	.170E-9	3.200	0.50	0.50	20.0	0.6	2.5	0.3
150	110	0.75	0.5	.250E-9	3.600	0.50	0.50	6.0	0.7	2.5	0.3
140	100	0.75	0.5	.400E-9	3.600	0.50	0.50	4.0	0.7	2.5	0.3
165	120	0.75	0.5	.100E-9	3.600	0.50	0.50	8.0	0.7	2.5	0.3
140	100	1.0	0.5	.450E-9	3.300	0.50	0.50	5.0	0.7	2.5	0.3
140	100	1.0	0.5	.658E-11	4.647	0.50	0.50	8.0	0.7	2.5	0.3
125	90	1.0	0.5	.484E-10	4.124	0.50	0.50	5.0	0.7	2.5	0.3
125	90	1.0	0.5	.623E-9	3.365	0.50	0.50	4.0	0.7	2.5	0.3
125	90	1.0	0.5	.475E-8	2.792	0.50	0.50	2.5	0.7	2.5	0.3
140	100	1.0	0.5	.954E-10	3.745	0.50	0.50	5.0	0.7	2.5	0.3
120	85	0.75	0.5	.343E-9	3.069	0.50	0.50	12.0	0.65	2.5	0.3
147	105	0.75	0.5	.150E-8	2.144	0.50	0.50	16.0	0.30	5.84	1.0
55	45	0.75	0.5	.500E-9	3.069	0.50	0.50	6.0	0.65	2.5	0.3
65	50	0.75	0.5	.225E-10	3.457	0.50	0.50	8.0	0.30	5.84	1.0
125	90	0.75	0.5	.332E-9	3.250	0.25	0.50	7.5	0.7	2.5	0.3
120	85	0.75	0.5	.265E-9	3.395	0.25	0.50	7.5	0.7	2.5	0.3
95	70	0.75	0.5	.210E-8	2.775	0.25	0.50	7.0	0.7	2.5	0.3
120	85	0.75	0.5	.225E-9	3.250	0.25	0.50	8.0	0.7	2.5	0.3
125	90	0.75	0.5	.225E-9	3.250	0.25	0.50	8.0	0.7	2.5	0.3
125	90	0.75	0.5	.248E-9	3.295	0.25	0.50	6.0	0.7	2.5	0.3
125	90	0.75	0.5	.400E-10	3.500	0.25	0.50	6.0	0.7	2.5	0.3
125	90	0.75	0.5	.158E-9	3.425	0.25	0.50	5.5	0.7	2.5	0.3
120	85	0.75	0.5	.376E-9	3.216	0.25	0.50	5.5	0.7	2.5	0.3
95	70	0.75	0.5	.329E-8	2.747	0.25	0.50	5.0	0.7	2.5	0.3
85	60	0.75	0.5	.199E-9	3.536	0.25	0.50	6.0	0.7	2.5	0.3
75	55	0.75	0.5	.629E-9	3.108	0.25	0.50	5.5	0.7	2.5	0.3
60	45	0.75	0.5	.916E-9	3.165	0.25	0.50	5.0	0.7	2.5	0.3
70	55	0.75	0.5	.392E-10	3.954	0.25	0.50	6.5	0.7	2.5	0.3
70	55	0.75	0.5	.392E-10	3.954	0.25	0.50	6.5	0.7	2.5	0.3
55	45	0.75	0.5	.109E-9	3.846	0.25	0.50	6.5	0.7	2.5	0.3
50	40	0.75	0.5	.109E-9	3.846	0.25	0.50	6.5	0.7	2.5	0.3

(*continued*)

Material; Condition; Environment*	Code	YS	UTS
[Q] Ni Alloys/superalloys			
Inconel 718; ST (1900F) + A (1400F/10h + 1200F/10h)			
Plt; −320F LN2	Q3LC10LA4	191	225
Plt; 600F air; >.6Hz	Q3LC11AA14	158	180
Plt	Q3LC11AB1	165	205
Plt; 1000F air; >.6Hz	Q3LC12AA18	145	170
Inconel 718; ST (2000F) + A (1325F/4h + 1150F/16h)			
Plt; 800F air; >.2Hz	Q3LE11AA16	133	170
Plt; 1000F air; >.6Hz	Q3LE13AA18	125	157
Plt	Q3LE13AB1	145	187
GTA weld-STA; 600F air, >.6Hz	Q3LEB3AA14	140	180
GTA weld-STA; 800F air, >.6Hz	Q3LEB3AA16	132	175
GTA weld-STA; 1000F air, >.6Hz	Q3LEB3AA18	129	167
GTA weld-STA	Q3LEB3AB1	147	193
Inconel 718, Bolt material			
185 ksi UTS Bolts	Q3LP18AB1	180	205
225 ksi UTS Bolts	Q3LQ18AB1	205	225
Inconel X-750; ST (2100F) + A (1550F/24h + 1300F/20h)			
Plt; 600F air, >.6Hz	Q3SD10AA14	87	147
Plt; 800F air, >.6Hz	Q3SD10AA16	83	143
Plt; 1000F air, >.6Hz	Q3SD10AA18	81	134
Plt & Forg	Q3SD26AB1	100	150
Forg; −452F LHe	Q3SD26LA2	125	220
Rene and Udimet Alloys			
Rene 41; ST (1950F) + A (1400F/16h)			
Plt & Forg	QQ7AD13AB1	138	184
Forg; 1100F air	Q7AD26AA19	112	157
Forg; 1200F air	Q7AD26AA20	110	155

* Unless noted, assume Lab Air (LA) environment and any orientation except S-T, S-L, C-R, C-L, and R-L.

(*continued*)

K_{Ie}	K_{Ic}	A_k	B_k	C	n	p	q	DK_0	R_{cl}	a	S.R.
125	90	0.75	0.5	.300E-9	2.900	0.25	0.50	8.0	0.7	2.5	0.3
125	90	0.75	0.5	.284E-9	3.234	0.25	0.50	7.5	0.7	2.5	0.3
125	90	0.75	0.5	.400E-9	2.900	0.25	0.50	8.0	0.7	2.5	0.3
100	70	0.75	0.5	.224E-8	2.829	0.25	0.50	7.0	0.7	2.5	0.3
120	85	0.75	0.5	.146E-8	2.733	0.25	0.50	7.5	0.7	2.5	0.3
100	70	0.75	0.5	.500E-8	2.251	0.25	0.50	7.0	0.7	2.5	0.3
125	90	0.75	0.5	.250E-9	2.900	0.25	0.50	8.0	0.7	2.5	0.3
85	60	0.75	0.5	.262E-8	2.386	0.25	0.50	6.0	0.7	2.5	0.3
75	55	0.75	0.5	.259E-8	2.452	0.25	0.50	5.5	0.7	2.5	0.3
70	50	0.75	0.5	.154E-7	1.984	0.25	0.50	5.0	0.7	2.5	0.3
85	60	0.75	0.5	.469E-9	2.844	0.25	0.50	6.5	0.7	2.5	0.3
100	70	0.75	0.5	.131E-9	3.626	0.25	0.50	4.0	0.7	2.5	0.3
70	55	0.75	0.5	.750E-9	2.838	0.25	0.50	3.0	0.7	2.5	0.3
70	50	0.75	0.5	.672E-10	3.705	0.25	0.50	6.0	0.7	2.5	0.3
70	50	0.75	0.5	.200E-9	3.200	0.25	0.50	5.0	0.7	2.5	0.3
70	50	0.75	0.5	.114E-8	2.727	0.25	0.50	4.0	0.7	2.5	0.3
85	60	0.75	0.5	.102E-9	3.421	0.25	0.50	10.0	0.7	2.5	0.3
100	70	0.75	0.5	.207E-15	7.228	0.25	0.50	12.0	0.7	2.5	0.3
105	75	0.75	0.5	.306E-10	3.847	0.25	0.50	8.0	0.7	2.5	0.3
84	60	0.75	0.5	.581E-9	3.384	0.25	0.50	6.0	0.7	2.5	0.3
77	55	0.75	0.5	.250E-7	2.150	0.25	0.50	6.0	0.7	2.5	0.3

(*continued*)

Material; Condition; Environment*	Code	YS	UTS
[R] Misc. superalloys			
Multiphase alloys			
MP35N Rnd Rod	R3AB18AB1	254	274
[S] Copper/Bronze alloys			
Be-Cu Alloys			
CDA 172			
Rnd Rod	S0BA13AB1	159	179
C17510			
Peak-aged Plt	S1LB11AB1	110	125
Peak-aged Plt; −320F LN2	S1LB11LA4	120	141
Overaged Plt	S1LC11AB1	86	99
Al-Bronze Alloys			
CDA 630 Al-Bronze Extr	S6JB36AB1	82	117
[T] Magnesium alloys			
AM 503 Plt	T1AA11AB1	16	29
AZ-31 B-H24 Plt	T1DA12AB1	26	39
ZK-60 A-T5 Plt	T1MA12AB1	39	50
ZW1 Plt	T1NA11AB1	24	36
QE22A-T6 Plt	T2LB13AB1	26	33
[U] Misc. non-ferrous alloys			
Beryllium			
Cross-rolled sht	U1CA90AB1	55	70
Hot-pressed blk	U1CA93AB1	35	60
Columbium Alloys			
C- 103 Plt	U2CA10AB1	43	50
Zinc Alloys			
Zn-4 Al-0.04Mg Die cast alloy No. 3			
As cast; 135-200F air0	U4BA50AA9	25	34
As cast	U4BA50AB1	30	41

* Unless noted, assume Lab Air (LA) environment and any orientation except S-T, S-L, C-R, C-L, and R-L.

(continued)

K_{Ie}	K_{Ic}	A_k	B_k	C	n	p	q	DK_0	R_{cI}	a	S.R.
105	80	1.0	0.5	.264E-8	2.280	0.25	0.25	3.0	0.7	2.5	0.3
27	26	0.35	0.5	.532E-10	4.595	0.5	1.0	8.0	0.7	2.0	0.3
110	75	1.0	0.5	.315E-7	1.950	0.75	0.50	8.0	0.7	2.0	0.3
150	110	1.0	0.5	.404E-8	2.591	0.75	0.50	8.0	0.7	2.0	0.3
105	70	1.0	0.5	.900E-8	2.500	0.75	0.50	8.0	0.7	2.0	0.3
74	53	1.0	0.5	.123E-8	3.190	0.50	0.50	8.5	0.7	2.5	0.3
17	12	1.0	0.5	.304E-6	3.183	0.25	0.25	0.75	0.7	1.5	0.3
28	20	1.0	0.5	.294E-6	2.717	0.25	0.25	1.4	0.7	1.5	0.3
28	20	1.0	0.5	.292E-6	2.576	0.25	0.25	1.4	0.7	1.5	0.3
18	13	1.0	0.5	.294E-6	3.210	0.25	0.25	0.75	0.7	1.5	0.3
18	13	1.0	0.5	.294E-6	3.210	0.25	0.25	0.75	0.7	1.5	0.3
11	9.5	1.0	1.5	.200E-9	3.750	0.15	0.15	7.0	0.7	1.75	0.3
12.5	10	1.0	1.5	.500E-10	5.000	0.15	0.15	8.5	0.7	1.75	0.3
42	30	1.0	1.0	.112E-7	2.322	0.5	0.5	10.0	0.7	2.0	0.3
20	14	1.0	0.5	.132E-6	2.812	0.25	0.5	1.5	0.7	1.25	0.3
20	14	1.0	0.5	.500E-7	3.700	0.25	0.5	2.5	0.7	1.25	0.3

Appendix B

Equations for Uniaxial True Stress and True Strain

Typical engineering and true stress-strain curves are shown in Figure B-1. The following equations describe the method of obtaining the true stress-strain parameters in terms of known quantities extracted from an engineering stress-starin curve generated in the laboratory in accordance with the ASTM Standards.

The engineering or conventional stress is defined as

$$\sigma = \frac{P}{A_0} \qquad \text{(B.1)}$$

where P is the tensile load on the specimen, and A_0 is the initial cross-sectional area of the test specimen. The true stress is defined as

$$\sigma_T = \frac{P}{A} \qquad \text{(B.2)}$$

where A is the cross-sectional area of the test specimen under the load P. After the load is removed the area is A_p.

From elasticity theory:

$$A = A_p \left(1 - v\frac{\sigma_T}{E}\right)^2 \qquad \text{(B.3)}$$

Substituting Eq. (B.3) in Eq. (B.2) gives

$$\sigma_T = \frac{P}{A_p \left(1 - v\frac{\sigma_T}{E}\right)^2} \qquad \text{(B.4)}$$

Appendix B Equations for Uniaxial True Stress and True Strain   359

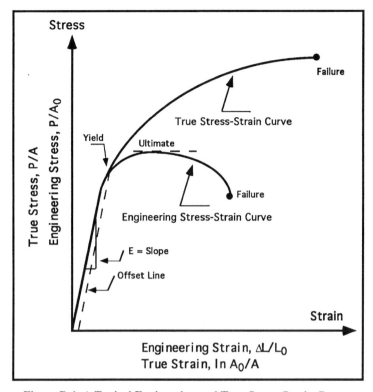

Figure B-1 A Typical Engineering and True Stress-Strain Curves

For plastic straining the volume remains constant. That is, the volume of the uniaxial test specimen remains constant (area $\times$ length) during uniform plastic deformation and as the specimen length increases ($L_0 + \Delta L_0$). Its cross-sectional area decreases uniformly up to the maximum load corresponding to the ultimate of the material.

$$A_0 L_0 = A_p L_p \tag{B.5}$$

From which

$$A_p = A_0 \frac{L_0}{L_p} \tag{B.6}$$

where L_0 and L_p are the lengths of the strained region before loading and after removal of the load, respectively. Substituting Eq. (B.6) in Equation (B.4) gives,

$$\sigma_T = \frac{PL_p/L_0}{A_0\left(1 - v\frac{\sigma_T}{E}\right)^2} \tag{B.7}$$

Substituting Eq. (B.1) in Eq. (B.7) gives

$$\sigma_T = \sigma \frac{L_p/L_0}{\left(1 - v\frac{\sigma_T}{E}\right)^2} \tag{B.8}$$

Plastic strain is defined as:

$$\varepsilon_p = \frac{L_p - L_0}{L_0} \tag{B.9}$$

From which

$$\varepsilon_p = \frac{L_p}{L_0} - 1 \tag{B.10}$$

and

$$\frac{L_p}{L_0} = 1 + \varepsilon_p \tag{B.11}$$

Substituting Eq. (B.11) in Eq. (B.8) gives

$$\sigma_T = \sigma \frac{1 + \varepsilon_p}{\left(1 - v\frac{\sigma_T}{E}\right)^2} \tag{B.12}$$

True strain is defined as:

$$\varepsilon_T = \int_{L_0}^{L} \frac{dL}{L} \tag{B.13}$$

Integrating and substituting limits gives

$$\varepsilon_T = \ln L/L_0 \tag{B.14}$$

Appendix B Equations for Uniaxial True Stress and True Strain

From Eq. (B.11), (B.14) becomes

$$\varepsilon_T = \ln(1 + \varepsilon) \tag{B.15}$$

or

$$e^{\varepsilon_T} = 1 + \varepsilon \tag{B.16}$$

The total true strain is the sum of the elastic plus plastic true strain.

$$\varepsilon_T = \frac{\sigma_T}{E} + \varepsilon_{TP} \tag{B.17}$$

Rearranging terms gives

$$\varepsilon_{TP} = \varepsilon_T - \frac{\sigma_T}{E} \tag{B.18}$$

From Eq. (B.16)

$$1 + \varepsilon_P = e^{\varepsilon_{TP}} \tag{B.19}$$

Substituting Eq. (B.18) in Eq. (B.19) gives

$$1 + \varepsilon_P = e^{\varepsilon_T - \sigma_T/E} \tag{B.20}$$

or

$$1 + \varepsilon_P = \frac{e^{\varepsilon_T}}{e^{\sigma_T/E}} \tag{B.21}$$

Substituting Eq. (B.16) in Eq. (B.21) gives

$$1 + \varepsilon_P = \frac{1 + \varepsilon}{e^{\sigma_T/E}} \tag{B.22}$$

Substituting Eq. (B.22) in Eq. (B.12) gives

$$\sigma_T = \sigma \frac{1 + \varepsilon}{\left(1 - v\frac{\sigma_T}{E}\right)^2 e^{\sigma_T/E}} \tag{B.23}$$

Appendix B Equations for Uniaxial True Stress and True Strain

The reader should note that true stress/true strain parameters can be obtained from the engineering stress/strain curve using equations B-16 and B-23 prior to onset of the necking strain[1]. Equations B-16 and B-23 are valid based on the assumption of uniform distribution of strain along the gage length, L_G, and the material volume due to the application of monotonic load remains constant as described by equation B-5. Thus, equations B-16 and B-23 should be used prior to necking up to the maximum load on the engineering stress-strain curve (points to the right of ultimate strength of the material). Beyond the maximum load the true stress, σ_T, should be calculated from the actual load and cross-sectional area ($\sigma_T = P/A$). Furthermore, the true strain, ε_T, beyond the maximum load should be calculated based on the actual diameter, D, (if the test specimen has a circular cross section) or area, A (i.e. $\varepsilon_T = \ln A/A_0$ or $2 \ln D_0/D$).

To find plastic necking strain, the total deflection of the tensile specimen is the sum of the deflections from uniform plastic straining, neck strain, and elastic straining.

$$\overline{\varepsilon_F} L_G = \varepsilon_{PU} L_G + \varepsilon_{PN} L_{N_0} + \varepsilon_{EF} L_G \tag{B.24}$$

where $\overline{\varepsilon_F}$ is the total deflection of the tensile specimen divided by the initial gauge length, L_G, ε_{PU} is the plastic strain at ultimate stress; ε_{PN} is the plastic necking deflection divided by the effective initial length of the necked region, L_{N_0}; and ε_{EF} is the elastic strain at fracture. It is assumed that for thin tensile specimens with a thickness, t_s:

$$L_{N_0} = t_s \tag{B.25a}$$

and for specimens thicker than the maximum thickness for plane stress, t_0,

$$L_{N_0} = t_0 \tag{B.25b}$$

Substituting Eq. (B.25) into Eq. (B.24) and solving for the plastic necking strain gives

$$\varepsilon_{PN} = (\overline{\varepsilon_F} - \varepsilon_{PU} - \varepsilon_{EF}) \frac{L_G}{t_s} \tag{B.26}$$

From Eq. (B.16)

$$\varepsilon_{EF} = e^{\varepsilon_{TEF}} - 1 \tag{B.27}$$

---

[1] Necking strain is defined as the strain required to deform the test specimen from maximum load to fracture.

Appendix B Equations for Uniaxial True Stress and True Strain

From elastic theory

$$\varepsilon_{TEF} = \frac{\sigma_{TF}}{E} \qquad (B.28)$$

Substituting Eq. (B.28) in Eq. (B.27) gives

$$\varepsilon_{EF} = e^{\sigma_{TF}/E} - 1 \qquad (B.29)$$

Substituting Eq. (B.29) in Eq. (B.26) gives

$$\varepsilon_{PN} = (\overline{\varepsilon_F} - \varepsilon_{PU} - e^{\sigma_{TF}/E} + 1)\frac{L_G}{t_s} \qquad (B.30)$$

Again, from Eq. (B.16)

$$\varepsilon_{PU} = e^{\varepsilon_{TPU}} - 1 \qquad (B.31)$$

From Eq. (B.17)

$$\varepsilon_{TPU} = \varepsilon_{TU} - \frac{\sigma_{TU}}{E} \qquad (B.32)$$

Substituting Eq. (B.32) in Eq. (B.31) gives

$$\varepsilon_{PU} = e^{(\varepsilon_{TU} - \sigma_{TU}/E)} - 1 \qquad (B.33)$$

or

$$\varepsilon_{PU} = \frac{e^{\varepsilon_{TU}}}{e^{\sigma_{TU}/E}} - 1 \qquad (B.34)$$

From Eq. (B.16)

$$e^{\varepsilon_{TU}} = 1 + \varepsilon_U \qquad (B.35)$$

Substituting Eq. (B.35) in Eq. (B.34) gives

$$\varepsilon_{PU} = \frac{1 + \varepsilon_U}{e^{\sigma_{TU}/E}} - 1 \qquad (B.36)$$

Substituting Eq. (B.36) in Eq. (B.30) gives

$$\varepsilon_{PN} = \left(\overline{\varepsilon_F} - \frac{1 + \varepsilon_U}{e^{\sigma_{TU}/E}} - e^{\sigma_{TF}/E} + 2 \right) \frac{L_G}{t_s} \qquad (B.37)$$

For the sum of the true plastic strain at ultimate plus that for necking the total true plastic strain

$$\varepsilon_{TPF} = \varepsilon_{TPU} + \varepsilon_{TPN} \qquad (B.38)$$

From Eq. (B.18)

$$\varepsilon_{TPU} = \varepsilon_{TU} - \frac{\sigma_{TU}}{E} \qquad (B.39)$$

Substituting Eq. (B.39) in Eq. (B.38) gives

$$\varepsilon_{TPF} = \varepsilon_{TU} - \frac{\sigma_{TU}}{E} + \varepsilon_{TPN} \qquad (B.40)$$

From Eq. (B.15)

$$\ln(1 + \varepsilon_{PF}) = \ln(1 + \varepsilon_U) - \frac{\sigma_{TU}}{E} + \ln(1 + \varepsilon_{PN}) \qquad (B.41)$$

From which

$$\ln(1 + \varepsilon_{PF}) = \ln\left[\frac{(1 + \varepsilon_U)(1 + \varepsilon_{PN})}{e^{\sigma_{TU}/E}} \right] \qquad (B.42)$$

or

$$1 + \varepsilon_{PF} = \frac{(1 + \varepsilon_U)(1 + \varepsilon_{PN})}{e^{\sigma_{TU}/E}} \qquad (B.43)$$

From Eq. (B.12)

$$\sigma_{TF} = \sigma_F \frac{1 + \varepsilon_{PF}}{\left(1 - v\frac{\sigma_{TU}}{E}\right)^2} \qquad (B.44)$$

Appendix B Equations for Uniaxial True Stress and True Strain

Substituting Eq. (B.43) in Eq. (B.44) gives

$$\sigma_{TF} = \sigma_F \frac{(1+\varepsilon_U)(1+\varepsilon_{PN})}{\left(1 - v\frac{\sigma_{TU}}{E}\right)^2 e^{\sigma_{TU}/E}} \quad (B.45)$$

Substituting Eq. (B.37) in Eq. (B.45) gives

$$\sigma_{TF} = \sigma_F \frac{1+\varepsilon_U}{\left(1 - v\frac{\sigma_{TF}}{E}\right)^2 e^{\sigma_{TU}/E}} \left[1 + \left(\overline{\varepsilon_F} - \frac{1+\varepsilon_U}{e^{\sigma_{TU}/E}} - e^{\sigma_{TF}/E} + 2\right)\frac{L_G}{t_s}\right] \quad (B.46)$$

where t_s is greater than t_0, from Eq. (B.25b), Eq. (B.37) becomes

$$\varepsilon_{PN} = \left(\overline{\varepsilon_F} - \frac{1+\varepsilon_U}{e^{\sigma_{TU}/E}} - e^{\sigma_{TF}/E} + 2\right)\frac{L_G}{t_0} \quad (B.47)$$

or

$$\varepsilon_{PN} = \left(\overline{\varepsilon_F} - \frac{1+\varepsilon_U}{e^{\sigma_{TU}/E}} - e^{\sigma_{TF}/E} + 2\right)\frac{L_G}{(t_0/\varepsilon_{PN})\varepsilon_{PN}} \quad (B.48)$$

From which

$$\varepsilon_{PN} = \left[\left(\overline{\varepsilon_F} - \frac{1+\varepsilon_U}{e^{\sigma_{TU}/E}} - e^{\sigma_{TF}/E} + 2\right)\frac{L_G}{(t_0/\varepsilon_{PN})}\right]^{1/2} \quad (B.49)$$

Again substituting Eq. (B.49) in Eq. (B.45) gives

$$\sigma_{TF} = \sigma_F \frac{1+\varepsilon_U}{\left(1 - v\frac{\sigma_{TF}}{E}\right)^2 e^{\sigma_{TU}/E}} \left\{1 + \left[\left(\overline{\varepsilon_F} - \frac{1+\varepsilon_U}{e^{\sigma_{TU}/E}} - e^{\sigma_{TF}/E} + 2\right)\frac{L_G}{(t_0/\varepsilon_{PN})}\right]^{1/2}\right\}$$

(B.50)

Substituting Eq. (B.40) in Eq. (B.17) gives

$$\varepsilon_{TF} = \frac{\sigma_{TF}}{E} + \varepsilon_{TU} - \frac{\sigma_{TU}}{E} + \varepsilon_{TPN} \quad (B.51)$$

From Eq. (B.15)

$$\ln(1 + \varepsilon_F) = \frac{\sigma_{TF}}{E} + \ln(1 + \varepsilon_U) - \frac{\sigma_{TU}}{E} + \ln(1 + \varepsilon_{PN}) \quad (B.52)$$

From which

$$\ln(1 + \varepsilon_F) = \ln\left[(1 + \varepsilon_U)(1 + \varepsilon_{PN})\frac{e^{\sigma_{TF}/E}}{e^{\sigma_{TU}/E}}\right] \quad (B.53)$$

or

$$1 + \varepsilon_F = (1 + \varepsilon_U)(1 + \varepsilon_{PN})\frac{e^{\sigma_{TF}/E}}{e^{\sigma_{TU}/E}} \quad (B.54)$$

or

$$\varepsilon_F = (1 + \varepsilon_U)(1 + \varepsilon_{PN})\frac{e^{\sigma_{TF}/E}}{e^{\sigma_{TU}/E}} - 1 \quad (B.55)$$

For a tensile specimen with a circular cross-section where reduction of area, RA, is given, reduction of area is defined as

$$RA = \frac{A_0 - A_p}{A_0} \quad (B.56)$$

or

$$RA = 1 - \frac{A_p}{A_0} \quad (B.57)$$

Substituting Eq. (B.6) in Eq. (B.57) gives

$$RA = 1 - \frac{L_0}{L_p} \quad (B.58)$$

Substituting Eq. (B.11) in Eq. (B.58) gives

$$RA = 1 - \frac{1}{1 + \varepsilon_p} \quad (B.59)$$

From which for the necked region

$$\varepsilon_{PF} = \frac{RA}{1 - RA} \quad (B.60)$$

Index

Accumulative damage, 12, 60–62, 97
Airy stress function, 108
Anisotropy
    beam analysis, 248–249
    material, 128–134
Annotation for drawing, 265–267
Anti buckling device, 207–208
Apparent fracture toughness, 138–139

Back surface correction, 155
Basquin exponent, 82
Bauschinger effect, 72
Beam analysis, 248–249
Biaxial stress state, 124–125
Bolts integrity assessment
    bearing stress, 239
    bolted joint, 9–10, 230–248
    fatigue crack growth, 231–248
    pad analysis, 237–248
    pad bending, 239
    preloaded bolt, 230–237
Brittle failure
    cleavage, 1, 4, 124–126, 132, 135
    ductile brittle transition, 132

Center crack, 117
Charpy test, 132–133
Cleavage, 1, 4, 124–126, 132, 135
Closure Phenomenon, 196–198
    Elber, 196–198
    Newman, 203–206
    Raju, 206
Coffin-Manson law, 82
Combined stress intensity, 114–117
Comments on
    Miner's rule, 50–51

Compact tension specimen (*see* testing)
Compliance method, 189–191
Constant amplitude fatigue test, 34–37
Constant life diagram, 41–47
Conventional fatigue
    High cycle fatigue, 13
    low cycle fatigue, 13
Conventional stress-strain curve, 80–81
Correction
    Back surface, 155, 156
    Plastic zone, 307–309
    Thickness, 126, 130, 169, 307–309, 317, 321
    Width, 302
Crack closure effect, 194–196
Crack extension force, 105
Crack initiation concept, 57–67
    PSB, 64
Crack growth analysis
    bolts, 231–236
    pads, 237–248
Crack growth rate, 179
    ASTM testing, (*see* testing)
    cycle by cycle, 227–229
    equations, 180–188
    NASA/FLAGRO, 178, 185–186, 207–216,
    stress ratio effect, 194–196
    Threshold value, 241–247
Crack initiation life, 68
Crack opening stress intensity, 206
Crack tip
    modes, 122–123
    plasticity, 146–152
Critical stress intensity factor, 6, 124–132
Crystal structures, 70–72

Curve
    Empirical representation of S-N, 38–40
    S-N, 7, 13–15, 32–34
    ε-N, 13–15, 32, 92–98
    cyclic stress-strain, 8, 86–88, 90–92
Cycle by cycle fatigue crack growth, 227–229
Cyclic or fluctuating load, 15–18
Cyclic stress-strain curve, 15, 74–82
Cyclic hardening and softening, 74

Damage tolerance, 10, 253–268
Deformation
    plastic (*see* plastic zone)
    Modes, 106
Design philosophy
    fail safe, 4, 256
    slow crack growth, 4, 254
Determination of
    $\frac{\partial U_F}{\partial c}$ energy term, 292
    $\frac{\partial U_U}{\partial c}$ energy term, 294
Development of
    R-curve, 161–164
    Constant life diagram, 42–44
Diagram
    Gerber, 42–43
    Goodman, 42–43
    S–N, 40–42, 48–51
    ε–N, 92–98
    residual strength capability, 134–146
    universal slope, 98–99
    Kitagawa, 202
Dimple shape, 57–59
Dislocation, 62
Ductile behavior
    brittle transition, 132
    fracture
Dugdale's model, 3, 248

Eddy current, 270–271
Effective crack size, 141
Effect of crack size
    threshold stress intensity, 199–203
    crack closure, 200
Effect of thickness
    fracture toughness, 130–131
Effect of temperature
    fracture toughness, 130–134
Effect of rate of loading
    fracture toughness, 130

Effect of yielding
    fracture toughness, 130
Effective crack length, 141
Effective stress intensity, 196, 198
Effective stress range, 198
Effective yield stress, 203
Elastic energy release rate, 3
Endurance limit, 42–43
Energy balance approach, 5, 6, 117, 118
Energy release rate, 105, 290
Engineering stress-strain curve, 69–70
Equiaxed dimples, 59
Examples
    chapter (2), 23, 24, 44, 48, 79, 87, 90, 95
    chapter (3), 113, 127, 142, 150, 192
    chapter (4), 186, 208, 212, 213, 221, 234, 239, 244
Extrusion, 63–66

Factors influencing
    endurance limit, 34–35
    fracture toughness, 130–134
    threshold stress intensity
    S–N, 38, 40
Fail safe design, 4, 256
Failure criterion
    Octohedral shear stress, 295–296, 303
    Von-Mises, 148–150
Fatigue ductility coefficient, 82
Fatigue strength coefficient, 82
Fatigue
    low cycle, 13
    high cycle, 13
    spectrum, 18–31
Fatigue notch factor, 88–93
Flat failure, (*see* cleavage)
Flaw
    elliptical, 153
    embedded, 153–154
    size inspection (*see* NDI)
Flight spectrum, 28–31
FMDM theory, 5, 289
Forman equation, 185
Fracture
    brittle, 64
    critical parts classification, 261–262
    monotonic load, 57–59
    plane strain (*see* plane strain)
    plane stress (*see* plane stress)

Index

surface examination, 51–57
toughness, 4, 124–128
Fracture control
    acceptability of hardware, 263–265
    activities, 260–261
    board, 267–268
    contents, 261
    development, 259
    parts classification, 261–262
    plan, 10, 258–268
Fracture toughness
    apparent, 138–139
    plane strain, 122–123
    plane stress, 122–123
    surface crack, 159–160
    testing (*see* testing)
    computation, 312–320

Griffith theory, 2, 3, 103–105, 289
Grip fixture apparatus, 172–174

High cycle fatigue, 13,
High energy components, 264, 265
Hysteresis loop, 15, 71–74

Intrusion, 63–66
Inspection (*see* NDI)
Intergranular fracture, 66
Irwin plastic zone, 170–172
Initiation
    fatigue crack, 57, 177–178

Kitagawa diagram, 202

Low released mass, 256
Leak-before -burst, 5, 160–164
LEFM, 103
Linear cumulative damage, 47–51
Low cycle fatigue, 13, 67–68

Magnetic particle inspection, 270
Mandrelizing, 17
Manson, 98
Material selection, 270 (*see* also appendix A)
Material anisotropy, 128–130
Mean stress effect, 88
Microvid coalescence, 57–59
Miner's rule, 7, 15, 47–49
Mixed mode fracture, 106, 306–309
Modes of deformation, 106

Monotonic failure, 57–59
Multiple load path design, 253–257

NASA/FLAGRO computer code, 8, 178
Natural stress-strain curve, 81–83
Net section yielding, 313, 317
Neuber rule, 8, 15, 68, 93–98
Newman crack closure, 247–252
Non destructive inspection (NDI)
    Eddy current, 270–271
    magnetic particle, 270
    penetrant, 269
    radiographic, 272
    special, 288
    ultrasonic, 271
Notional crack, 148

Opening mode, 106
Opening stress, 198, 203
Overload, 216–218

Paris Law, 8, 183
Part through cracks
    emanating from a hole, 158–159
    fracture toughness, 188–190
    stress intensity factor, 179–184
Penetrant inspection, (*see* NDI)
Persistent slip bands (PSB), 64
Plane strain fracture toughness, 122, 123
Plane stress fracture toughness, 122–123
Plastic zone shape
    Irwin, 146–148
    Tresca, 150–152
    Von-Mises, 148–150
Power law representation
    cyclic stress-strain curve, 78–79
    stress-strain curve, 71
Plate bending, 239
Pre-cracking, 166, 189
Preloaded bolts, 230–237
Probability of flaw size detection, 276–278

Radiographic inspection, 272–274
Rain flow technique, 21–23
Range, 21
Residual strength capability, 4, 7, 156–160
    abrupt failure, 158–160
Retardation effect, 9
    Wheeler model, 9
    Willenborg model, 9

Reversal to failure, 82
R-curve, 5, 8, 139–141
Rolled threads, 9

Safe-life, 8, 155, 202, 206
Shuttle, 38–39
Simulative law, 8
Slip mechanism
    monotonic load, 61–62
    cyclic load, 63–66
Slope method
    polynomial, 182
    secant, 180–181
Small cracks, 199–202
Smith-Watson-Topper, 88
Special level inspection, (see NDI)
Stage I and stage II, 7
Strain ratio, 26
Strain-controlled, 76–78
Strain-life prediction, 8, 82–86
Stress and strain at notch, 104–112
Stress concentration factor, 153–157
Stress intensity factor, 105–122
    derivation, 107–111
    equation, 117–122
    range, 179
    surface crack, 153–157
Stress function
    Airy, (see Airy)
Stress ratio, 18
Stress-strain curve, 69–71, 309
Striation (beach) marking, 51
Superposition principle, 114–116
Surface cracks, 152–160
    from a hole, 158–159
    longitudinal in a pipe, 157–158
    fracture toughness, 159–160
Surface energy, 104

Tearing failure, 135, 138–141
Testing
    crack growth test, 188–193
    fracture toughness, 164–173
    K_{Ic} test, 164–168
    K_c test, 169–174
    S–N test, 34–38
    specimen, 165, 171–174
Theoretical stress concentration
    factor, 89, 93–94
Threaded bolt, 230–237
Threshold stress intensity factor, 199–203
Through cracks, 137–142
Tractability, 265
Transition point, 86–88
Tresca yield criterion, 150–152
Triaxial stress state, 124, 125
True stress-strain curve, 70–71 (see also appendix B)

Ultrasonic inspection, 271
Universal slope method, 98–99

Variables affecting NDI, 278–286
Variable amplitude stress, 216–218
Verification of FMDM result, 309–312
Von-Mises yield criterion, 148–150

Walker equation, 185
Welded joints failure, 2–3
Westergaard stress function, 109
Wheeler model, 15, 218–219
Willenborg model, 15, 219–221
Wöhler, 7, 13

X-Ray inspection, 272–274

Yield criterion
    Tresca, 150–152
    Von-Mises, 148–150
Yield strength effect, 130, 203